9

Springer Series on Chemical Sensors and Biosensors

Methods and Applications

Series Editor: G. Urban

Springer Series on Chemical Sensors and Biosensors

Series Editor: G. Urban

Recently Published and Forthcoming Volumes

Optical Guided-wave Chemical and Biosensors II
Volume Editors: Mohammed Z., Lakhtakia A.
Vol. 8, 2009

Optical Guided-wave Chemical and Biosensors I
Volume Editors: Mohammed Z., Lakhtakia A.
Vol. 7, 2009

Hydrogel Sensors and Actuators
Volume Editors: Gerlach G., Arndt K. -F.
Vol. 6, 2009

Piezoelectric Sensors
Volume Editors: Steinem C., Janshoff A.
Vol. 5, 2006

Surface Plasmon Resonance Based Sensors
Volume Editor: Homola J.
Vol. 4, 2006

Frontiers in Chemical Sensors
Novel Principles and Techniques
Volume Editors: Orellana G., Moreno-Bondi M. C.
Vol. 3, 2005

Ultrathin Electrochemical Chemo- and Biosensors
Technology and Performance
Volume Editor: Mirsky V. M.
Vol. 2, 2004

Optical Sensors
Industrial, Environmental and Diagnostic Applications
Volume Editors: Narayanaswamy R., Wolfbeis O. S.
Vol. 1, 2003

Mathematical Modeling of Biosensors

An Introduction for Chemists and Mathematicians

Volume Authors: Romas Baronas · Feliksas Ivanauskas
Juozas Kulys

Springer

Chemical sensors and biosensors are becoming more and more indispensable tools in life science, medicine, chemistry and biotechnology. The series covers exciting sensor-related aspects of chemistry, biochemistry, thin film and interface techniques, physics, including opto-electronics, measurement sciences and signal processing. The single volumes of the series focus on selected topics and will be edited by selected volume editors. The Springer Series on Chemical Sensors and Biosensors aims to publish state-of-the-art articles that can serve as invaluable tools for both practitioners and researchers active in this highly interdisciplinary field. The carefully edited collection of papers in each volume will give continuous inspiration for new research and will point to existing new trends and brand new applications.

ISSN 1612-7617
ISBN 978-94-007-3090-8 e-ISBN 978-90-481-3243-0
DOI: 10.1007/978-90-481-3243-0

Springer Dordrecht Heidelberg London New York

Softcover reprint of the hardcover 1st edition 2009

Cover design: SPi Publisher Services

Printed on acid-free paper

Springer is part of Springer Science+Business Media (www.springer.com)

Series Editor

Prof. Dr. Gerald Urban

IMTEK - Laboratory for Sensors
Institute for Microsystems Engineering
Albert-Ludwigs-University
Georges-Köhler-Allee 103
79110 Freiburg
Germany
urban@imtek.de

Volume Authors

Romas Baronas
Feliksas Ivanauskas

Vilnius University
Dept. Mathematics & Informatics
Naugarduko 24
LT-03225 Vilnius
Lithuania
romas.baronas@mif.vu.lt
feliksas.ivanauskas@mif.vu.lt

Juozas Kulys

Vilnius Gediminas Technical University
Fac. Fundamental Sciences
Dept. Chemistry & Bioengineering
Sauletekio Ave. 11
LT-10223 Vilnius
Lithuania
JKulys@bchi.lt

Preface

Biosensors are analytical devices in which specific recognition of the chemical substances is performed by biological material. The biological material that serves as recognition element is used in combination with a transducer. The transducer transforms concentration of substrate or product to electrical signal that is amplified and further processed. The biosensors may utilize enzymes, antibodies, nucleic acids, organelles, plant and animal tissue, whole organism or organs. Biosensors containing biological catalysts (enzymes) are called catalytical biosensors. These type of biosensors are the most abundant, and they found the largest application in medicine, ecology, and environmental monitoring.

The action of catalytical biosensors is associated with substrate diffusion into biocatalytical membrane and it conversion to a product. The modeling of biosensors involves solving the diffusion equations for substrate and product with a term containing a rate of biocatalytical transformation of substrate. The complications of modeling arise due to solving of partially differential equations with non-linear biocatalytical term and with complex boundary and initial conditions.

The book starts with the modeling biosensors by analytical solution of partial differential equations. Historically this method was used to describe fundamental features of biosensors action though it is limited by substrate concentration, and is applicable for simple biocatalytical processes. Using this method the action of biosensors was analyzed at critical concentrations of substrate and enzyme activity. The substrates conversion in single and multienzyme membranes was studied. The different schemes of substrates conversion which found practical application for biosensors construction were analyzed. The biosensors dynamics was considered at the simplest scheme of biocatalyzer action.

The other part of the book covers digital modeling of biosensors. The biosensors based on amperometric as well as potentiometric transducers are considered. The action of biosensors containing single and multienzymes were modeled using the finite difference technique at nonstationary and steady state. Special emphasis was placed to model biosensors utilizing a complex biocatalytical conversion and biosensors with multipart transducers geometry and biocatalytical membranes structure.

The final part of the book is dedicated to the basic concepts of the theory of the difference schemes for the digital solving of linear diffusion equations which are basis for biosensors modeling.

The book can be recommended for the master and doctoral studies as well as for special studies of biosensors modeling. The Part 3 can also be used for independent study of digital solution of differential equations.

The book was prepared for the period of students teaching by R. Baronas and F. Ivanauskas at Vilnius University and by J. Kulys at Vilnius Gediminas Technical University. The authors acknowledge particular universities for the support of the manuscript preparation. The contribution of the coauthors of the cited publications is highly appreciated.

Vilnius,
February 2009

Romas Baronas
Feliksas Ivanauskas
Juozas Kulys

Acknowledgements

The authors acknowledge Vilnius University (Romas Baronas and Feliksas Ivanauskas) and Vilnius Gediminas Technical University (Juozas Kulys) for the support of monograph preparation.

Contents

Introduction xvii

Part I Analytical Modeling of Biosensors

Biosensor Action 3
1 Kinetics of Biocatalytical Reactions 3
2 Transducer Function 5
3 Scheme of Biosensor Action 6

Modeling Biosensors at Steady State and Internal Diffusion Limitations .. 9
1 Biosensors Containing Single Enzyme 9
2 Biosensors Containing Multienzymes 10
2.1 Consecutive Substrates Conversion 10
2.2 Parallel Substrates Conversion 14
2.3 Biosensors Utilizing Cyclic Substrates Conversion 15
3 Biosensors Utilizing Synergistic Substrates Conversion 16
4 Biosensors Based on Chemically Modified Electrodes 18

Modeling Biosensors at Steady State and External Diffusion Limitations 21
1 Biosensor Using Single Enzyme 21
2 Biosensors with Multienzymes 22
3 Biosensor Utilizing Non Michaelis–Menten Enzyme 23

Modeling Biosensors Utilizing Microbial Cells 27
1 Metabolite Biosensor 27
2 BOD Biosensor 30

Modeling Nonstationary State of Biosensors 33
1 Potentiometric Biosensors 33
2 Amperometric Biosensors 34

Part II Numerical Modeling of Biosensors

Mono-Layer Mono-Enzyme Models of Biosensors ... 43
1 Mathematical Model of an Amperometric Biosensor ... 44
1.1 Governing Equations ... 44
1.2 Initial and Boundary Conditions ... 45
1.3 Dimensionless Model ... 46
2 Characteristics of the Biosensor Response ... 47
2.1 Biosensor Current ... 47
2.2 Biosensor Sensitivity ... 48
2.3 Maximal Gradient of the Current ... 48
2.4 Response Time ... 49
3 Finite Difference Solution ... 50
3.1 Numerical Approximation of Equations ... 50
3.2 Calculation Procedure ... 51
3.3 Validation of Numerical Solution ... 53
3.4 Numerical Error Analysis ... 57
4 Peculiarities of the Biosensor Response ... 60
4.1 Effect of the Enzyme Membrane Thickness ... 60
4.2 Stability of the Response ... 63
4.3 The Response Versus the Substrate Concentration ... 64
4.4 The Response Versus the Maximal Enzymatic Rate ... 66
4.5 Choosing the Enzyme Membrane Thickness ... 68
4.6 Biosensor Resistance ... 70
4.7 Maximal Gradient of the Current ... 71
5 Flow Injection Analysis ... 72
5.1 Mathematical Model ... 73
5.2 Numerical Solution ... 73
5.3 Biosensor Response ... 75
5.4 Peculiarities of the Biosensor Response ... 77
5.5 Sequential Injection Analysis ... 80
6 Biosensors with Chemical Amplification ... 81
6.1 Mathematical Model ... 82
6.2 Finite Difference Solution ... 84
6.3 Concentration Profiles ... 85
6.4 Peculiarities of the Biosensor Response ... 86
7 Potentiometric Biosensors ... 91
7.1 Mathematical Model ... 91
7.2 Biosensor Response ... 92
7.3 Finite Difference Solution ... 93
7.4 Validation of Numerical Solution ... 94
7.5 Simulated Biosensor Response ... 95
7.6 Peculiarities of the Biosensor Response ... 98

8 Enzyme Inhibition ... 103
8.1 Substrate Inhibition ... 103
8.2 Effect of Substrate Inhibition ... 105
8.3 Product Inhibition ... 108
8.4 Effect of Product Inhibition ... 109

One-Layer Multi-Enzyme Models of Biosensors ... 113
1 Biosensors Response to Mixture of Compounds ... 114
1.1 Mathematical Model ... 114
1.2 Solution of the Problem ... 116
1.3 Generation of Data Sets ... 117
1.4 Concluding Remarks ... 120
2 Biosensors Acting in Trigger Mode ... 120
2.1 Mathematical Models ... 121
2.2 Finite Difference Solution ... 126
2.3 Simulated Response ... 128
2.4 Peculiarities of the Response ... 129
2.5 Concluding Remarks ... 136

Multi-Layer Models of Biosensors ... 139
1 Multi-Layer Approach ... 140
1.1 Mathematical Model of Multi-Layer System ... 141
1.2 Numerical Approximation ... 143
1.3 Three-Layer Model ... 145
2 Two-Compartment Model ... 147
2.1 Mathematical Model ... 147
2.2 Transient Numerical Solution ... 150
2.3 Validation of Numerical Solution ... 152
2.4 Simulated Biosensor Responses ... 153
2.5 Effect of the Diffusion Layer ... 155
2.6 The Nernst Diffusion Layer ... 157
2.7 Dimensionless Model ... 160
2.8 Impact of the Diffusion Module ... 162
3 Biosensors with Outer Porous Membrane ... 163
3.1 Mathematical Model ... 164
3.2 Numerical Simulation ... 166
3.3 Effect of the Porous Membrane ... 168
3.4 Concluding Remarks ... 171
4 Biosensors with Selective and Outer Perforated Membranes ... 172
4.1 Mathematical Model ... 173
4.2 Numerical Simulation ... 176
5 Biosensors Based on Chemically Modified Electrode ... 178
5.1 Mathematical Model ... 178
5.2 Numerical Simulation ... 182
5.3 Dimensionless Model ... 184

5.4 Simulated Biosensor Action 186
5.5 Impact of the Diffusion Module 189
5.6 Impact of the Substrate Concentration 191
5.7 Concluding Remarks 192
6 Optical and Fluorescence Biosensors 193
6.1 Mathematical Model 193
6.2 Numerical Simulation 197
6.3 Simulated Biosensor Action 198
6.4 Impact of the Substrate Concentration 201

Modeling Biosensors of Complex Geometry 203
1 Biosensor Based on Heterogeneous Microreactor 204
1.1 Structure of Modeling Biosensor 204
1.2 Mathematical Model 206
1.3 Numerical Simulation 211
1.4 Effect of the Tortuosity of the Microreactor Matrix 214
1.5 Effect of the Porosity of the Microreactor Matrix 214
1.6 Concluding Remarks 216
2 Biosensor Based on Array of Microreactors 217
2.1 Principal Structure of Biosensor 217
2.2 Mathematical Model 219
2.3 Numerical Simulation 222
2.4 Effect of the Electrode Coverage with Enzyme 225
2.5 Concluding Remarks 227
3 Plate-Gap Biosensor 228
3.1 Principal Structure of Biosensor 228
3.2 Mathematical Model 229
3.3 Governing Equations 230
3.4 Initial Conditions 231
3.5 Boundary and Matching Conditions 231
3.6 Biosensor Response 232
3.7 Numerical Simulation 233
3.8 Effect of the Gaps Geometry 236
3.9 Concluding Remarks 236
4 Biosensors with Selective and Perforated Membranes 237
4.1 Principal Structure of Biosensor 238
4.2 Mathematical Model 239
4.3 Numerical Simulation 242
4.4 Effect of the Perforation Topology 245
4.5 Concluding Remarks 246

Part III Numerical Methods for Reaction–Diffusion Equations

The Difference Schemes for the Diffusion Equation 249
1 The Grids 250
1.1 An Equidistant Grid in the Straight 251
1.2 A Non-equidistant Grid in a Straight 251
1.3 The Equidistant Grid in a Plane 252
1.4 The Non-equidistant Grid in a Plane 253
1.5 The Grid in a Multidimensional Case 253
2 The Approximation of the Function Derivatives 254
2.1 The Derivative of the First Order 254
2.2 The Approximation of the Second Order Derivative 256
2.3 The Approximation of the Second Order Derivative on a Non-equidistant Grid 256
3 The Explicit Difference Scheme 258
3.1 The Calculation of a Solution 261
3.2 The Convergence and the Stability 262
4 The Implicit Difference Scheme 264
4.1 The Convergence and the Stability 266
4.2 The Calculation of a Solution 266
5 The Elimination Method for the System of Linear Equations 268
5.1 Stability of the Elimination Method 271
6 The Crank–Nicolson Difference Scheme 271
6.1 The Calculation of a Solution 273
6.2 The Convergence and the Stability 274
7 The Difference Scheme with the Weights 274
7.1 The Convergence and the Stability 276
8 The Crank–Nicolson Difference Scheme on Non-equidistance Grid 276
8.1 The Calculation of a Solution 278
8.2 The Convergence and the Stability 279
9 The Explicit Difference Scheme in the Cylindrical Coordinates 280
9.1 The Calculation of a Solution 281
9.2 The Convergence and Stability 282
10 The Crank–Nicolson Difference Scheme in the Cylindrical Coordinates 282
10.1 The Calculation of a Solution 283
10.2 The Convergence and Stability 285
11 The Discontinuous Diffusion Coefficient 285
12 The Explicit Difference Scheme 287
12.1 The Calculation of a Solution 288
13 The Crank–Nicolson Difference Scheme 288
13.1 The Calculation of a Solution 289

The Difference Schemes for the Reaction–Diffusion Equations 293
1 The Boundary-Value Problem for the System of Reaction–Diffusion Equations 293
2 The Explicit Difference Scheme 295
2.1 The Calculation of a Solution 296
2.2 The Convergence and the Stability 297
3 The Non-linear Crank–Nicolson Type Difference Scheme 298
3.1 Calculation of a Solution 299
3.2 The Convergence and the Stability 303
4 The Linear Crank–Nicolson Type Difference Scheme 303
4.1 The Calculation of a Solution 304
4.2 The Convergence and the Stability 307
5 Law of Conservation of Mass 307
6 The Alternating Directions Method 309
6.1 Calculation of a Solution 311
6.2 The Convergence and the Stability 312
7 The Explicit Method for the Multidimensional Problems 312
7.1 The Calculation of a Solution 314
7.2 The Convergence and the Stability 315

References 317

Index 329

About Authors 333

Introduction

The action of biocatalytical biosensors can be modeled with partial differential equations (PDE) of substrates and products diffusion and conversion in biocatalytical membranes. This book deals with biosensors modeling using analytical and digital solution of the PDE. The intrinsic logics of the book is to evaluate critical parameters and conditions that determinate the biosensors response. Since the analytical solutions of PDE describing biosensors action is possible at limited conditions the modeling of complex biosensor action are performed using digital solution of PDE.

The first part of the book is dedicated to the modeling biosensors by analytical solution of partial differential equations. First chapter of Part I contains tutorial introduction of kinetics of biocatalytical reactions, transducer function of biosensors and a general scheme of biosensor action. In second chapter of Part I the modeling biosensors at steady state and internal diffusion limitation is considered with special contribution to varies schemes of enzymes action. Third chapter of Part I concerns the modeling of biosensors at steady state and external diffusion limitations. The action of biosensor containing single enzyme, biosensors with multienzyme and biosensor utilizing non Michaelis–Menten enzyme kinetics was analyzed. Fourth chapter of Part I contains results of modeling biosensors utilizing microbial cells acting as specific biocatalytical or unspecific biochemical oxygen demand microreactor. The main task of fifth chapter of Part I is to analyze limited cases of biosensors modeling at nonstationary state at some critical concentrations of substrate when analytical solution of PDE was performed. The non stationary response of amperometric as well as potentiometric biosensor was analyzed.

At the end of the first part advantages and disadvantages of analytical modeling of biosensors are shown. The largest advantage of aproximal analytical solution is a possibility to get analytical solution of PDE. The disadvantages include limited concentration interval of reactive components, not applicable to biosensors with complex biocatalytical schemes, very complex solution of non stationary state, lack of analytical solution for complex initial and boundary conditions.

In the second part of the book the corresponding reaction–diffusion problems are solved using digital modeling. The solving PDE was performed using the finite difference technique. First chapter of Part II covers mathematical models with nonlinear reaction kinetics. The biosensors are assumed to be flat electrodes having a mono-layer of an enzyme applied onto the electrode surface. Coupling the enzyme-catalyzed reaction in the enzyme layer (enzyme membrane) with the one-dimensional-in-space diffusion, the mathematical models are described by the non-stationary reaction-diffusion equations. The biosensors based on amperometric as well as potentiometric transducers are considered. The batch and the injection modes of the biosensor operation are modeled in this chapter. The biosensors utilizing the amplification by the conjugated electrochemical and the enzymatic substrates conversion are also investigated. This chapter ends with the modeling of the biosensors with the substrate as well as the product inhibition. The initial boundary value problems are solved numerically by applying the finite difference technique.

Second chapter of Part II deals with the mathematical models of two types of amperometric multi-enzyme biosensors. One type of the biosensors utilizes enzymatic reactions assuming no interaction between the analyzed substrates and the reaction products. The mathematical model of such biosensors is to simulate the biosensor response to a mixture of compounds (substrates). The second type of the biosensors utilizes the enzymatic reaction followed by a cyclic product conversion. Two kinds of the product regeneration in the two-enzyme biosensors are analyzed: enzymatic and electrochemical.

Third chapter of Part II covers multi-layer mathematical models. The biosensors acting in slightly-stirred buffer solutions are described by two-compartment mathematical models. The biosensor operation is analyzed with a special emphasis to the Nernst diffusion layer. This chapter also discusses the multienzyme systems, were the enzymes are immobilized separately in different active layers packed in a sandwich like multi-layer arrangement. The effect of the diffusion limitation to the substrate is investigated when inert outer membranes are applied to stabilize the enzyme layer and to prolong the calibration curve of the biosensor. This chapter also presents the mathematical models of the amperometric biosensor based on the chemically modified electrode as well as of the peroxidase-based optical biosensor.

Fourth chapter of Part II considers modeling of biosensors for which a two-dimensional-in space domain is required to describe mathematically the biosensor action. Firstly, an amperometric biosensor based on a carbon paste electrode encrusted with a single microreactor is considered. Then, an analytical system based on an array of enzyme microreactors immobilized on a single electrode is investigated. Carbon paste porous electrodes are also investigated by applying a plate-gap model. The last section of the this chapter focuses on the modeling of a practical amperometric biosensor containing the selective and the perforated membranes. The perforated membrane is analyzed with a special emphasis to the geometry of the membrane perforation.

Contemporary numerical methods for solving problems of the mathematical chemistry are gaining increasing popularity. The aim of first chapter of Part III is to introduce the reader with the relevant facts about the basic concepts of the theory of the difference schemes for the linear diffusion equations. The linear diffusion equations play an important and crucial role in most models of a biosensor theory. The most popular simple and together effective difference schemes for the linear diffusion equations are presented here. This method is being frequently used in solving applied problems not only by professional mathematicians, but also by laymen. The concepts presented below are of a primary nature and are sufficient for the solution of the problems of the biosensor. In this book the notations of [222] are mainly applied. The many aspects of the numerical methods for the solution of the partial differential equations are presented in [5, 12, 187, 216].

The difference schemes are extensively applied to the solution of a biosensor problems in second chapter of Part III. This chapter is devoted to various difference approximations of the reaction–diffusion equations. The difference technique, developed in a previous chapter, is employed for the construction of the difference schemes. The main subject of investigation is the system of two nonlinear reaction–diffusion equations in one and two dimensional in space cases.

Part I
Analytical Modeling of Biosensors

Abstract This is part one of the book Mathematical Modeling of Biosensors. The part is dedicated to the modeling biosensors by analytical solution of partial differential equations. This part contains tutorial introduction to kinetics of biocatalytical reactions, transducer function of biosensors and a general scheme of biosensor action. A special emphasis is placed to the modeling biosensors at steady state and internal or external diffusion limitation with special contribution to varies schemes of enzymes action, the modeling of biosensors utilizing microbial cells acting as specific biocatalytical or unspecific biochemical oxygen demand microreactor and the modeling biosensors at nonstationary state at some critical concentrations of substrate when analytical solution of PDE is performed.

Keywords Biocatalyzer · Biosensor · Diffusion · Kinetics · PDE

Biosensor Action

Contents

1 Kinetics of Biocatalytical Reactions 3
2 Transducer Function 5
3 Scheme of Biosensor Action 6

1 Kinetics of Biocatalytical Reactions

The biosensors contain immobilized enzymes or other biological catalysts [128, 131, 258]. The biocatalyst catalyzes the conversion of the substrate to the product. Biological catalysts (enzymes) show high activity and specificity. The activity of enzymes may exceed the rate of chemically catalyzed reaction by a factor $4.6 \times 10^5 - 1.4 \times 10^{17}$ [58]. The enzymatic activity of the enzymes depends on many factors, i.e. the free energy of reaction, the substrate docking in the active center of enzyme, the proton tunneling and other factors [88, 124, 132, 143].

The general principles of catalytic activity of enzymes are known, but particular factors that determine high enzyme activity are often not established [178]. The specificity of enzymes depends on the enzyme type [65, 81]. There are enzymes which catalyze the conversion of just one substrate. Other enzymes show broad substrates specificity. Oxidoreductases, i.e. enzymes that catalyze electron transfer, may catalyze, for example, the oxidation or reduction of many substrates. To characterize the substrates with diverse activity a slang expression "good substrate" and "bad substrate" is used.

The following scheme of biocatalyzer action was postulated by Henri in 1902 [109]:

$$\mathrm{E} + \mathrm{S} \rightleftarrows \mathrm{ES} \rightarrow \mathrm{E} + \mathrm{P}\,, \tag{1}$$

where E, S, ES and P correspond to the enzyme, the substrate, the enzyme–substrates complex and the product, respectively. In biochemistry the concentrations are expressed as $\mathrm{mol/dm^3} \equiv \mathrm{M}$, whereas in models the concentrations of components are typically expressed in $\mathrm{mol/cm^3}$.

R. Baronas et al., *Mathematical Modeling of Biosensors*, Springer Series on Chemical Sensors and Biosensors 9, DOI 10.1007/978-90-481-3243-0_1,

Michaelis and Menten confirmed this scheme of enzymes action using acetate to keep pH of solution [185]. Following the scheme (1) the change of concentration of each component can be expressed by ordinary differential equation (ODE):

$$\begin{aligned}\frac{\mathrm{d}E}{\mathrm{d}t} &= -k_1 ES + k_{-1} E_S + k_2 E_S,\\ \frac{\mathrm{d}S}{\mathrm{d}t} &= -k_1 ES + k_{-1} E_S,\\ \frac{\mathrm{d}E_S}{\mathrm{d}t} &= k_1 ES - k_{-1} E_S - k_2 E_S,\\ \frac{\mathrm{d}P}{\mathrm{d}t} &= k_2 E_S,\end{aligned} \tag{2}$$

where t is time, E, S, E_S, P correspond to the concentrations of the enzyme, the substrate, the enzyme–substrates complex and the product, respectively, and kinetic constants k_1, k_{-1} and k_2 correspond to the respective reactions: the enzyme substrate interaction, the reverse enzyme substrate decomposition and the product formation.

To solve the system of ODE (2), Briggs and Haldane applied quasi-steady state approach (QSS) to E_S which means that $\mathrm{d}E_S/\mathrm{d}t \approx 0$ [55]. The calculated "initial rate" of the steady state reaction rate was expressed (S is equal to the initial concentration S_0):

$$V(S) = -\frac{\mathrm{d}S}{\mathrm{d}t} = \frac{V_{max} S}{K_M + S}, \tag{3}$$

where $V_{max} = k_2 E$ is the maximal enzymatic rate, and $K_M = (k_{-1} + k_2)/k_1$, and it is called the Michaelis constant. The Michaelis constant is the concentration of the substrate at which half the maximum velocity of an enzyme-catalyzed reaction is achieved [55,185]. Typical values of constants of the enzymes which are used for the biosensor preparation are: $k_1 = 10^6 - 10^8\,\mathrm{M^{-1}s^{-1}}$ and $k_{-1} \approx k_2 = 100 - 1{,}000\,\mathrm{s^{-1}}$.

Calculations show that during the enzymes action the quasi-steady state is established during $4.0 - 0.1$ ms at enzyme and substrate concentration 10^{-8} M and 10^{-3} M, respectively. It is sufficient to establish a quasi-steady state in the membranes of biosensors with the thickness more than 2×10^{-4} cm since the thickness δ_d of the effective diffusion layer calculated using the Cottrell equation is [19]:

$$\delta_d = \sqrt{\pi D t}, \tag{4}$$

where the diffusion coefficient D for low molecular weight molecules is about $3 \times 10^{-6}\,\mathrm{cm^2/s}$.

For a more complex biocatalytical process, the establishment of the quasi-steady state requires much longer period of time. It was shown, for example, that for the synergistic reactions, involving cyclic mediators conversion, the time of the QSS establishing can be as large as 180 s [147]. Therefore the expression for the "initial rate" is no longer valid, and the modeling should include the rates of all individual reactions.

2 Transducer Function

The purpose of the transducer is to convert the biochemical recognition into an electronic signal. The transducer is a device that responds selectively to the substrate, the product, the mediator or other compound the concentration of which is related to the analyte under determination [128, 131]. The transducer should show high selectivity since the biosensor selectivity depends on the specificity of the biocatalytical process and the selectivity of the transducer.

The transducers include amperometric and ion-selective electrodes, optical systems and other physical devices realizing different physical phenomena. The biocatalytical membrane is located at close proximity to transducer. There are two fundamental categories of transducers in respect of their response. The transducer of the first type, i.e. the amperometric electrode, is monitoring faradaic current which arises when the electrons are transferred between the substrates, the product or the enzyme active center and an electrode. As a result of electrochemical reaction the concentration of oxidized (reduced) compound at surface of the transducer drops down. The transducers of the second type, i.e. ion-selective electrodes, optical fibber, do not perturb the concentration of the determining compound at the surface. The difference between the transducer types produces different boundary conditions for the modeling of the biosensors.

The boundary condition for the first category of transducer can be written

$$P = 0 \quad \text{or} \quad S = 0 \quad \text{at} \quad x = 0, \tag{5}$$

where x stands for space, P and S are the concentrations of the product and the substrate at the transducer surface, respectively. This boundary condition means that the kinetics of electron transfer is fast, and the potential of the transducer is high enough to keep a current at diffusion limiting condition. If the kinetics of electron transfer is slow then the transducer current depends on the electrode potential and is obtained from the Butler–Volmer expression [56]. The modeling of biosensors at this type of boundary conditions has not been performed due to this uncommon state for the real biosensors.

The boundary condition of the transducer of the second category is

$$\frac{\mathrm{d}P}{\mathrm{d}x} = 0 \quad \text{or} \quad \frac{\mathrm{d}S}{\mathrm{d}x} = 0 \quad \text{at} \quad x = 0, \tag{6}$$

For the ion-selective electrodes this corresponds to the Nerstian boundary condition [56]. For the optical transducer and other transducers this condition means non-leakage (zero flux) of the product or the substrate on the boundary between the transducer and the biocatalytical membrane.

3 Scheme of Biosensor Action

The biosensor produces a signal when the analyte under determination diffuses from the bulk solution into the biocatalytical membrane. The biocatalyst catalyzes the substrate conversion to the product, which is determined by the transducer. The concentration change of S and P is associated with the diffusion and the enzymatic reaction. Following Fick the compounds concentration change in the biocatalytical membrane can be written

$$\begin{aligned} \frac{\partial S}{\partial t} &= D_e \frac{\partial^2 S}{\partial x^2} - V(S), \\ \frac{\partial P}{\partial t} &= D_e \frac{\partial^2 P}{\partial x^2} + V(S), \quad x \in (0, d), \quad t > 0. \end{aligned} \tag{7}$$

where x and t stand for space and time, respectively, $S(x,t)$ is the concentration of the substrate, $P(x,t)$ is the concentration of the reaction product, d is the thickness of the enzyme membrane, D_e is the diffusion coefficient of compounds in the enzyme membrane, that is typically used the same for the substrate, the product and the mediator.

The solution of (7) at corresponding initial and boundary conditions produces the concentration change of S and P in time and membrane thickness.

For the first type of transducers the response R of biosensor can be written

$$R(t) = C_1 \left. \frac{\partial P}{\partial x} \right|_{x=0}, \tag{8}$$

and for the second type of transducers

$$R(t) = C_2 P(0,t), \tag{9}$$

or

$$R(t) = C_3 \log P(0,t), \tag{10}$$

where C_1, C_2, C_3 are the appropriate constants.

The logarithmic dependence is characteristic of ion-selective electrodes, whereas for optical and other transducers linear dependence between the response and the concentration is realized.

Simple analytical solution of (7) is impossible even for the simplest initial and boundary conditions due to the hyperbolic function of the enzymatic rate dependence on the substrate concentration (3). Therefore the description of biosensors action is divided into the simplest cases for which analytical solutions still exist. This approach was used widely, especially at the beginning of the development of biosensors, to recognize the principles of the biosensors action. The aproximal analytical solution gives information about the critical cases. They are also useful to test the correctness of numerical calculations found at initial and boundary limiting conditions.

When the concentration S_0 to be measured is very small in comparison with the Michaelis constant K_M,

$$\forall x, t : x \in [0, d], t > 0 : \quad 0 < S(x,t) < S_0 \ll K_M, \tag{11}$$

the nonlinear function $V(S)$ simplifies to that of the first order,

$$V(S) = \frac{V_{max} S}{S + K_M} \approx \frac{V_{max}}{K_M} S. \tag{12}$$

Practically, the enzyme reaction can be considered first-order when the concentration of the detected species is below one-fifth of K_M, i.e. $S_0 < 0.25 K_M$, [99]. This case is rather typical for the biosensors with a high enzyme loading factor.

The nonlinear reaction–diffusion system (7) reduces to a linear one,

$$\begin{aligned} \frac{\partial S}{\partial t} &= D_e \frac{\partial^2 S}{\partial x^2} - kS, \\ \frac{\partial P}{\partial t} &= D_e \frac{\partial^2 P}{\partial x^2} + kS, \quad x \in (0, d), \quad t > 0. \end{aligned} \tag{13}$$

where k is the first-order reaction constant (linear enzyme kinetic coefficient),

$$k = \frac{V_{max}}{K_M}. \tag{14}$$

Analytical solutions are typically made at steady state and external and internal diffusion limiting conditions.

The steady state (stationary) conditions mean that

$$\frac{\partial S}{\partial t} = 0, \quad \frac{\partial P}{\partial t} = 0. \tag{15}$$

The external diffusion limitation indicates that the substrates transport through the diffusion (stagnant) layer [134] is a rate limiting process. At internal diffusion limitation the substrates diffusion through external diffusion layer is fast and the process is limited by the diffusion inside an enzyme membrane. The disadvantage of these approximate solutions is an error at the boundaries between the different approximate treatments. It is helpful to illustrate this approach by reference to a trivial problem of the substrate conversion in the biocatalytical membrane of the biosensor and at the concentration of the substrate less than K_M. The calculated profile of the substrate concentration at the steady state or stationary conditions is shown in Fig. 1.

It is possible to identify an abrupt of concentration change of the substrate at the boundary of biocatalytical membrane/stagnant layer as well as at the boundary stagnant layer/bulk solution. This comes from approximate solutions at the boundaries

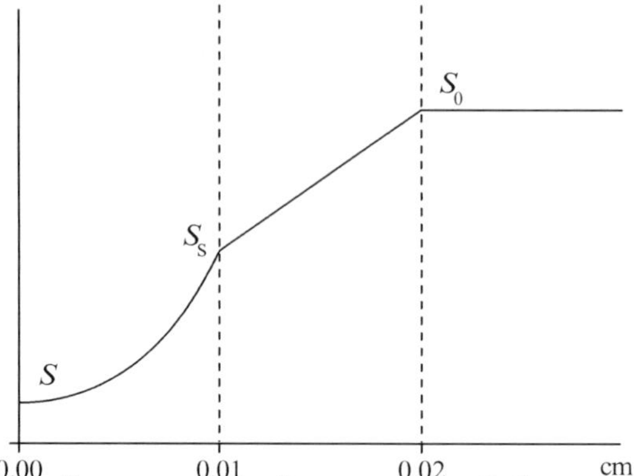

Fig. 1 The substrate concentration profile in a biosensor at steady state conditions. The concentration profile was calculated with the boundary conditions $\partial S/\partial x = 0$ at $x = 0$ and $S = S_0$ at $x \geq d + \delta$. The diffusion coefficient (D_0) in a stagnant layer is $3 \times 10^{-6}\,\mathrm{cm^2/s}$, in membrane is $D_e = 10^{-6}\,\mathrm{cm^2/s}$, $V_{\max} = 5 \times 10^{-7}\,\mathrm{mol/cm^3 s}$, $K_M = 10^{-5}\,\mathrm{mol/cm^3}$, $d = \delta = 0.01$ cm, $S_0 = 10^{-6}\,\mathrm{mol/cm^3}$

during different approximate treatments. The change of the steady state concentration of the substrate in membrane can be calculated as

$$\frac{S}{S_s} = \frac{\cosh(\alpha x)}{\cosh(\alpha d)}, \tag{16}$$

where S, S_s is substrate concentration at transducer surface and at the boundary of membrane and stagnant solution, respectively,

$$\alpha^2 = \frac{V_{max}}{K_M D_e}. \tag{17}$$

On the other hand, at the steady state a substrate flux through the boundary of stagnant layer/bulk solution is equal to the flux through the boundary of biocatalytical membrane/stagnant layer:

$$D_0 \frac{S_0 - S_s}{\delta} = D_e \frac{\partial S}{\partial x}\bigg|_{x=d} = D_e \alpha \tanh(\alpha d) S_s. \tag{18}$$

A combination of these two solutions (16), (18) produces the concentration profile of the substrate in the biocatalytical membrane and the stagnant layer (Fig. 1). It is possible to notice that the greatest error of calculations is at $x = d$ and $x = d + \delta$. However, at the limiting (the internal or the external diffusion limitation) cases the two expressions produce very good approximations to the full equation. Therefore, the modeling of the biosensors at two limiting cases was used to solve different biosensors problems.

Modeling Biosensors at Steady State and Internal Diffusion Limitations

Contents

1 Biosensors Containing Single Enzyme 9
2 Biosensors Containing Multienzymes 10
2.1 Consecutive Substrates Conversion 10
2.2 Parallel Substrates Conversion 14
2.3 Biosensors Utilizing Cyclic Substrates Conversion 15
3 Biosensors Utilizing Synergistic Substrates Conversion 16
4 Biosensors Based on Chemically Modified Electrodes 18

1 Biosensors Containing Single Enzyme

The most popular glucose biosensor is based on glucose oxidase (GO) that catalyzes β-D-glucose oxidation with oxygen [261, 273],

$$\beta\text{-D-glucose} + O_2 \overset{\text{D-glucose oxidase}}{\longrightarrow} \text{D-glucono-}\delta\text{-lactone} + H_2O_2 \tag{1}$$

The hydrogen peroxide produced is oxidized on platinum electrode acting as a transducer. One of the first tasks of modeling of this type of the biosensors was devoted to evaluate the dependence of biosensors response on enzymatic parameters [128]. The action of the biosensors was analyzed at the internal diffusion limiting conditions and at the steady state conditions.

The biosensor response (the current density) was calculated as

$$i(t) = n_e D_e F \left.\frac{\partial P}{\partial x}\right|_{x=0}, \tag{2}$$

where n_e – the number of electrons (for hydrogen peroxide $n = 2$), F – the Faraday number, D_e – the diffusion coefficient of the substrate and the product in the biocatalytical membrane.

R. Baronas et al., *Mathematical Modeling of Biosensors*, Springer Series on Chemical Sensors and Biosensors 9, DOI 10.1007/978-90-481-3243-0_2,

To calculate the biosensor response the change of the product concentration at the transducer surface as indicated in (2) was evaluated. Solving (Chapter 1, eq. 13) with the boundary conditions $\partial S/\partial x = 0$, $P = 0$ at $x = 0$, $S = S_0$, $P = 0$ at $x \geq d$ and at $S_0 \ll K_M$ (the concentration of oxygen is taken in access) produces the stationary biosensor response (the steady state current density) as:

$$I = n_e F D_e \frac{S_0}{d}\left(1 - \frac{1}{\cosh(\alpha d)}\right), \tag{3}$$

The solution shows that the biosensor response is a linear function of the substrate concentration. The sensitivity of the biosensors expressed as $\mathrm{d}I/\mathrm{d}S_0$ does not depend on the enzyme activity if the diffusion module (αd) is larger than 1 since

$$1 - \frac{1}{\cosh(\alpha d)} \approx 1. \tag{4}$$

At $\alpha d < 1$ the approximal solution of (13) is

$$I \approx n_e F D_e S_0 \frac{\alpha^2 d}{2} = n_e F S_0 \frac{V_{max} d}{2K_M}. \tag{5}$$

In this case the biosensor sensitivity is determined by the enzyme parameters V_{max} and K_M.

The inactivation of an enzyme following the diffusion module change from $(\alpha d > 1)$ to $(\alpha d < 1)$ produces a wrong interpretation of an enzyme stability in the biocatalytical membranes. Simple calculations show that if the enzyme inactivates in a solution following the first order reaction with a rate constant $k_{in} = 0.1\ \mathrm{h}^{-1}$ the half-time (τ) of enzyme inactivation is 6.9 h ($\tau = \ln(2)/k_{in}$). If the same enzyme is used for the biocatalytical membranes preparation, and at the beginning of activation the diffusion module of the biosensors is, for example 100 ($d = 0.01$ cm), the response of biosensor will decrease two times only after 87 h.

2 Biosensors Containing Multienzymes

2.1 Consecutive Substrates Conversion

The analysis of the biosensor action containing consecutive substrate conversion with two enzymes has also been analyzed at internal diffusion limitation and steady state conditions [130]. These consecutive reactions occur in the bienzyme electrode based on immobilized D-glucose oxidase and peroxidase. Under the action of D-glucose oxidase (1), D-glucose is oxidized with the production of hydrogen peroxide. During the second stage, hydrogen peroxide is reduced by ferrocyanide ion (6). This reaction is catalyzed by peroxidase

$$H_2O_2 + 2Fe(CN)_6^{4-} + 2H^+ \longrightarrow 2Fe(CN)_6^{3-} + 2H_2O \tag{6}$$

Under the stationary conditions at excess concentrations of oxygen and ferrocyanide, when reactions (1) and (6) are the first order, the change of concentration within the bienzyme membrane is described by the system of the following equations:

$$\begin{aligned}
\frac{d^2 S}{dx^2} &= \frac{V_{max} S}{K_M D_e} = \alpha_1^2 S\,, \\
\frac{d^2 P_1}{dx^2} &= -\frac{V_{max} S}{K_M D_e} + \frac{V'_{max} P_1}{K'_M D_e} = -\alpha_1^2 S + \alpha_2^2 P_1\,, \\
\frac{d^2 P_2}{dx^2} &= -2\frac{V'_{max} P_1}{K'_M D_e} = -2\alpha_2^2 P_1\,,
\end{aligned} \tag{7}$$

where S, P_1 and P_2 are the concentrations of glucose, hydrogen peroxide and ferricyanide, respectively; D_e – the diffusion coefficients of S, P_1 and P_2, which are taken equal; V_{max}, V'_{max}, K_M and K'_M are the corresponding parameters of the enzyme reactions (1) and (6).

The solution of the system (7) taking into consideration the boundary conditions $S = S_0$, $P_1 = 0$, $P_2 = 0$ at $x \geq d$ and $dS/dx = dP_1/dx = 0$, $P_2 = 0$ at $x = 0$ gives the dependence of electrode current density I on the kinetic and diffusive parameters ($\alpha_1 \neq \alpha_2$):

$$\begin{aligned}
I &= FD_e \frac{dP_2}{dx}\bigg|_{x=0} \\
&= \frac{2FD_e\alpha_1^2}{d(\alpha_2^2 - \alpha_1^2)}\left[\left(1 - \frac{1}{\cosh \alpha_2 d}\right) - \frac{\alpha_2^2}{\alpha_1^2}\left(1 - \frac{1}{\cosh \alpha_1 d}\right)\right] S_0\,.
\end{aligned} \tag{8}$$

Hence, it follows that the current of the bienzyme electrode is proportional to the substrate (glucose) concentration. The current is determined by means of diffusion module $\alpha_1 d$ and $\alpha_2 d$. When the rate of enzyme reaction is great ($\alpha_1 d > 1$ and $\alpha_2 d > 1$) the response reaches its maximal value and is determined by the substrate diffusion,

$$I = \frac{2FD_e}{d} S_0 \tag{9}$$

When the activity of peroxidase is considerably greater than the activity of D-glucose oxidase ($\alpha_2 d > \alpha_1 d > 1$) the biosensor response is determined by the D-glucose oxidase parameters

$$I = \frac{2FD_e}{d}\left(1 - \frac{1}{\cosh \alpha_1 d}\right) S_0\,. \tag{10}$$

Under the kinetic control ($\alpha_1 d \ll 1$), (10) is transformed to:

$$I = \frac{F V_{max} d}{K_M} S_0 \,. \tag{11}$$

Due to high molecular activity of the peroxidase, these bienzyme biosensors operate in the mode controlled by the D-glucose oxidase reaction. Sensitivity, as well as the stability of electrodes are close to that of mono-enzyme D-glucose electrode.

The modeling of trienzyme biosensor utilizing consecutive substrates conversion with three enzymes was completed at internal diffusion limitation and steady state conditions [145]. The example of successful application of three enzymes might be sensitive to the creatinine biosensor [248]. In the membrane of this biosensor creatininase (E_1) hydrolyses creatinine (S) to creatine (P_1). The creatine is further hydrolyzed with creatinase (E_2) to sarcosine (P_2). The oxidation of sarcosine with sarcosine oxidase (E_3) produces hydrogen peroxide (P_3) that is determined amperometrically:

$$\mathrm{S} \xrightarrow{\mathrm{E_1}} \mathrm{P_1} \xrightarrow{\mathrm{E_2}} \mathrm{P_2} \xrightarrow{\mathrm{E_3}} \mathrm{P_3} \tag{12}$$

The rate of each reaction ($V_i(S)$) can be characterized by the standard enzyme parameters $V_{max}^{(i)}$ and $K_M^{(i)}$, where $i = 1, 2$ and 3, for E_1, E_2 and E_3 catalyed process, respectively. At concentration of S, P_1 and P_2 less than the Michaelis–Menten constants ($K_M^{(i)}$), $V_1 = V_{max}^{(1)} S / K_M^{(1)}$, $V_2 = V_{max}^{(2)} P_1 / K_M^{(2)}$, $V_3 = V_{max}^{(3)} P_2 / K_M^{(3)}$.

At substrates concentration less than $K_M^{(i)}$ and at a constant diffusion coefficients the diffusion equations and the enzymatic conversions take a form

$$\begin{aligned}
\frac{1}{D_e}\frac{\partial S}{\partial t} &= \frac{\partial^2 S}{\partial x^2} - \alpha_1^2 S \,, \\
\frac{1}{D_e}\frac{\partial P_1}{\partial t} &= \frac{\partial^2 P_1}{\partial x^2} + \alpha_1^2 S - \alpha_2^2 P_1 \,, \\
\frac{1}{D_e}\frac{\partial P_2}{\partial t} &= \frac{\partial^2 P_2}{\partial x^2} + \alpha_2^2 P_1 - \alpha_3^2 P_2 \,, \\
\frac{1}{D_e}\frac{\partial P_3}{\partial t} &= \frac{\partial^2 P_3}{\partial x^2} + \alpha_3^2 P_2 \,,
\end{aligned} \tag{13}$$

where D_e – the diffusion coefficient of all compounds in the enzyme membrane, $\alpha_i = (V_{max(i)}^{(i)} / K_M^{(i)} D_e)^{1/2}$, $i = 1, 2, 3$.

The biosensor response (the current density) was calculated as

$$i(t) = 2FD_e \left.\frac{\partial P_3}{\partial x}\right|_{x=0} \,. \tag{14}$$

The solution of (13) was found at the steady state conditions ($\partial S/\partial t = \partial P_1/\partial t = \partial P_2/\partial t = \partial P_3/\partial t = 0$) with the boundary conditions: $S = S_0$, $P_1 = 0$, $P_2 = 0$, $P_3 = 0$ at $x \geq d$, $\partial S/\partial x = 0$, $\partial P_1/\partial x = 0$, $\partial P_2/\partial x = 0$, $P_3 = 0$ at $x = 0$, where d – the membrane thickness.

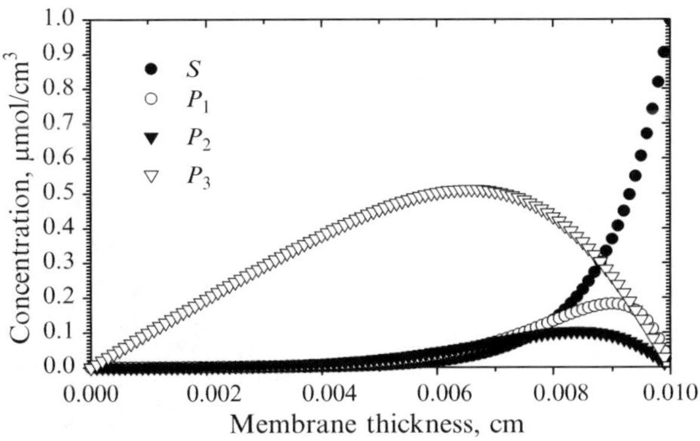

Fig. 1 The profile of compounds concentration in trienzyme membrane of the biosensor. For calculations $S_0 = 10^{-6}$ mol/cm^3, $\alpha_1 d = 10.0$, $\alpha_2 d = 10.1$, $\alpha_3 d = 10.3$ and $d = 0.01$ cm were used

Calculations show that a significant concentration of products in a membrane is produced if all the diffusion modules ($\alpha_i d$) are larger than 1 (Fig. 1).

To prove the correctness of the calculations the distribution of compounds in a membrane was also determined at the boundary condition $\partial P_3/\partial x = 0 (x = 0)$. In this case the sum of the compounds was equal to S_0 at all x values. At $\alpha_1 \neq \alpha_2 \neq \alpha_3$ the expression of three enzyme biosensors response (the current density) is

$$I = 2FD_e\left[\frac{\alpha_2^2\alpha_3^2}{(\alpha_2^2-\alpha_1^2)(\alpha_3^2-\alpha_1^2)}(1-\cosh(\alpha_1 d)) - \frac{\alpha_1^2\alpha_3^2}{(\alpha_2^2-\alpha_1^2)(\alpha_3^2-\alpha_2^2)}(1-\cosh(\alpha_2 d)) + \frac{\alpha_1^2\alpha_2^2}{(\alpha_3^2-\alpha_1^2)(\alpha_3^2-\alpha_2^2)}(1-\cosh(\alpha_3 d))\right]\frac{S_0}{d}. \quad (15)$$

It is impossible in practice to achieve equal values of the diffusion modules for all enzymes. Therefore the biosensor response has not been derived at $\alpha_1 = \alpha_2 = \alpha_3$.

The dependence of the response of the biosensor on the diffusion module of the least active enzymes E_1 and E_2 is shown in Fig. 2. It is easy to notice that the response is very small, still diffusion modules are less than 1. The maximal biosensor response of 6×10^{-5}A/cm^2 is achieved when the diffusion modules are greater than 10.

Experiments show that among three immobilized enzymes the lowest stability demonstrates creatininase (E_1). The model permits to predict sensitivity change of the biosensor during the enzyme inactivation. If the inactivation follows exponential decay, for example, with half-time 2 days, and at the beginning the biosensor contains large catalytic activities ($\alpha_i d \approx 10$), the response decreases just 34.7 % during 10 days (Fig. 3). The apparent half-time of biosensor inactivation increases up to 11.6 days. In fact, this biosensor can be used even longer, i.e. during 15 days with permanent calibration.

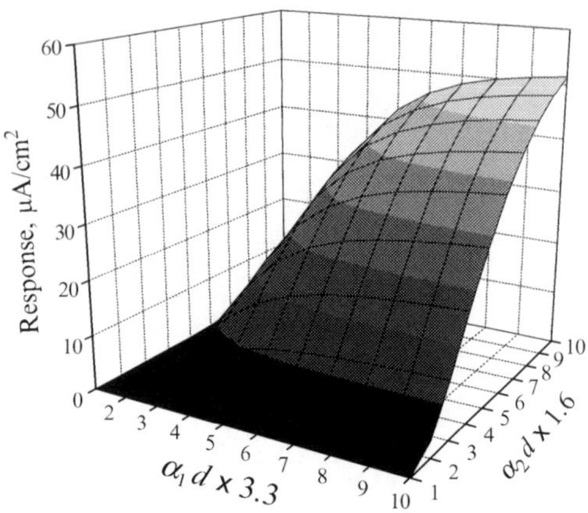

Fig. 2 The dependence of the biosensor response on the diffusion modules $\alpha_1 d$ and $\alpha_2 d$. For calculations $S_0 = 10^{-6}$ mol/cm^3 and $\alpha_3 d = 10.3$ were used

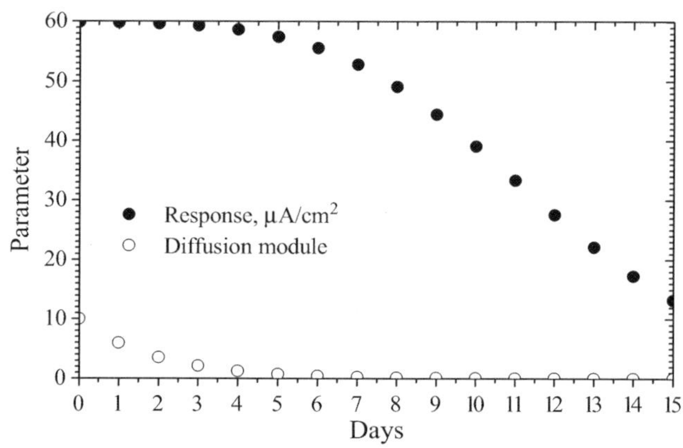

Fig. 3 The changes of the biosensor response and the diffusion module of trienzyme biosensor during the enzyme inactivation. The half-time of enzyme inactivation is 2 days, $S_0 = 10^{-6}$ mol/cm^3, $\alpha_1 d = 10.0$, $\alpha_2 d = 10.1$, $\alpha_3 d = 10.3$ and $d = 0.01$ cm were used

2.2 *Parallel Substrates Conversion*

In the presence of adenosine triphosphate (ATP), D-glucose oxidation is paralleled by the reaction of glucose phosphorylation in the bienzyme biosensors, the catalytic membrane of which is made of D-glucose oxidase and hexokinase [130],

$$\text{D-glucose} + \text{ATP} \xrightarrow{\text{hexokinase}} \text{D-glucose-6-phosphate} + \text{ADP} \tag{16}$$

The increase in ATP concentration leads to the decrease of the response occurring as a result of the D-glucose oxidase reaction (1). For the calculation of the dependence of the biosensor response it is assumed that the oxidation and phosphorylation are first-order for D-glucose and ATP, respectively. The solution of equations of diffusion and enzyme reactions with boundary conditions: $S = S_0$, $S_1 = S_{10}$, $P_1 = 0$ at $x \geq d$, and $\mathrm{d}S/\mathrm{d}x = \mathrm{d}S_1/\mathrm{d}x = 0$, $P_1 = 0$ at $x = 0$ gives the dependence of the density I of the biosensor current on the D-glucose (S_0) and ATP (S_{10}) concentration ($\alpha_1 \neq \alpha_2$),

$$I = 2FD_e \frac{\mathrm{d}P_1}{\mathrm{d}x}\bigg|_{x=0} = \frac{2FD_e}{d}\left[S_0\left(1 - \frac{1}{\cosh \alpha_1 d}\right) - S_{10}\left(\frac{\alpha_2^2}{\alpha_2^2 - \alpha_1^2}\left(1 - \frac{1}{\cosh \alpha_1 d}\right) - \frac{\alpha_1^2}{\alpha_2^2 - \alpha_1^2}\left(1 - \frac{1}{\cosh \alpha_2 d}\right)\right)\right], \quad (17)$$

where $\alpha_2 = (V_{max}/K_M D_e)^{1/2}$ is related to hexokinase reaction.

Thus, it follows that at a high rate of both enzyme reactions ($\alpha_1 d > 1$ and $\alpha_2 d > 1$) the density of the biosensor current is determined by the difference between the D-glucose and ATP concentrations,

$$I = \frac{2FD_e}{d}(S_0 - S_{10}), \quad (18)$$

i.e. the decrease in the biosensors current is proportional to the concentration of coenzyme (ATP),

$$\Delta I = \frac{2FD}{d} S_{10}. \quad (19)$$

If the reaction of D-glucose oxidase proceeds rapidly ($\alpha_1 d > 1$), and the phosphorylation is at low rate ($\alpha_2 d < 1$), then

$$\Delta I = \frac{FD_e V'_{max} d}{K'_M} S_{10}. \quad (20)$$

The sensitivity of such a biosensor is directly proportional to the activity of hexokinase. The experimental results indicate that the action of the ATP electrode is determined by the activity of this enzyme [130]. The hexokinase inactivation results in a quick loss of the biosensors sensitivity.

2.3 *Biosensors Utilizing Cyclic Substrates Conversion*

Rich biocatalytical possibilities permit to construct different systems utilizing the cyclic substrates conversion. The cyclic conversion substrates in an enzyme membrane may considerably increase the sensitivity of the biosensor.

$$\mathrm{S} \xrightarrow{\alpha_1} \mathrm{P}_1 \xrightarrow{\alpha_2} \mathrm{S} + \mathrm{P}_2 \quad (21)$$

where $\alpha_1 d$ and $\alpha_2 d$ are the diffusion modules of the corresponding reactions.

If P_2 is considered to be an electrode-active compound, the biosensors response is

$$I = n_e F D_e \left.\frac{\mathrm{d}P_2}{\mathrm{d}x}\right|_{x=0} = \frac{n_e F D_e}{d} \frac{\alpha_1^2 \alpha_2^2}{\alpha_1^2 + \alpha_2^2} \left[\frac{d^2}{2} - \frac{1}{\alpha_1^2 + \alpha_2^2} \left(1 - \frac{1}{\cosh(\sqrt{\alpha_1^2 + \alpha_2^2} d)} \right) \right] S_0 \quad (22)$$

where the boundary conditions are: $S = S_0$, $P_1 = 0$, $P_2 = 0$, when $x \geq d$; $\mathrm{d}S/\mathrm{d}x = \mathrm{d}P_1/\mathrm{d}x = 0$, $P_2 = 0$, when $x = 0$.

Two important conclusions can be drawn from Eq. 22:

(i) Under the kinetic control of the first or second reaction ($\alpha_1 d < 1$ or $\alpha_2 d < 1$) the amplification of the signal does not take place.
(ii) At a high enzymatic activity ($\alpha_1 d > 1$ and $\alpha_2 d > 1$) the response of the biosensor increases by a value which is directly proportional to the square of the membrane thickness:

$$I = I_d \frac{\alpha_1^2 \alpha_2^2 d^2}{2(\alpha_1^2 + \alpha_2^2)}, \quad (23)$$

where I_d corresponds to the diffusion controlled response of the biosensors containing single enzyme

$$I_d = n_e F D_e \frac{S_0}{d}. \quad (24)$$

The amplification is rapidly increased by the rise of the enzymatic activity. For example, at $\alpha_1 d = \alpha_2 d = 4$, the 4-fold increase of the sensitivity occurs. At $\alpha_1 d = \alpha_2 d = 10$, the amplification enlarges 25-fold.

The possibility of a considerable increase in the sensitivity of the biosensors by means of chemical amplification was demonstrated using alcohol dehydrogenase with cyclic coenzyme (NAD) conversion and in other biocatalytical systems [136].

Biocatalytical systems may utilize other substrates conversion in addition to cyclic conversion. A biosensor showing submicromolar sensitivity to hydrogen peroxide was developed utilizing fungal peroxidase and pyrroloquinoline quinone-dependent glucose dehydrogenase. High sensitivity of the biosensor was achieved by triggering the initiator conversion with peroxidase following the signal amplification by cyclic conversion of the mediator formed [149].

3 Biosensors Utilizing Synergistic Substrates Conversion

Biocatalytical reactions in biocatalytical membranes may be conjugated with chemical conversions. These synergistic reactions allow to generate high sensitive biosensors. High sensitive biosensors for heterocyclic compounds determination were built using the oxidases-catalyzed hexacyanoferrate(III) reduction [146]. The detection limit of some heterocyclic compounds determination was 2×10^{-10} mol/cm^3. The sensitivity of the biosensors was 300–10,000 times larger in comparison to the determination of hexacyanoferrate(III).

The steady state current of the biosensor was calculated using the synergistic scheme of oxidases action. Following the scheme the oxidized glucose oxidase (GO_{ox}) is reduced with glucose ant reduction of hexacyanoferrate(III) (Fer) is catalyzed by reduced glucose oxidase (GO_{red}),

$$\mathrm{GO_{ox}} + \text{D-glucose} \xrightarrow{k_{red}} \mathrm{GO_{red}} + \mathrm{P} \tag{25}$$

$$\mathrm{GO_{red}} + 2\mathrm{Fer} \xrightarrow{k_f} \mathrm{GO_{ox}} + 2\mathrm{Fer_{red}} \tag{26}$$

In the presence of heterocyclic compounds that act as mediators (M) they are oxidized with hexacyanoferrate(III) to cation radicals. The cation radical (M_{ox}) formed reacts with reduced oxidase. The reduced mediator (M_{red}) is further oxidized with hexacyanoferrate(III),

$$\mathrm{GO_{red}} + 2\mathrm{M}_{ox} \xrightarrow{k_{ox}} \mathrm{GO_{ox}} + 2\mathrm{M_{red}} \tag{27}$$

$$\mathrm{M_{red}} + \mathrm{Fer} \xrightarrow{k_{exc}} \mathrm{M_{ox}} + \mathrm{Fer_{red}} \tag{28}$$

Chemical reaction (28) increases Fer_{red} production rate, therefore the rate of overall process is larger than the reactions (26) and (27).

Since the electrochemically active compound is hexacyanoferrate(II) (Fer_{red}), the steady state response can be calculated like the current density of the biosensor with a chemical amplification [136]:

$$I = \frac{FD_e}{d}\frac{\alpha_1^2\alpha_2^2}{\beta^2}\left(\frac{d^2}{2} - \frac{1}{\beta^2}\left(1 - \frac{1}{\cosh(\beta d)}\right)\right) M_0 , \tag{29}$$

where $\alpha_1 d$ and $\alpha_2 d$ are the diffusion modules, $\beta^2 = \alpha_1^2 + \alpha_2^2$, d is the enzyme membrane thickness, M_0 corresponds to the total mediator. The diffusion modules were calculated as $\alpha_1 d = (k_{ox}E_0/D_e)^{1/2}d$, $\alpha_2 d = (k_{exc}Fer/D_e)^{1/2}d$, where E_0 and Fer stand for the total enzyme and the hexacyanoferrate(III) concentrations, respectively, k_{exc} is the bimolecular electron exchange constant between the mediator and hexacyanoferrate (III).

The density I_0 of the steady state current of the biosensor in the absence of the mediator was calculated as

$$I_0 = \frac{FD_e}{d}\left(1 - \frac{1}{\cosh(\gamma d)}\right) Fer , \tag{30}$$

where $\gamma = (k_f E_0/D_e)^{1/2}$, k_f is the constant of the reaction of the hexacyanoferrate(III) with the reduced glucose oxidase.

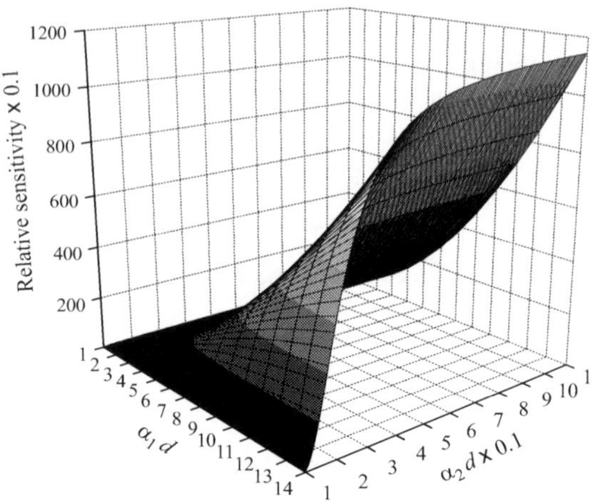

Fig. 4 The dependence of the biosensor relative sensitivity S_r on the diffusion modules $\alpha_1 d$ and $\alpha_2 d$

The analysis of the dependence of the relative sensitivity ($S_r = I/I_0$) on the diffusion modules reveals that the S_r is larger than 1 if $\alpha_1 d$ and $\alpha_2 d$ are larger than 0.5 (Fig. 4).

At $\alpha_1 d = \alpha_2 d = 1$, the relative sensitivity S_r of the biosensor is 12.9. It increases if both diffusion modules are larger than 1. The calculations show that for the biosensor containing 1.3×10^{-7} mol/cm^3 of glucose oxidase $\alpha_1 d$ is 14.5 and $\alpha_2 d$ is 113.5. In contrast, the diffusion module (γd) of the biosensor acting with the pure hexacyanoferrate(III) is 0.13 due to the low constant of the reduced enzyme. Since γd is less than 1 it means that the biosensor acts in a kinetic regime. It is easy to notice that at $\alpha_2 d > \alpha_1 d > 1$ and $\gamma d < 1$ the relative sensitivity equals

$$\frac{I}{I_0} = \frac{k_{ox}}{k_f}. \tag{31}$$

The comparison of calculated and experimentally determined values reveals that the calculated relative sensitivity of the biosensors is about three times larger than the experimentally determined. This deference can be caused by the limited stability of the oxidized heterocyclic compounds, uncounted parallel reaction of reduced enzyme with oxygen and external diffusion limitation of hexacyanoferrate(III) and glucose.

4 Biosensors Based on Chemically Modified Electrodes

Chemically modified electrodes (CME) are produced by modifying carbon electrodes with redox active component (mediator) which reacts with an enzyme [138]. For modification of electrodes adsorption or covalent immobilization of the mediator is used. The peculiarities of CME based biosensors modeling arise due to the mediator location on the electrode that produces special boundary conditions.

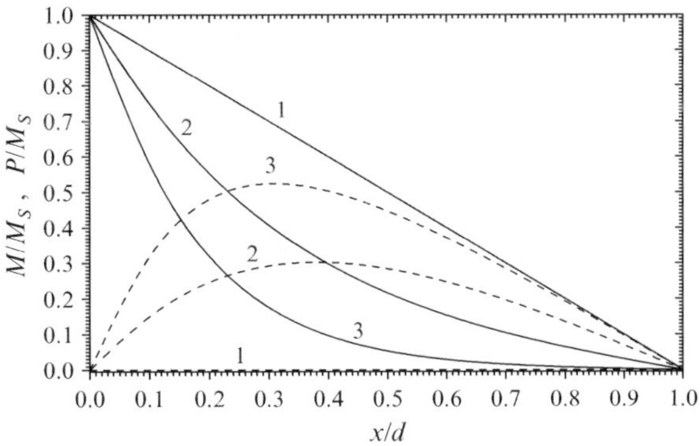

Fig. 5 Concentration profiles of the mediator (*solid lines*) and the product (*dashed lines*) in the enzyme membrane. $\alpha d = 0\,(1)$, $\alpha d = 3\,(2)$, $\alpha d = 6\,(3)$

It was assumed that the concentration of the mediator (M_s) on the electrode is constant due to the desorption of the mediator. The enzyme membrane with thickness d is fixed on CME (Fig. 5).

At the steady state and in the absence of a substrate the concentration of the mediator decreases uniformly in the membrane. In the presence of the substrate a part of the mediator is reduced (oxidized) and the reaction product is oxidized (reduced) on the electrode. For calculations the concentration of the mediator was assumed to be much less than K_M and the enzymatic reaction was described as a bimolecular process of the mediator interaction with an enzyme. The constant of reaction rate is k_b:

$$V(M) = k_b M E_0 \tag{32}$$

The steady state diffusion and enzymatic equation (33) was solved with the boundary conditions $M = M_s$, $P = 0$ at $x = 0$ and $M = 0$, $P = 0$ at $x \geq d$

$$\begin{aligned} D_e \frac{\mathrm{d}^2 M}{\mathrm{d}x^2} - k_b E_0 M &= 0\,, \\ D_e \frac{\mathrm{d}^2 P}{\mathrm{d}x^2} + k_b E_0 M &= 0\,, \end{aligned} \tag{33}$$

where M and P are the concentrations of the mediator and the product in the enzyme membrane of thickness d, respectively.

The mediator (M) and the product (P) concentrations in the enzyme membrane were calculated form the following equations:

$$\begin{aligned} \frac{M}{M_s} &= \exp(\alpha x) - \frac{\exp(\alpha d)\sinh(\alpha x)}{\sinh(\alpha d)}\,, \\ \frac{P}{M_s} &= \frac{\exp(\alpha d)\sinh(\alpha x)}{\sinh(\alpha d)} - \exp(\alpha x) - \frac{x}{d} + 1\,, \end{aligned} \tag{34}$$

where $\alpha^2 = k_b E_0 / D_e$.

The density I of the stationary biosensor current is

$$I = n_e F D_e \left.\frac{\mathrm{d}P}{\mathrm{d}x}\right|_{x=0} = n_e F D_e \left(\alpha \coth(\alpha d) - \frac{1}{d}\right) M_s \,. \tag{35}$$

When $\alpha d \approx 1$, the response is

$$I = 0.33 n_e F d k_{ox} E_0 M_s \,. \tag{36}$$

For the first time the model was used to describe the action of glucose biosensors based on organic metal or on carbon electrodes modified with organic metal components [64]. Later the biosensors of this type have found one of the largest practical applications [172, 258].

Modeling Biosensors at Steady State and External Diffusion Limitations

Contents

1 Biosensor Using Single Enzyme 21
2 Biosensors with Multienzymes 22
3 Biosensor Utilizing Non Michaelis–Menten Enzyme 23

1 Biosensor Using Single Enzyme

The modeling of the biosensors action at an external diffusion limitation is much easer due to the linear gradient of the substrates concentration in a stagnant layer. The analysis of such systems, however, did not receive a lot of attention since the internal diffusion problems are intrinsic for the catalytical biosensors. For a biosensor acting at the external diffusion limitation and at the steady state conditions the flux of the substrate through a stagnant layer is equal to the enzyme reaction rate on the surface of the transducer:

$$D_0 \frac{S_0 - S_s}{\delta} = \frac{V_{max,s} S_s}{K_M + S_s}, \tag{1}$$

where $V_{max,s}$ corresponds to the maximal enzyme rate on the surface expressed as $\mathrm{mol/cm^2s}$.

The solution of this equation is

$$S_s = \frac{1}{2}S_0 - \frac{1}{2}K_M - \frac{1}{2}\rho K_M + \frac{1}{2}\sqrt{S_0^2 + 2S_0K_M - 2S_0\rho K_M + K_M^2 + 2\rho K_M^2 + \rho^2 K_M^2}, \tag{2}$$

where

$$\rho = \frac{V_{max,s}\delta}{K_M D_0}. \tag{3}$$

R. Baronas et al., *Mathematical Modeling of Biosensors*, Springer Series on Chemical Sensors and Biosensors 9, DOI 10.1007/978-90-481-3243-0_3,

The biosensor response is

$$I = n_e F \frac{V_{max,s} S_s}{K_M + S_s}, \tag{4}$$

where S_s comes from (2).

At $S_0 < K_M$ the expression for the response is much simpler:

$$I = n_e F \frac{V_{max,s} S_0}{K_M(1 + \rho)}. \tag{5}$$

At this substrate concentration the sensitivity of the biosensor depends on the value of the diffusion module. At $\rho < 1$ the sensitivity depends on the kinetic parameters of an enzyme:

$$I = n_e F \frac{V_{max,s} S_0}{K_M}. \tag{6}$$

At $\rho > 1$ the biosensor acts in the diffusion controlled regime, and the sensitivity is determined by the diffusion parameter and the thickness of a stagnant layer,

$$I = n_e F \frac{S_0 D_0}{\delta}. \tag{7}$$

In this case the temperature inactivation of an enzyme or the other factors influencing the enzyme activity no longer perturb the sensitivity of the biosensor.

2 Biosensors with Multienzymes

The conversion of the substrate in, for example, adsorbed polyenzyme system, is catalyzed by the corresponding enzyme with the matching kinetic parameters $V_{max,s}^{(i)}$ and $K_M^{(i)}$:

$$\mathrm{S}_1 \longrightarrow \mathrm{S}_2 \longrightarrow \ldots \mathrm{S}_i \longrightarrow \mathrm{S}_n \tag{8}$$

At the steady state conditions, the substrate flux from the bulk solution, where the concentration is S_0, is equal to the rate of the product (the substrate of the following reactions) generation:

$$\begin{aligned}
D_0 \frac{S_0 - S_1}{\delta} &= \frac{V_{max,s}^{(1)} S_1}{K_M^{(1)} + S_1}, \\
D_0 \frac{S_2}{\delta} &= \frac{V_{max,s}^{(1)} S_1}{K_M^{(1)} + S_1} - \frac{V_{max,s}^{(2)} S_2}{K_M^{(2)} + S_2}, \\
D_0 \frac{S_i}{\delta} &= \frac{V_{max,s}^{(i-1)} S_{i-1}}{K_M^{(i-1)} + S_{i-1}} - \frac{V_{max,s}^{(i)} S_i}{K_M^{(i)} + S_i}, \qquad i = 3, \ldots, n-1,
\end{aligned} \tag{9}$$

$$D_0 \frac{S_n}{\delta} = \frac{V_{max,s}^{(n-1)} S_{n-1}}{K_M^{(n-1)} + S_{n-1}} .$$

At $S_i < K_M^{(i)}$ and $\rho_i = V_{max,s}^{(i)} \delta_i / (K_M^{(i)} D_0)$

$$S_{n-1} = \frac{\prod^{n-2} \rho_i}{\prod^{n-1} (1 + \rho_i)} S_0 , \tag{10}$$

and

$$S_n = \frac{\prod^{n-1} \rho_i}{\prod^{n-1} (1 + \rho_i)} S_0 . \tag{11}$$

It follows that the degree of the substrate conversion depends on the stadium for that diffusion module is less than 1. The longer the chain (larger enzyme numbers n) the less the product formation. Even at $n = 3$, the significant yield of the product is generated if all diffusion modules are greater than 1.

If the rate of the last product formation is proportional to the current, the biosensors response can be expressed

$$I = n_e F \frac{V_{max,s}^{(n-1)} \prod^{n-2} \rho_i}{K_M^{n-1} \prod^{n-1} (1 + \rho_i)} S_0 . \tag{12}$$

At $\rho_i > 1$ the sensitivity of the biosensor is determined by the kinetic parameters of the last reaction. At $\rho_i < 1$ the sensitivity is a product of $V_{max,s}^{(n-1)} / K_M^{(n-1)}$ and the diffusion modules.

3 Biosensor Utilizing Non Michaelis–Menten Enzyme

Very often the kinetics of an enzyme action is much more complicated in comparison with the simplest scheme of an enzyme action (Chapter 1, eq. 3). Inhibition, activation, allostery and other types of non Michaelis–Menten kinetics determinate the diversity of the enzymes [98]. In addition to the scheme (Chapter 1, eq. 1), the interaction of the enzyme substrate complex (ES) with other substrate molecule (S) following the generation of non-active complex (ES_2) may produce one of the simplest non-Michaelis–Menten scheme of an enzyme action

$$\mathrm{ES} + \mathrm{S} \leftrightarrow \mathrm{ES}_2 \tag{13}$$

The dependence of the steady state "initial" rate on the substrate concentration shows the maximum at $S = \sqrt{K_M K_I}$ in contrast to the typical Michaelis–Menten kinetics:

$$V(S) = -\frac{dS}{dt} = \frac{V_{max} S}{K_M + S + S^2/K_I}, \tag{14}$$

where K_I is the constant of ES_2 dissociation.

The biosensor is considered as an infinite nontransparent plate covered by a thin (molecular) layer of an enzyme. The substrate flux to the layer is perpendicular to the surface of the transducer. The thickness of the stagnant (the Nernst) layer covering the enzyme layer is δ. At the steady state conditions the concentration of the substrate on the transducer can be calculated from the equality of the substrate flux and the enzyme conversion on the surface:

$$D\frac{S_0 - S_s}{\delta} = \frac{V_{max} S_s}{K_M + S_s - S_s^2/K_I}. \tag{15}$$

At the steady state conditions the biosensor response I can be expressed as follows:

$$I = n_e F \frac{V_{max} S_s}{K_M + S_s - S_s^2/K_I}. \tag{16}$$

Assuming for simplicity $K_I = K_M$ and using dimensionless parameters, $c_b = S_0/K_M$ and $c_s = S_s/K_M$, the expression (15) simplifies,

$$c_b - c_s = \frac{\rho c_s}{1 + c_s + c_s^2}, \tag{17}$$

where ρ is the diffusion module, $\rho = V_{max}\delta/(K_M D)$.

Eq. 17 can be solved using the Cardano formula or graphically. For the calculations the enzyme concentration 10^{-11} mol/cm^2 that corresponds to monolayer of the enzyme molecules adsorbed on the geometrically flat surface was used. The values of the other parameters used for the calculations were: the catalytic constant ($k_2 = 10^3$ 1/s) corresponds to the moderately active enzyme, the Michaelis constant (10^{-5} mol/cm^3) is typical for many enzymes, the diffusion coefficient of the substrate 10^{-6} cm^2/s, the thickness of the stagnant layer 0.03 cm. For these parameters the calculated dimensionless parameter ρ equals 30.

The solution of (17) at $0 < c_b < 9.714$ gave a single value of surface concentration that was less in comparison with c_b. At $c_b = 9.714$ two values of $c_s = 0.589$ and $c_s = 4.063$ were calculated. At $11.091 > c_b > 9.714$ three values of c_s were found. At $c_b = 11.091$ two values of c_s (1.201 and 7.688) were found again. At $c_b > 11.091$ a single value was calculated. To verify the correctness of the calculations a graphical solution of (17) was found. A function $Y = c_s/(1 + c_s + c_s^2)$ was plotted to show the enzymatic rate, and the function $y = (c_b - c_s)/\rho$ at fixed c_b concentration – to show the diffusion rate. The crossing of these functions gave c_s. In Fig. 1, three approximate values of $c_s = 0.64$, $c_s = 2.8$ and $c_s = 5.5$ gave the crossing of the functions at $c_b = 10$ and $\rho = 30$. These values fitted the calculated c_s (0.636, 2.833 and 5.529).

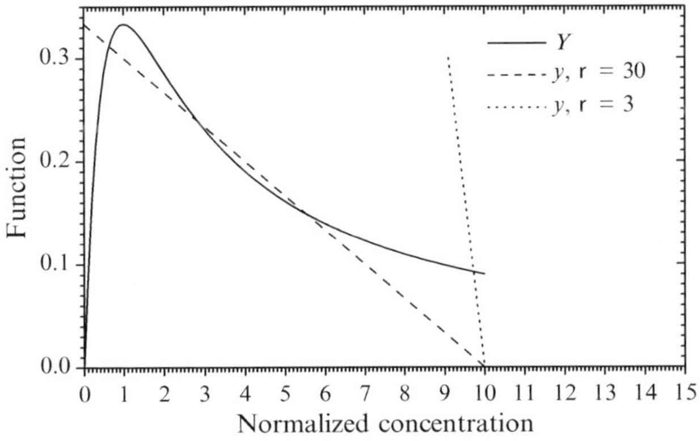

Fig. 1 Graphical surface concentration calculation. The parameters of calculations are presented in the text

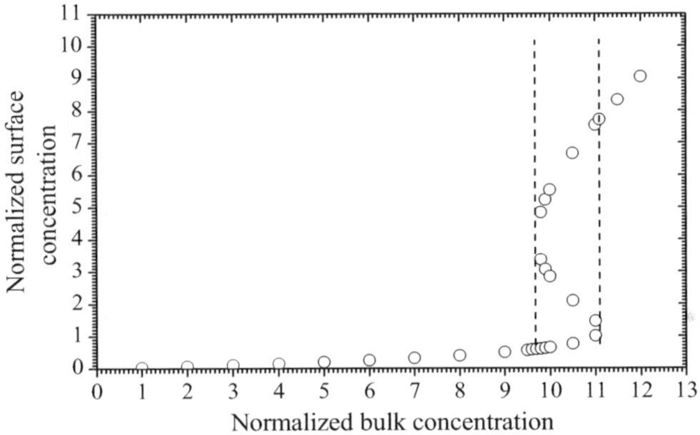

Fig. 2 The dependence of normalized surface concentration on normalized bulk concentration at $\rho = 30$. Vertical dot lines mark the zone of multi-steady state concentrations

The dependence of the substrate surface concentration on the bulk concentration is shown in Fig. 2. The presented results demonstrate that a surface concentration is less than a bulk concentration. At critical bulk concentrations 9.714 and 11.091, two steady state concentrations are available whereas at the intermediate three concentrations are possible. However, an intermediate concentration is not stable, since any perturbation of parameters produces a low or high concentration.

The biosensor response (16) is related to the surface concentration of the substrate. The generation of different surface concentrations produces multi-response of biosensor (Fig. 3). The multi-response can be achieved at $11.091 \geq c_b \geq 9.714$.

The modeling shows that multi-surface concentration is possible at large diffusion parameter (ρ). The decrease of the ρ value up to 3, generates just 1 concentration $c_s = 9.7$ at $c_b = 10$ (Fig. 3). It is worth noticing that many interfaces

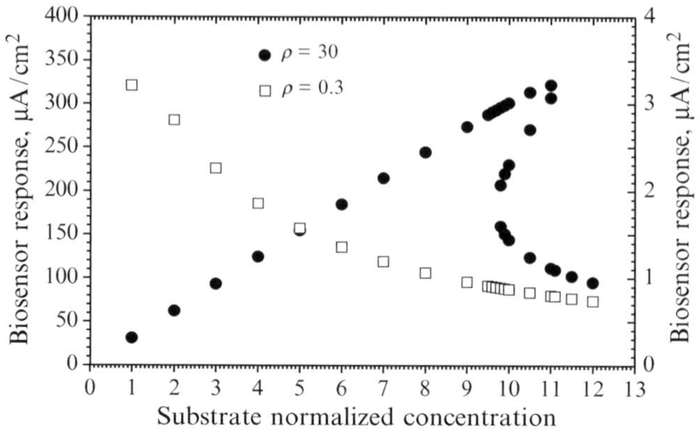

Fig. 3 The dependence of the biosensor response on normalized bulk concentration of the substrate. The data calculated with $\rho = 30$ belongs to the *left hand y* axis, with $\rho = 0.3$ – to the *right hand y* axis

used for the enzyme adsorption have a surface being much more than geometrical. Therefore the enzyme concentration and consequently ρ may increase many orders of magnitude.

At $\rho = 0.3$ the biosensor acts in a kinetic regime. The surface concentration of the substrate is little less than the bulk concentration. The response of the biosensor drops down almost 10–100 times, and the decrease of the response in the concentration range 1–12 is associated with the enzyme activity decrease (Fig. 3). From the calculations it follows that all parameters which change the diffusion module (ρ) perturb the multi-steady state zone. The thickness of the stagnant layer (δ) is the most difficult controllable parameter. Using a rotating disk electrode or the precious flow rate may help to control the thickness of this layer.

Multi-steady state surface concentration may have far-reaching consequences for the stability of the biosensors response. It can generate oscillations of the concentration and the response of the biosensor if the negligible perturbation of an enzyme activity or mass transport occurs.

Modeling Biosensors Utilizing Microbial Cells

Contents

1 Metabolite Biosensor 27
2 BOD Biosensor 30

1 Metabolite Biosensor

The biocatalytical system of microbial cells can be used as biocatalyzers for the biosensor preparation [127, 130]. They can show very high specificity for some substrates. For example, yeast cells *Hansenula anomala* grown in lactate reach breeding media induce cytochrome b_2, and shows high specificity to L-lactate [125]. The scheme of substrates distribution in a microbial biosensor is depicted in Fig. 1.

The peculiarity of the modeling of the microbial biosensors is a slow substrate and product transport through the microbial cell wall. If the substrate transport is slower than the diffusion through the bulk solution and the semipermeable membrane the substrate and the product concentration change in the cell can be written:

$$\begin{aligned} \frac{\mathrm{d}S_c}{\mathrm{d}t} &= k(S_0 - S_c) - V_c\,, \\ \frac{\mathrm{d}P_c}{\mathrm{d}t} &= V_c - k' P_c\,, \end{aligned} \tag{1}$$

where S_c and P_c are the concentrations of the substrate and the product in the cell, respectively, k and k' are constants of substrate transport into the cell and the product transport from the cell, respectively, and V_c is the rate of the enzymatic process in the cell. The constants k and k' are related to permeability (h) of the cell wall, that can be expressed as

$$\begin{aligned} k &= h_s/l\,, \\ k' &= h_p sur_c/vol_c \end{aligned} \tag{2}$$

R. Baronas et al., *Mathematical Modeling of Biosensors*, Springer Series on Chemical Sensors and Biosensors 9, DOI 10.1007/978-90-481-3243-0_4,

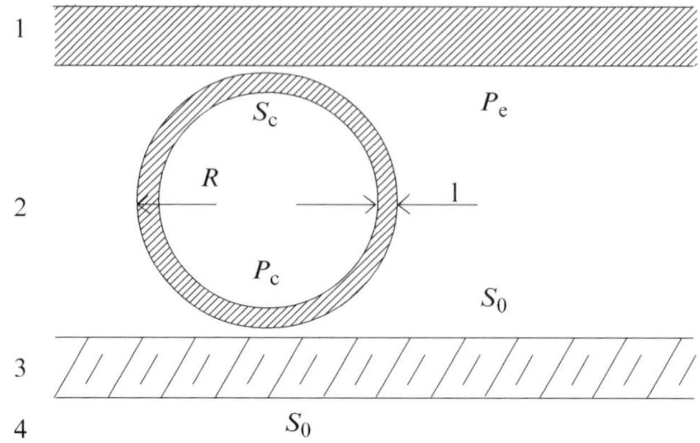

Fig. 1 The scheme of the substrates distribution in a microbial biosensor. Transducer (**1**), biocatalytical membrane with microbial cell (**2**), semipermeable membrane (**3**), bulk solution (**4**), radius of cell (R), thickness of cell wall (l)

where sur_c and vol_c are the surface and the volume of the microbial cell, respectively.

For permeability calculations the equilibrium distribution of the substrate and the product between the cell wall and the solution is used,

$$\begin{aligned} h_s &= \frac{D_s}{l}\frac{S_{m,eq}}{S_{0,eq}}, \\ h_p &= \frac{D_p}{l}\frac{P_{m,eq}}{P_{c,eq}}. \end{aligned} \tag{3}$$

If the kinetics of the substrate conversion in the cell obeys the Michaelis–Menten scheme (Chapter 1, eq. 1) the substrate concentration change in the cell can be rewritten

$$\frac{\mathrm{d}S_c}{\mathrm{d}t} = k(S_0 - S_c) - \frac{V_{max,c}S_c}{K_{M,c} + S_c}. \tag{4}$$

This equation can be integrated with the initial conditions $S_c = 0$ at $t = 0$. A more simple solution can be produced at $S_c \ll K_{M,c}$:

$$S_c = \frac{-\exp\left(\dfrac{-tkK_{M,c} - tV_{max,c} + \ln(kS_0)K_{M,c}}{K_{M,c}}\right) + kS_0}{k + \dfrac{V_{max,c}}{K_{M,c}}}. \tag{5}$$

Inserting S_c solution into (1) gives a very bulky expression of P_c. Therefore, it is omitted.

To simplify further analysis the biosensor action was analyzed at the steady state conditions when $\mathrm{d}S_c/\mathrm{d}t$ and $\mathrm{d}P_c/\mathrm{d}t$ equals zero. In this case

$$P_c = \frac{\rho' S_0}{1+\rho}, \tag{6}$$

where dimensionless parameters ρ and ρ' represent the ratios of the enzymatic reaction in the cell and the compounds transfer,

$$\begin{aligned} \rho &= \frac{V_{max,c}}{K_{M,c}k}, \\ \rho' &= \frac{V_{max,c}}{K_{M,c}k'}. \end{aligned} \tag{7}$$

The concentration P_c of the product produced in the cell is the basis for the calculation of the product concentration P_e at the transducer since

$$\frac{\mathrm{d}P_e}{\mathrm{d}t} = k' P_c - \frac{t_p P_e}{\delta}, \tag{8}$$

where t_p and δ are the transfer coefficient and the thickness of the semipermeable membrane.

At the steady state conditions, P_e is equal to

$$P_e = \frac{V_{max,c}}{K_{M,c}} \frac{\delta}{(1+\rho)t_p} S_0. \tag{9}$$

The response E of the biosensor measured as an electrode potential in the case of the potentiometric transducer is

$$E = E_0 + \frac{RT}{n_e F} \ln P_e = \frac{RT}{n_e F} \ln S_0 + C, \tag{10}$$

where E_0 is a characteristic constant for the ion-selective electrode, R is the universal gas constant, T is the absolute temperature (K), and C is a constant.

In the case of the amperometric transducer the expression of the response is more complex due to the product consumption at the electrode. This can be accounted for by adding a transfer rate (t_p/δ) and a constant of the electrochemical conversion (k_{el}). Therefore, the response of the biosensor can be calculated as

$$I = n_e F \frac{k_{el} k'}{t_p/\delta + k_{el}} P_c = n_e F \frac{k_{el} k'}{t_p/\delta + k_{el}} \frac{\rho'}{1+\rho} S_0. \tag{11}$$

2 BOD Biosensor

The special case of the biosensor based on microorganisms is the so called BOD (biochemical oxygen demand) biosensor. The BOD biosensor determinates organic compounds that can be oxidized by a microbial cell. These biosensors are used for water quality management, ecology and environmental science.

The BOD biosensor action was modeled assuming that the rate of oxygen consumption by the cells in the layer near the transducer (oxygen electrode) is determined by the substrate concentration [127]. If the rate of the substrate conversion in the cells is limited by the transport through the cell wall, the substrate concentration in the layer near the electrode surface is determined by the diffusion from the external solution and the transport into the cell. At a steady state

$$\frac{D_s}{d}\frac{S_0 - S}{d} = \frac{V_s S}{K_s + S}, \tag{12}$$

where D_s is the diffusion coefficient of the substrate through the semipermeable membrane, d is the thickness of the enzyme membrane, S_0 and S are the concentrations of the substrate in the external solution and in the layer near the oxygen electrode, respectively, V_s is the rate of transport through the cell membrane at saturating concentrations of the substrate, and K_s is the saturation constant.

The solution of the equation gives the expression for the relative concentration of the substrate in the layer near the oxygen electrode:

$$\frac{S}{S_0} = \frac{1 - K_s/S_0(1+\rho)}{2} + \left(\left(\frac{1 - K_s/S_0(1+\rho)}{2}\right)^2 + \frac{K_s}{S_0}\right)^{1/2} \tag{13}$$

where the dimensionless module $\rho = V_s d^2/D_s K_s$ quantitatively characterizes the diffusion limitations.

In the case of two extreme concentrations of the substrate the solution of (6) is simpler:

$$S_0 - S = \rho K_s \quad \text{at} \quad S_0 \gg K_s + \frac{V_s d^2}{D_s}, \tag{14}$$

and

$$S = \frac{S_0}{1+\rho} \quad \text{at} \quad S_0 \ll K_s + \frac{V_s d^2}{D_s}. \tag{15}$$

To calculate the stationary concentration O_2 of the oxygen it was assumed that the substrate flow through the membrane is equal to the oxygen flow:

$$\frac{D_s}{d}(S_0 - S) = \frac{1}{N}\frac{D_{O_2}}{d}(O_{2_0} - O_2), \tag{16}$$

where D_{O_2} is the diffusion coefficient of oxygen, N is the stoichiometric coefficient of the substrate conversion.

The decrease in the oxygen electrode current is proportional to the oxygen concentration change:

$$\frac{I_0 - I}{I_0} O_{2_0} = O_{2_0} - O_2 = N \frac{D_s}{D_{O_2}} (S_0 - S) . \tag{17}$$

At saturating substrate concentration

$$\frac{I_0 - I}{I_0} O_{2_0} = N \frac{V_s d^2}{D_{O_2}} . \tag{18}$$

At a low concentration of S_0

$$\frac{I_0 - I}{I_0} O_{2_0} = N \frac{D_s}{D_{O_2}} \frac{\rho}{1 + \rho} S_0 . \tag{19}$$

Thus, at a high substrate concentration the maximal decrease in the current is determined by the rate of transport into the cell and diffusion ratio. At low concentrations, the response of the sensor is directly proportional to S_0. The sensitivity varies with the value of the module. When $\rho \gg l$ the sensor operates under diffusive conditions and the sensitivity is maximal. At $\rho \ll l$ the action of the sensor is limited by the kinetic parameters and the sensitivity is determined by the rate of the transport of the substances into the cell,

$$\frac{I_0 - I}{I_0} O_{2_0} = N \frac{V_s d^2}{D_{O_2} K_s} S_0 . \tag{20}$$

The examination of the calibration curves for the microbial biosensor showed that a 50% consumption of oxygen took place at substrate (L-lactate) concentration much higher than K_s indicating an important role of diffusion [130].

Modeling Nonstationary State of Biosensors

Contents

1 Potentiometric Biosensors 33
2 Amperometric Biosensors 34

1 Potentiometric Biosensors

The biosensors response at a transition state can be modeled solving partial differential equations (PDE) of the substrates diffusion and the biocatalytical conversion with the initial and the boundary conditions. The analytical solutions, however, exist at very limited cases. The Laplace transformation that is typically used for solving the diffusion equations is no longer applicable for the solution of such problems. Therefore, for the modeling of the diffusion and enzymatic reactions the other methods of PDE solving are used.

Carr [61] used the Fourier method to solve (Chapter 1, eq. 7) at $S \ll K_M$ and $S \gg K_M$ with the initial and the boundary conditions: $S = P = 0$ at $0 < x < d$ and $t = 0$; $\partial S/\partial x = \partial P/\partial x = 0$ at $x = 0$; $S = S_0$, $P = 0$ at $x = d$.

At $S \ll K_M$ the kinetics of the substrate concentration change at the transducer surface is

$$P(0,t) = S_0\left(\left(1 - \frac{1}{\cosh(\alpha d)}\right) - \frac{4}{\pi}\exp\left(-\frac{\pi^2 D_e t}{4d^2}\right)\left(\frac{4 + (\alpha d\pi)^2\left(1 - \exp\left(-\frac{V_{max}}{K_M}t\right)\right)}{4 + (\alpha d\pi)^2}\right)\right), \quad (1)$$

where d is the thickness of the diffusion layer, $\alpha = \sqrt{V_{max}/K_M D_e}$.

R. Baronas et al., *Mathematical Modeling of Biosensors*, Springer Series on Chemical Sensors and Biosensors 9, DOI 10.1007/978-90-481-3243-0_5,

In the case of a very high substrate concentration (zero-order reaction rate), $S_0 \gg K_M$, the product concentration P_e at the electrode surface was found to be [61, 128]

$$P_e = P(0,t) = \frac{V_{max}d^2}{D_e}\left(\frac{1}{2} - \frac{16}{\pi^3}\right)\exp\left(-\frac{\pi^2 D_e t}{4d^2}\right). \tag{2}$$

The analysis of (1) reveals that if enzyme activity is high the exponential expression containing the V_{max} term is approaching zero. Therefore, (1) can be simplified,

$$P_e = S_0\left(\left(1 - \frac{1}{\cosh(\alpha d)}\right) - \frac{4}{\pi}\exp\left(-\frac{\pi^2 D_e t}{4d^2}\right)\right). \tag{3}$$

Introducing this expression into the Nernst equation produces the potentiometric biosensor response (25°C):

$$E = E_0 + 0.0591\log P_e = E_0 + 0.0591\log S_0\left(\left(1 - \frac{1}{\cosh(\alpha d)}\right)\right.$$
$$\left.-\frac{4}{\pi}\exp\left(-\frac{\pi^2 D_e t}{4d^2}\right)\right). \tag{4}$$

This shows that the dynamics of the biosensor response weekly depends on the enzyme activity. If the diffusion module $\alpha d \gg 1$ the biosensor will achieve the difference of 0.1 or 1.0 mV to the steady state response during a time ($\tau_{0.1}$ or $\tau_{1.0}$):

$$\tau_{0.1} = \frac{2.35d^2}{D_e}, \qquad \tau_{1.0} = \frac{1.42d^2}{D_e}. \tag{5}$$

As an example, the response of potentiometric biosensor that is different by 0.1 mV from the steady state will be achieved during 78 s at the membrane thickness of 0.01 cm and $D_e = 3 \times 10^{-6}\,\mathrm{cm}^2/\mathrm{s}$. The double increase of the membrane thickness will increase this time up to 313 s. The response time of the real biosensors, with the diffusion module about 1 is 18–26% longer.

2 Amperometric Biosensors

The dynamics of the amperometric biosensors was modeled assuming that there exist no outer diffusion limitations and the conversion of the compound in the biocatalytic membrane follows the first-order reaction kinetics [133]. The dynamics of three types of biosensors in which the consecutive, cyclic and parallel conversion of the substrates proceeds was calculated. The consecutive conversion of the substrates in the biocatalytic membranes takes place according to the following scheme:

$$\mathrm{S} \xrightarrow{k_1} \mathrm{Z} \xrightarrow{k_2} \mathrm{P} \tag{6}$$

where S is the substrate, Z is the intermediate, P is the electrochemically active product, and k_1 and k_2 are the first order rate constants (V_{max}/K_M) of the enzymatic conversion of the compounds.

The scheme for the cyclic substrate conversion is

$$\mathrm{S} \xrightarrow{k_1} \mathrm{Z} \xrightarrow{k_2} \mathrm{S} + \mathrm{P} \tag{7}$$

The parallel conversion proceeds according to

$$\mathrm{P}_1 \xleftarrow{k_1} \mathrm{S} \xrightarrow{k_2} \mathrm{P}_2 \tag{8}$$

where P_1 is the only electrochemically active product.

The mass transport and the consecutive substrates conversion (6) in the biocatalytical membrane are described by the following system of the reaction–diffusion equations:

$$\begin{aligned}
\frac{\partial S}{\partial t} &= D_e \frac{\partial^2 S}{\partial x^2} - k_1 S\,, \\
\frac{\partial Z}{\partial t} &= D_e \frac{\partial^2 Z}{\partial x^2} - k_2 Z + k_1 S\,, \\
\frac{\partial P}{\partial t} &= D_e \frac{\partial^2 P}{\partial x^2} + k_2 Z\,,
\end{aligned} \tag{9}$$

where S, Z, P correspond to the concentrations of the substrate, the intermediate and the product, respectively. Equations 9 together with the initial ($S = 0, Z = 0, P = 0$ when $0 \le x < d$, $t = 0$) and the boundary ($S = S_0, Z = 0, P = 0$ when $x \ge d$, $t \ge 0$; $\partial S/\partial x = 0$, $\partial Z/\partial x = 0$, $P = 0$ when $x = 0, t \ge 0$) conditions form together the initial boundary value problem.

The cyclic substrate conversion (7) is described by

$$\begin{aligned}
\frac{\partial S}{\partial t} &= D_e \frac{\partial^2 S}{\partial x^2} - k_1 S + k_2 Z\,, \\
\frac{\partial Z}{\partial t} &= D_e \frac{\partial^2 Z}{\partial x^2} + k_1 S - k_2 Z\,, \\
\frac{\partial P}{\partial t} &= D_e \frac{\partial^2 P}{\partial x^2} + k_2 Z\,.
\end{aligned} \tag{10}$$

The initial and the boundary conditions for this problem were assumed to be identical to those used in case of the consecutive substrates conversion.

The following equations describe the parallel substrate conversion (9):

$$\begin{aligned}\frac{\partial S}{\partial t} &= D_e\frac{\partial^2 S}{\partial x^2} - (k_1 + k_2)S\,,\\ \frac{\partial P_1}{\partial t} &= D_e\frac{\partial^2 P_1}{\partial x^2} + k_1 S\,,\\ \frac{\partial P_2}{\partial t} &= D_e\frac{\partial^2 P_2}{\partial x^2} + k_2 S\,,\end{aligned} \tag{11}$$

where P_1 and P_2 stand for the concentrations of the products P_1 and P_2, respectively. The following initial and the boundary conditions were assumed for the system of the parallel substrates conversion: $S = 0, P_1 = 0, P_2 = 0$ when $0 \le x < d, t = 0$; $\partial S/\partial x = 0$, $P_1 = 0$, $\partial P_2/\partial x = 0$ when $x = 0, t \ge 0$; $S = S_0, P_1 = 0, P_2 = 0$ when $x \ge d, t \ge 0$.

The systems (9)–(11) together with the initial and the boundary conditions were solved using the Green function.

The evaluation of the system (6) for the consecutive conversion gives the following expression ($k_1 \ne k_2$) for the biosensor response (the current density):

$$\begin{aligned} i(t) &= n_e F D_e S_0 \frac{4k_1k_2}{d^2}\\ &\times \sum_{n=1}^{\infty}\sum_{m=1}^{\infty} \frac{(-1)^{m+1}\gamma_{2m-1}\Phi(t)}{(k_1+\gamma_{2m-1}^2 D_e)(k_2+\gamma_{2m-1}^2 D_e)(4\gamma_n^2-\gamma_{2m-1}^2)}\,,\end{aligned} \tag{12}$$

where $\gamma_{2m-1} = (2m-1)\pi/2d$, $\gamma_n = n\pi/2d$, and function $\Phi(t)$ takes the following form:

$$\begin{aligned}\Phi(t) = 1 &- \frac{(k_1+\gamma_{2m-1}^2 D_e)(k_2+\gamma_{2m-1}^2 D_e)\exp(-4\gamma_n^2 D_e t)}{(4\gamma_n^2 D_e - \gamma_{2m-1}^2 D_e - k_1)(4\gamma_n^2 D_e - \gamma_{2m-1}^2 D_e - k_2)}\\ &- \frac{4\gamma_n^2 D_e(k_2+\gamma_{2m-1}^2 D_e)\exp(-(k_1+\gamma_{2m-1}^2 D_e)t)}{(k_2-k_1)(4\gamma_n^2 D_e - \gamma_{2m-1}^2 D_e - k_1)}\\ &+ \frac{4\gamma_n^2 D_e(k_1+\gamma_{2m-1}^2 D_e)\exp(-(k_2+\gamma_{2m-1}^2 D_e)t)}{(k_2-k_1)(4\gamma_n^2 D_e - \gamma_{2m-1}^2 D_e - k_2)}\,.\end{aligned}$$

It is possibly to notice a kinetic indistinguishability of the constants k_1 and k_2 for the consecutive conversion. At a high value of k_1 the dynamics of the response is identical to that of a monoenzyme electrode characterized by the rate constant k_2. Generally, the kinetic behavior of the electrode is determined by the diffusion modules $\alpha_1 d = (k_1/D_e)^{1/2}d$ and $\alpha_2 d = (k_2/D_e)^{1/2}d$, and at high values the processes proceed at the same rates as the substance transfer through the inert membrane (Table 1). At a low catalytic activity of one of the enzyme the response decreases more than four times.

Table 1 Proportionality coefficient a in the equation of 95% response ($t_{95} = ad^2/D_e$) of biosensors containing dual enzyme system [133]

Consecutive Conversion		Cyclic Conversion		Parallel Conversion	
$\alpha_1 d(\alpha_2 d \to \infty)$ or $\alpha_2 d(\alpha_1 d \to \infty)$	a	βd	a	βd	a
→0	1.577	→0	1.986	→0	1.577
0.1	1.571	0.1	1.984	0.1	1.571
0.3	1.554	0.3	1.937	0.3	1.554
0.5	1.443	0.5	1.885	0.5	1.443
0.7	1.336	0.7	1.819	0.7	1.336
1	1.158	1	1.728	1	1.158
2	0.707	2	1.497	2	0.707
4	0.393	4	1.364	4	0.393
5	0.379	5	1.345	5	0.379
7	0.374	7	1.329	7	0.374
10	0.374	10	1.320	10	0.374
20	0.372	20	1.314	20	0.372
∞	0.371	∞	1.312	∞	0.371

The transient behavior of the cyclic conversion is described as

$$i(t) = n_e F D_e \frac{4k_1 k_2 S_0}{d^2} \times \sum_{n=1}^{\infty} \sum_{m=1}^{\infty} \frac{(-1)^{m+1}\Phi(t)}{\gamma_{2m-1}(4\gamma_n^2 D_e - \gamma_{2m-1}^2 D_e)(\gamma_{2m-1}^2 D_e + k_1 + k_2)}, \quad (13)$$

where

$$\begin{aligned}\Phi(t) = 1 &- \frac{(\gamma_{2m-1}^2 D_e + k_1 + k_2) 4\gamma_n^2 D_e \exp(-\gamma_{2m-1}^2 D_e t)}{(k_1 + k_2)(4\gamma_n^2 D_e - \gamma_{2m-1}^2 D_e)} \\ &+ \frac{(\gamma_{2m-1}^2 D_e + k_1 + k_2)\gamma_{2m-1}^2 D_e \exp(-4\gamma_n^2 D_e t)}{(k_1 + k_2 - 4\gamma_n^2 D_e + \gamma_{2m-1}^2 D_e)(4\gamma_n^2 D_e - \gamma_{2m-1}^2 D_e)} \\ &- \frac{4\gamma_n^2 D_e \gamma_{2m-1}^2 D_e \exp(-(\gamma_{2m-1}^2 D_e + k_1 + k_2)t)}{(k_1 + k_2)(k_1 + k_2 - 4\gamma_n^2 D_e + \gamma_{2m-1}^2 D_e)}.\end{aligned}$$

In limiting cases, the solution of (13) becomes simpler. When $k_1 + k_2 > D_e/d^2$ and $m = 1$, the current is expressed as

$$i(t) = I - n_e F D_e \frac{k_1 k_2 S_0}{(k_1 + k_2)d} \sum_{n=1}^{\infty} \frac{32 \exp\left(-\left(\frac{\pi}{2d}\right)^2 D_e t\right) \gamma_n^2 D_e}{\pi(4\gamma_n^2 D_e - (\pi/2d)^2 D_e)^2}, \quad (14)$$

where I is the steady state response of the biosensor,

$$I = n_e F D_e \frac{S_0}{d} \frac{k_1 k_2}{k_1 + k_2} \left(\frac{d^2}{2D_e} - \frac{1}{k_1 + k_2} \left(1 - \frac{1}{\cosh \beta d} \right) \right), \tag{15}$$

where $\beta = ((k_1 + k_2)/D_e)^{1/2}$.

So, it follows that at $\beta d \geq 20$ a 95% response is practically constant and has a minimal value of $1.31d^2/D_e$ (Table 1), i.e. during the cyclic substrate conversion even at high enzyme rate the kinetics of biosensor response is 3.5–5.4 times slower than the substance diffusion through the membrane.

In the second limiting case when $k_1 + k_2 < D_e/d^2$, (13) takes the following form:

$$i(t) = I - n_e F D_e \frac{k_1 k_2 S_0}{(k_1 + k_2)d} \sum_{n=1}^{\infty} \frac{32\gamma_n^2 D_e \Phi(t)}{\pi[4\gamma_n^2 D_e - (\pi/2d)^2 D_e]} \tag{16}$$

where

$$\Phi(t) = \frac{\exp(-(\pi/2d)^2 D_e t)}{4\gamma_n^2 D_e - (\pi/2d)^2 D_e} + \frac{(\pi/2d)^2 D_e \exp(-(\pi/2d)^2 D_e t + k_1 t + k_2 t)}{((\pi/2d)^2 D_e + k_1 + k_2)((\pi/2d)^2 D_e + k_1 + k_2 - 4\gamma_n^2 D_e)}.$$

Summing over the first 10 terms of the series in this equation indicates that when $\beta d < 0.1$ the response is reached more slowly, $t_{95} = 1.99d^2/D_e$ (Table 1).

The solution of the transient behavior of the biosensor with a parallel substrate conversion is :

$$i(t) = n_e F D_e \frac{S_0 4 k_1}{d^2} \sum_{n=1}^{\infty} \sum_{m=1}^{\infty} \frac{(-1)^{m+1} \gamma_{2m-1} \Phi(t)}{(k_1 + k_2 + \gamma_{2m-1}^2 D_e)(4\gamma_n^2 - \gamma_{2m-1}^2)}, \tag{17}$$

where

$$\Phi(t) = 1 + \frac{(k_1 + k_2 + \gamma_{2m-1}^2 D_e) \exp(-4\gamma_n^2 D_e t)}{4\gamma_n^2 D_e - \gamma_{2m-1}^2 D_e - k_1 - k_2} - \frac{4\gamma_n^2 D_e \exp(-(k_1 + k_2 + \gamma_{2m-1}^2 D_e)t)}{4\gamma_n^2 D_e - \gamma_{2m-1}^2 D_e - k_1 - k_2}$$

As it follows from this equation, the response time of the biosensor is equal to that of the monoenzyme electrodes characterized by the constant value $(k_1 + k_2)$ (Table 1), i.e. the conversion of a part of the substrate into the electrochemically inactive product results in the same decrease of the stationary and non-stationary current. Therefore the response time of the biosensor remains unchanged.

The results of mathematical modeling were compared with the experimental data [133]. The comparison showed that the kinetics of the biosensor response depends on the biocatalytical schemes and rates of the biocatalytical conversion as predicted by the models.

Limitations of Analytical Solutions

The examples of the biosensors modeling by analytical solution of differential equations illustrate the possibilities and limitations of the method. Additional examples of the application of this approach to the biosensors modeling can be found in reviews [44, 168, 227, 235]. The greatest advantage of an aproximal analytical solution is a possibility of getting the analytical solution of PDE in some limited cases. The disadvantages of the method are

- Limited concentration interval of reactive components.
- Not applicable to biosensors with complex biocatalytical schemes.
- The solution of non stationary state is very complex.
- Practically it is impossible to get the solution for complex initial and boundary conditions.

The practical biosensors, however, contain multiplex geometry. The substrates are converted following the complex schemes. For this reason the quasi-steady state kinetics is no longer applicable. Modern modeling of the biosensors should cover all concentration intervals and transition state. These problems can be solved by the numerical modeling of the biosensors.

Part II
Numerical Modeling of Biosensors

Abstract This is the part two of the book Mathematical Modeling of Biosensors. The part is dedicated to the modeling of biosensors by numerical solution of non-linear reaction-diffusion problems using the finite difference technique. This part contains mathematical models of amperometric, potentiometric and optical biosensors acting in the batch as well as the injection modes. A special emphasis is placed to modeling biosensors utilizing a complex biocatalytical conversion and biosensors with multipart transducers geometry and biocatalytical membranes structure. By changing model parameters the output results are numerically analyzed with a special emphasis to the influence of the biosensor geometry and the catalytical parameters on the sensitivity of the biosensor response at transition and steady state conditions.

Keywords Biosensor · Diffusion · Modeling · Reaction · Simulation

Analytical solutions for reaction–diffusion problems involving nonlinear terms are known only in very limited cases [164, 165, 168, 210, 235]. Mell and Maloy (1975) postulated a numerical approach to simulate the steady state amperometric measurements for an enzyme electrode [180, 181]. Since then, the reaction–diffusion problems describing biochemical processes are often solved numerically [1, 3, 4, 25, 26, 30, 49, 56, 122, 151, 162, 192, 208, 262, 269, 276, 285]. The analytical solutions are often applied to validation of the corresponding numerical solutions.

The finite difference technique is a widely used numerical method for solving the reaction–diffusion problems. That technique is elaborated in many references [5, 222] and in the next part of this book.

In this part of the book the nonlinear modeling of the biosensor action. The corresponding reaction–diffusion problems are solved using the finite difference technique.

Mono-Layer Mono-Enzyme Models of Biosensors

Contents

1 Mathematical Model of an Amperometric Biosensor 44
1.1 Governing Equations 44
1.2 Initial and Boundary Conditions 45
1.3 Dimensionless Model 46
2 Characteristics of the Biosensor Response 47
2.1 Biosensor Current 47
2.2 Biosensor Sensitivity 48
2.3 Maximal Gradient of the Current 48
2.4 Response Time 49
3 Finite Difference Solution 50
3.1 Numerical Approximation of Equations 50
3.2 Calculation Procedure 51
3.3 Validation of Numerical Solution 53
3.4 Numerical Error Analysis 57
4 Peculiarities of the Biosensor Response 60
4.1 Effect of the Enzyme Membrane Thickness 60
4.2 Stability of the Response 63
4.3 The Response Versus the Substrate Concentration 64
4.4 The Response Versus the Maximal Enzymatic Rate 66
4.5 Choosing the Enzyme Membrane Thickness 68
4.6 Biosensor Resistance 70
4.7 Maximal Gradient of the Current 71
5 Flow Injection Analysis 72
5.1 Mathematical Model 73
5.2 Numerical Solution 73
5.3 Biosensor Response 75
5.4 Peculiarities of the Biosensor Response 77
5.5 Sequential Injection Analysis 80
6 Biosensors with Chemical Amplification 81
6.1 Mathematical Model 82
6.2 Finite Difference Solution 84
6.3 Concentration Profiles 85
6.4 Peculiarities of the Biosensor Response 86
7 Potentiometric Biosensors 91
7.1 Mathematical Model 91
7.2 Biosensor Response 92
7.3 Finite Difference Solution 93

R. Baronas et al., *Mathematical Modeling of Biosensors*, Springer Series on Chemical Sensors and Biosensors 9, DOI 10.1007/978-90-481-3243-0_6,

7.4 Validation of Numerical Solution 94
7.5 Simulated Biosensor Response 95
7.6 Peculiarities of the Biosensor Response 98
8 Enzyme Inhibition 103
8.1 Substrate Inhibition 103
8.2 Effect of Substrate Inhibition 105
8.3 Product Inhibition 108
8.4 Effect of Product Inhibition 109

A membrane biosensor may be considered as an electrode, having a layer of an enzyme applied onto the electrode surface [69]. Consider a scheme where the substrate (S) combines reversibly with the enzyme (E) to form a complex (ES). The complex then dissociates into the reaction product (P) and the enzyme is released [65, 78, 229, 258],

$$\mathrm{S + E \rightleftarrows ES \rightarrow E + P} \tag{1}$$

Assuming the quasi steady state approximation, the concentration of the intermediate complex (ES) does not change and may be neglected when simulating the biochemical behaviour of the biosensors [229, 235, 258]. According to the simplified scheme, the substrate (S) binds to the enzyme (E) and is converted to the product (P),

$$\mathrm{S} \xrightarrow{\mathrm{E}} \mathrm{P} \tag{2}$$

1 Mathematical Model of an Amperometric Biosensor

Consider the biosensor as the enzyme electrode having a layer of the enzyme immobilized onto the surface of the probe [85, 229, 258]. Assuming the symmetrical geometry of the electrode and homogeneous distribution of the immobilized enzyme in the enzyme layer of a uniform thickness, a mathematical model of the biosensor action can be expressed by a system of the reaction–diffusion equations.

1.1 Governing Equations

Coupling the enzyme-catalyzed reaction in the enzyme layer (enzyme membrane) with the one-dimensional-in-space diffusion, described by Fick's law, leads to the following equations [43, 235]:

$$\begin{aligned}\frac{\partial S}{\partial t} &= D_S\frac{\partial^2 S}{\partial x^2} - \frac{V_{max}S}{K_M + S},\\ \frac{\partial P}{\partial t} &= D_P\frac{\partial^2 P}{\partial x^2} + \frac{V_{max}S}{K_M + S}, \quad x \in (0, d), \quad t > 0,\end{aligned} \tag{3}$$

where x and t stand for space and time, respectively, $S(x,t)$ is the concentration of the substrate, $P(x,t)$ is the concentration of the reaction product, d is the thickness of the enzyme layer, D_S, D_P are the diffusion coefficients, V_{max} is the maximal enzymatic rate, and K_M is the Michaelis constant. The Michaelis constant K_M is the concentration of the substrate at which half the maximum velocity of the enzyme-catalyzed reaction is achieved [55, 185].

1.2 Initial and Boundary Conditions

Let $x = 0$ represent the electrode surface, while $x = d$ is the boundary between the analyzed solution and the enzyme membrane. Initially, no substrate as well as product appears inside the enzyme layer. The operation of the biosensor starts when the substrate appears over the surface of the enzyme membrane. This is expressed in the initial conditions ($t = 0$)

$$\begin{aligned} S(x,0) &= 0, \quad x \in [0,\ d), \\ S(d,0) &= S_0\,, \end{aligned} \tag{4}$$

$$\begin{aligned} P(x,0) &= 0, \quad x \in [0,\ d), \\ P(d,0) &= P_0\,, \end{aligned} \tag{5}$$

where S_0 and P_0 are the concentrations of the substrate and the product in the bulk solution, respectively. Usually, the zero concentration of the reaction product in the bulk is assumed, $P_0 = 0$.

In the scheme (2), the product (P) is an electro-active substance. In the case of the amperometric biosensors, due to the electrode polarization the concentration of the reaction product at the electrode surface ($x = 0$) is being permanently reduced to zero,

$$P(0,t) = 0, \quad t > 0. \tag{6}$$

At the electrode surface the substrate does not react. Because of this, at the electrode surface, the non-leakage (zero flux) boundary condition is defined for the substrate,

$$D_S \frac{\partial S}{\partial x}\bigg|_{x=0} = 0, \quad t > 0. \tag{7}$$

Assuming the finite diffusivity D_S of the substrate, the boundary condition (7) reduces to the following one:

$$\frac{\partial S}{\partial x}\bigg|_{x=0} = 0, \quad t > 0. \tag{8}$$

If the bulk solution is well-stirred and in a powerful motion then the diffusion layer ($0 < x < d$) remains at a constant thickness. Consequently, the concentration of the substrate as well as of the product over the enzyme surface (bulk solution/membrane interface) is assumed constant during the biosensor operation.

$$S(d,t) = S_0, \quad t > 0, \tag{9}$$

$$P(d,t) = P_0, \quad t > 0. \tag{10}$$

1.3 Dimensionless Model

In order to define the main governing parameters of the mathematical model (3)–(10), the dimensional variables (x and t) and unknown concentrations (S and P) are replaced with the following dimensionless parameters:

$$\hat{x} = \frac{x}{d}, \quad \hat{t} = \frac{D_S t}{d^2}, \quad \hat{S} = \frac{S}{K_M}, \quad \hat{P} = \frac{P}{K_M}, \tag{11}$$

where $\hat{x}$ stands for the dimensionless distance from the electrode surface, $\hat{t}$ is the dimensionless time, $\hat{S}$ and $\hat{P}$ are the dimensionless concentrations of the substrate and the product, respectively.

The governing equations (3) in the dimensionless coordinates $\hat{x}$ and $\hat{t}$ are expressed as follows:

$$\begin{aligned} \frac{\partial \hat{S}}{\partial \hat{t}} &= \frac{\partial^2 \hat{S}}{\partial \hat{x}^2} - \sigma^2 \frac{\hat{S}}{1+\hat{S}}, \\ \frac{\partial \hat{P}}{\partial \hat{t}} &= \frac{D_P}{D_S}\frac{\partial^2 \hat{P}}{\partial \hat{x}^2} + \sigma^2 \frac{\hat{S}}{1+\hat{S}}, \quad \hat{x} \in (0,\ 1), \quad \hat{t} > 0, \end{aligned} \tag{12}$$

where σ^2 is the dimensionless diffusion module which is also known as the Damköhler number [10],

$$\sigma^2 = \frac{V_{max} d^2}{K_M D_S}. \tag{13}$$

The diffusion module σ^2 compares the rate of enzyme reaction (V_{max}/K_M) with the mass transport through the enzyme layer (D_S/d^2).

Assuming (11), the initial conditions (4) and (5) transform to the following conditions:

$$\begin{aligned} &\hat{S}(\hat{x},0) = 0, \quad \hat{P}(\hat{x},0) = 0, \quad \hat{x} \in [0,\ 1), \\ &\hat{S}(1,\ 0) = \hat{S}_0, \quad \hat{P}(1,0) = \hat{P}_0, \end{aligned} \tag{14}$$

where $\hat{S}_0$ and $\hat{P}_0$ are the dimensionless concentrations of the substrate and the product in the bulk, respectively, $\hat{S}_0 = S_0/K_M$, $\hat{P}_0 = P_0/K_M$.

The boundary conditions (6)–(10) are rewritten as follows ($\hat{t} > 0$):

$$\hat{P}(0,\hat{t}) = 0\,, \quad \left.\frac{\partial \hat{S}}{\partial \hat{x}}\right|_{\hat{x}=0} = 0\,, \qquad\qquad (15)$$
$$\hat{S}(1,\hat{t}) = \hat{S}_0\,, \quad \hat{P}(1,\hat{t}) = \hat{P}_0\,.$$

Assuming the same diffusion coefficients for both species (the substrate and the product, $D_S = D_P$) and the zero concentration of the product in the bulk ($\hat{P}_0 = 0$), only the following two dimensionless parameters remain in the dimensionless mathematical model (12)–(15): the dimensionless substrate concentration $\hat{S}_0$ in the bulk solution and the diffusion module σ^2. The advent of the diffusion module σ^2 is one of the most important outcomes of defining a dimensionless model of the biosensor action. The diffusion module is the main parameter expressing all internal characteristics of the biosensor.

2 Characteristics of the Biosensor Response

2.1 Biosensor Current

The measured current is accepted as a response of the biosensor in physical experiments. The anodic current depends upon the flux of the electro-active substance (product) at the electrode surface, i.e. on the border $x = 0$. In the case of the amperometry, the biosensor current is also directly proportional to the area of the electrode surface. The anodic current $i_A(t)$ of the amperometric biosensor at time t can be obtained explicitly from the Faraday and the Fick laws,

$$i_A(t) = n_e F A D_P \left.\frac{\partial P}{\partial x}\right|_{x=0}, \qquad (16)$$

where n_e is a number of electrons involved in a charge transfer, A is the electrode surface area and F is the Faraday constant , $F = 96{,}485\,\mathrm{C/mol}$.

Due to the direct proportionality, the current is normalized with the area of that surface. Consequently, the density $i(t)$ of the biosensor current at time t is expressed as follows:

$$i(t) = \frac{i_A}{A} = n_e F D_P \left.\frac{\partial P}{\partial x}\right|_{x=0}. \qquad (17)$$

The system (3)–(10) approaches the steady state as $t \to \infty$,

$$I = \lim_{t\to\infty} i(t). \qquad (18)$$

I is taken as the density of the steady state biosensor current.

Assuming (11) the dimensionless density $\hat{i}$ of the current (flux) and the corresponding dimensionless density $\hat{I}$ of the steady state current are defined as follows:

$$\hat{i}(\hat{t}) = \frac{\partial \hat{P}_e}{\partial \hat{x}}\bigg|_{\hat{x}=0} = \frac{i(t)\,d}{n_e F D_P K_M}, \quad \hat{I} = \lim_{\hat{t}\to\infty} \hat{i}(\hat{t}). \tag{19}$$

2.2 Biosensor Sensitivity

The sensitivity is also one of the most important characteristics of the biosensors [66, 100, 123, 229, 258, 275]. The biosensor sensitivity can be expressed as the gradient of the steady state current with respect to the substrate concentration. Since the biosensor current as well as the substrate concentration varies even in orders of magnitude, especially when comparing different sensors, another useful parameter to consider is a dimensionless sensitivity. The dimensionless sensitivity for the substrate concentration S_0 is given by

$$B_S(S_0) = \frac{S_0}{I(S_0)} \times \frac{\mathrm{d}I(S_0)}{\mathrm{d}S_0}, \tag{20}$$

where B_S stands for the dimensionless sensitivity of the amperometric biosensor and $I(S_0)$ is the density of the steady state biosensor current calculated at the substrate concentration S_0. B_S varies between 0 and 1. Let us remind the reader that S_0 denotes the substrate concentration in the bulk solution, see (9).

2.3 Maximal Gradient of the Current

The maximal gradient of the biosensor current calculated with respect to the time is another common characteristic of the biosensor action [229]. This type of measurement is rather often used in the biosensors operating in the transient mode. Since the biosensor current as well as the time varies even in orders of magnitude, the dimensionless maximal gradient is used to compare different sensors. The dimensionless maximal gradient that varies between 0 and 1 is given by

$$B_G = \max_{i(t)>0} \left\{ \frac{t}{i(t)} \times \frac{\mathrm{d}i(t)}{\mathrm{d}t} \right\}, \tag{21}$$

where B_G is the dimensionless maximal gradient of the biosensor current with respect to the time. Due to a possible delay in rising the current, the condition $i(t) > 0$ avoids a possible indeterminacy when calculating the maximal gradient B_G.

2.4 Response Time

The time interval from the beginning of the biosensor action up to the moment of the current measured is called the biosensor response time. The moment of the measurement depends on the type of the device.

Devices operating in the stationary mode usually use the time when the absolute current slope value falls below a given small value. Since the biosensor current varies even in orders of magnitude, the current is usually normalized with the current value. In other words, the time T needed to achieve a given dimensionless decay rate ε is accepted as the response time,

$$T = \min_{i(t)>0} \left\{ t : \frac{t}{i(t)} \left| \frac{\mathrm{d}i(t)}{\mathrm{d}t} \right| < \varepsilon \right\}. \tag{22}$$

However, the response time T as the approximate time of the steady state is very sensitive to the decay rate ε, $T \to \infty$ when $\varepsilon \to 0$. Because of this, the time of the partial stationary current is usually used. The resultant relative output signal function $i^*(t)$ is expressed as follows:

$$i^*(t) = \frac{i(t)}{I}, \tag{23}$$

Let us notice, that for the amperometric biosensors

$$\forall t : t > 0 : \quad 0 \le i^*(t) \le 1; \quad i^*(0) = 0, \quad i^*(T) \approx 1. \tag{24}$$

The half-time $T_{0.5}$ of the steady state or the half of the time moment of occurrence of the steady state current is often used instead of the time T to evaluate the dynamics of the biosensor operation,

$$T_{0.5} = \{t : i^*(t) = 0.5\}. \tag{25}$$

The times $T_{0.9}$ and $T_{0.95}$ needed to reach 90% and 95% of the stationary current, respectively, are also rather often used as the response times, $i^*(T_{0.9}) = 0.9$, $i^*(T_{0.95}) = 0.95$.

In the case of the transient operational mode, the slope $\mathrm{d}i(t)/\mathrm{d}t$ is used to find out the maximal gradient (21) of the current [229]. Then, the time T_G of the maximal derivative of $i(t)$ is accepted as the response time,

$$T_G = \left\{ t : \frac{t}{i(t)} \times \frac{\mathrm{d}i(t)}{\mathrm{d}t} = B_G \right\}, \tag{26}$$

where B_G is the maximal gradient defined by (21).

3 Finite Difference Solution

For many reaction–diffusion problems involving nonlinear terms, the exact solutions are impossible. Therefore, the problem (3)–(10) is often solved numerically. The finite difference technique is the most widely used numerical method for solving the reaction–diffusion problems [5, 222].

To find a numerical solution of the problem in the domain $[0, d]\times[0, T]$ a discrete grid is introduced. The easiest way is to use a uniform discrete grid

$$\begin{aligned}\omega_h &= \{x_i : x_i = ih,\ i = 0, 1, \dots, N,\ hN = d\},\\ \omega_\tau &= \{t_j : t_j = j\tau,\ j = 0, 1, \dots, M;\ \tau M = T\}.\end{aligned} \tag{27}$$

where ω_h and ω_τ are the uniform discrete grids for space and time, respectively.

The following notation is used in the finite difference approximations presented below:

$$\begin{aligned}&S_i^j = S(x_i, t_j), \quad P_i^j = P(x_i, t_j), \quad i_j = i(t_j),\\ &i = 0, \dots, N, \quad j = 0, \dots, M.\end{aligned} \tag{28}$$

3.1 *Numerical Approximation of Equations*

There are several kinds of the finite difference schemes available for the approximation of the reaction–diffusion equations. The degree of the implicity is the main difference between them. The implicit scheme and the explicit one are two widely used schemes in practice [49, 56]. The next part of this book considers in detail the finite difference approximation of the model (3)–(10). In this chapter only two simple finite difference schemes are shortly introduced.

3.1.1 Explicit Scheme

The partial differential equations (3) can be approximated with the following explicit scheme:

$$\begin{aligned}&\frac{S_i^{j+1} - S_i^j}{\tau} = D_S \frac{S_{i+1}^j - 2S_i^j + S_{i-1}^j}{h^2} - \frac{V_{max} S_i^j}{K_M + S_i^j},\\ &\frac{P_i^{j+1} - S_i^j}{\tau} = D_P \frac{S_{i+1}^j - 2P_i^j + P_{i-1}^j}{h^2} + \frac{V_{max} S_i^j}{K_M + S_i^j},\\ &i = 1, \dots, N-1, \quad j = 0, \dots, M-1.\end{aligned} \tag{29}$$

3.1.2 Implicit Scheme

When approximating the reaction–diffusion equations (3) with an implicit scheme, the following finite difference equations are obtained:

$$\begin{aligned}
&\frac{S_i^{j+1}-S_i^j}{\tau}=D\frac{S_{i+1}^{j+1}-2S_i^{j+1}+S_{i-1}^{j+1}}{h^2}-\frac{V_{max}S_i^j}{K_M+S_i^j},\\
&\frac{P_i^{j+1}-P_i^j}{\tau}=D\frac{P_{i+1}^{j+1}-2P_i^{j+1}+P_{i-1}^{j+1}}{h^2}+\frac{V_{max}S_i^j}{K_M+S_i^j},\\
&i=1,\dots,N-1,\quad j=0,\dots,M-1\,.
\end{aligned} \tag{30}$$

3.1.3 Approximation of Initial and Boundary Conditions

The initial conditions (4) and (5) are approximated as follows:

$$\begin{aligned}
S_i^0=0,\quad i=0,\dots,N-1,\\
S_N^0=S_0,
\end{aligned} \tag{31}$$

$$\begin{aligned}
P_i^0=0,\quad i=0,\dots,N-1,\\
P_N^0=P_0,
\end{aligned} \tag{32}$$

The boundary conditions (9)–(10) are approximated as follows:

$$S_0^j=S_1^j,\quad S_N^j=S_0,\quad j=1,\dots,M, \tag{33}$$

$$P_0^j=P_1^j,\quad P_N^j=P_0,\quad j=1,\dots,M. \tag{34}$$

3.2 Calculation Procedure

The calculation of the numerical solution starts at the layer $t=t_0=0$ with applying Eqs. 31–32. When the solution on the layer t_j $(j=0,1,\dots,M-1)$ has been calculated, the solution on the next layer $t=t_{j+1}$ can be calculated by using (29) or (30). When using the implicit scheme (30), the systems of linear algebraic equations can be efficiently solved in both steps because of the tridiagonality of the matrices of the systems.

Having the numerical solution of the problem, the biosensor current at time $t=t_j$ is calculated by

$$i(t_j) \approx i_j = n_e F D_P P_1^j / h, \quad j = 1, \dots, M. \tag{35}$$

In the numerical simulation, the response time T was assumed as the time when the absolute biosensor current slope value falls below a given small dimensionless decay rate ε normalized with the current value,

$$T \approx \min_{i_j > 0} \left\{ t_j : \frac{t}{i_j} \times \frac{|i_j - i_{j-1}|}{\tau} < \varepsilon \right\}. \tag{36}$$

In calculations, $\varepsilon = 10^{-2}$ can be successfully used.

The half-time $T_{0.5}$ of the steady state or a half of the time moment of occurrence of the steady state current, can be calculated after the approximate steady state was reached. Having the calculated stationary current I and the transitional currents at all the times t_j, $t_j < T$, $j = 1, \dots, M$, $T_{0.5}$ is calculated as follows:

$$T_{0.5} \approx \min_{1 \le j \le M} \{ t_j : i_j \ge 0.5 I \}. \tag{37}$$

$T_{0.9}$, $T_{0.95}$ and any other time T_α are calculated very similarly,

$$T_\alpha \approx \min_{1 \le j \le M} \{ t_j : i_j \ge \alpha I \}. \tag{38}$$

where α is between 0 and 1.

The maximal gradient B_G of the current is calculated as

$$B_G \approx \max_{j=1,\dots,M,\, i_j > 0} \left\{ \frac{t_j}{i_j} \times \frac{i_j - i_{j-1}}{\tau} \right\} = \max_{j=1,\dots,M,\, i_j > 0} \left\{ j \times \frac{i_j - i_{j-1}}{i_j} \right\}. \tag{39}$$

Having the maximal gradient B_G, the corresponding time T_G can be obtained from

$$T_G \approx \left\{ t_j : \frac{t_j}{i_j} \times \frac{i_j - i_{j-1}}{\tau} = B_G, \quad i_j > 0, \quad j = 1, \dots, M \right\}. \tag{40}$$

The calculation of the biosensor sensitivity by (20) requires the simulation of the biosensor at least two values of the substrate concentration S_0. To find out the biosensor sensitivity at appropriate value of S_0, the response has to calculated for the other concentration close to the first one,

$$B_S(S_0) \approx \frac{S_0}{I(S_0)} \times \frac{I(S_0 + o(S_0)) - I(S_0)}{o(S_0)}, \tag{41}$$

where $o(S_0)$ is a relatively small part of S_0, e.g. 5% from the concentration to be analyzed, i.e $o(S_0) = 0.05 S_0$.

The calculation of the other characteristics of the biosensor, such as the steady state current (18), the maximal gradient (21), the response times (25) and (26) is rather easy.

3.3 Validation of Numerical Solution

Before applying the numerical solution to proceed with an investigation of the modeling process, the solution has to be validated. The usual method of validating the numerical solution is the use of some special limits of the input parameters for which analytic solutions are available. When applying this method, the numerical solution in the limits is tested against the associated analytical solutions.

Exact analytical solutions of the reaction–diffusion problems exist only for the linear cases of the reaction term. Considerable attention is given in the literature on the development analytical as well as numerical methods of solving the reaction–diffusion problems [201, 235].

The nonlinear term in the system (3) of the reaction–diffusion equations is called the Michaelis–Menten function. In certain cases the Michaelis–Menten function approaches the linear functions.

3.3.1 First-Order Reaction Rate

When the concentration S_0 to be measured is very small in comparison with the Michaelis constant K_M,

$$\forall x, t : x \in [0, d],\ t > 0 : \quad 0 < S(x,t) < S_0 \ll K_M, \tag{42}$$

the nonlinear Michaelis–Menten function $V(S)$ simplifies to that of the first order,

$$V(S) = \frac{V_{max} S}{S + K_M} \approx \frac{V_{max} S}{K_M} = kS, \tag{43}$$

where k is the linear enzyme kinetic coefficient. Practically, the enzyme reaction can be considered of the first order when the concentration of the detected species is below one-fifth of K_M, i.e. $S_0 < 0.2K_M$, [99]. This case is rather typical for biosensors with high enzyme loading factor.

Assuming the approximation (43), the initial boundary value problem (3)–(10) can be solved analytically [73,232,235]. The concentration profiles for both species, the substrate and the product, are expressed as follows:

$$S(x,t) = S_0 \left(1 - \frac{4}{\pi} \sum_{n=0}^{\infty} \frac{\sin((2n+1)(d-x)\pi/(2d))}{2n+1} \times \frac{k + ue^{-(k+u)t}}{k+u} \right), \tag{44}$$

with $u = D_S(2n+1)^2\pi^2/(4d^2)$,

$$
\begin{aligned}
P(x,t) = S_0 \frac{2k}{\pi} \sum_{m=1}^{\infty} \Bigg[& \sin(m(d-x)\pi/d) \Bigg(\frac{1-(-1)^m}{mw} (1-e^{wt}) \\
& + \frac{4(-1)^m}{\pi} \sum_{n=0}^{\infty} \Bigg(\frac{(-1)^n}{(2n+1)(k+u)} \\
& \times \left(\frac{k(1-e^{-wt})}{w} + \frac{u(e^{-(k+u)t} - e^{-wt})}{w-k-u} \right) \times \frac{4m^2}{4m^2-(2n+1)^2} \Bigg) \Bigg) \Bigg],
\end{aligned}
\tag{45}
$$

with $w = D_P m^2 \pi^2 / d^2$.

The corresponding steady state concentrations are [10, 62, 128]

$$
\begin{aligned}
S_{ss}(x) &= \lim_{t\to\infty} S(x,t) = S_0 \times \frac{\cosh(\alpha x)}{\cosh(\alpha d)}, \\
P_{ss}(x) &= \lim_{t\to\infty} P(x,t) \\
&= S_0 \times \frac{D_S}{D_P} \times \frac{(1/\cosh(\alpha d)(1-x/d) + x/d - \cosh(\alpha x)}{\cosh(\alpha d)},
\end{aligned}
\tag{46}
$$

where

$$
\alpha = \sqrt{\frac{k}{D_S}} = \sqrt{\frac{V_{max}}{K_M D_S}}. \tag{47}
$$

Having the concentration of the product, the density of the biosensor current has been calculated as follows [232]:

$$
\begin{aligned}
i(t) = & -n_e F D_P 2k \frac{S_0}{d} \times \sum_{m=1}^{\infty} \Bigg[\frac{(-1)^m - 1}{w} \left(1 - e^{wt}\right) \\
& + \frac{4m}{\pi} \sum_{n=0}^{\infty} \Bigg(\frac{(-1)^n}{(2n+1)(k+u)} \times \frac{4m}{4m^2-(2n+1)^2} \\
& \times \left(\frac{k(1-e^{-wt})}{w} + \frac{u(e^{-(k+u)t} - e^{-wt})}{w-k-u} \right) \Bigg) \Bigg].
\end{aligned}
\tag{48}
$$

Assuming $t \to \infty$, the corresponding density of the steady state current of the amperometric biosensors is calculated from $i(t)$ [128],

$$
I = n_e F D_P \frac{S_0}{d} \left(1 - \frac{1}{\cosh(\alpha d)} \right). \tag{49}
$$

Figures 1 and 2 show some results of the numerical simulation obtained in the case of the first-order reaction rate. The numerical results were obtained by applying the implicit finite difference scheme (30) and were compared with the

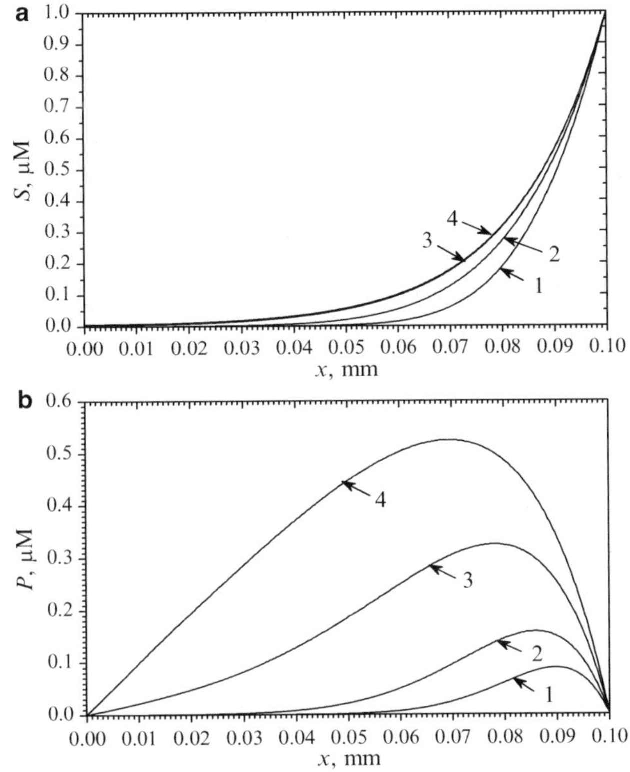

Fig. 1 Concentration profiles of the substrate (**a**) and the product (**b**) at the following values of time t: 0.5 (1), 1 (2), 3 (3) s and the steady state (4), $S_0 = 1\,\mu M = 0.01K_M$, $D_S = D_P = 300\,\mu m^2/s$, $d = 0.1$ mm, $K_M = 100\,\mu M$, $V_{max} = 100\,\mu M/s$

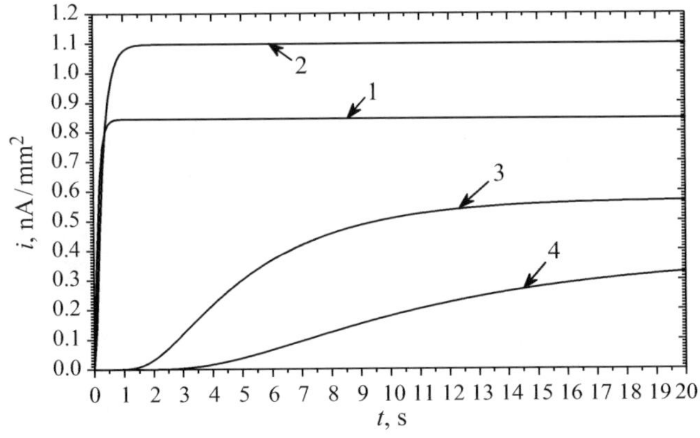

Fig. 2 Dynamics of the biosensor current at four values of the thickness d: 0.01 (1), 0.015 (2), 0.1 (3), 0.15 (4) mm. The other parameters are the same as in Fig. 1

corresponding analytical ones. No noticeable difference between the associated results was observed. So, Figs. 1 and 2 also show the analytical solutions (44)–(46) and (48), respectively.

One can see in Fig. 1 that at relatively large time $t = 3$ s (curve 3) the profile of the substrate concentration practically coincides with that at the steady state conditions (curve 4), while the corresponding concentrations of the product differ notably.

The evolution of the density $i(t)$ of the biosensor current is presented in Fig. 2. The biosensor response was modeled for the biosensors having different enzyme membrane thickness d: 0.01, 0.015, 0.1 and 0.15 mm. One can see in Fig. 2 that the biosensor current appears with some delay at relatively thick enzyme layers. This delay increases with the increase of the enzyme membrane thickness. When comparing the evolution of the biosensor current in two cases of relatively thin ($d = 0.01$ and 0.015 mm) membranes, one can see that the biosensor response is noticeably higher at the thicker membrane ($d = 0.015$ mm) than at the thinner one ($d = 0.01$ mm). However, when comparing the biosensor responses in the other two cases of the ten times thicker ($d = 0.1$ and 0.15 mm) membranes, one can see the opposite tendency, i.e. the biosensor of the thicker ($d = 0.15$ mm) membrane generates lower response than the thinner one ($d = 0.1$ mm).

3.3.2 Zero-Order Reaction Rate

In the case of a very high substrate concentration,

$$\forall x, t: \; x \in [0,\, d],\; t > 0: \quad K_M \| S_0 > S(x,t) > 0, \tag{50}$$

the nonlinear Michaelis–Menten function $V(S)$ reduces to that of the zero order,

$$V(S) = \frac{V_{max} S}{S + K_M} \approx V_{max}. \tag{51}$$

The inequalities (50) are valid for a low enzyme loading factor and large substrate concentrations. In these cases, the analytical solution of the initial boundary value problem (3)–(10) is [62, 235],

$$S(x,t) = S_0 - \frac{4}{\pi} \sum_{n=0}^{\infty} \left(\frac{\sin((2n+1)(d-x)\pi/(2d))}{2n+1} \right. \\ \left. \times \left(V_{max} \left(\frac{1 - e^{-wt}}{w} \right) + S_0 e^{-wt} \right) \right), \tag{52}$$

$$P(x,t) = \frac{4 V_{max}}{\pi} \sum_{n=0}^{\infty} \left(\frac{\sin((2n+1)(d-x)\pi/(2d))}{2n+1} \times \frac{1 - e^{-vt}}{v} \right), \tag{53}$$

where

$$w = D_S(2n+1)^2\pi^2/(4d^2), \tag{54}$$

$$v = D_P(2n+1)^2\pi^2/d^2. \tag{55}$$

The corresponding stationary concentrations are:

$$S_{ss}(x) = S_0 + \frac{V_{max}\left(x^2 - d^2\right)}{2D_S}, \tag{56}$$

$$P_{ss}(x) = V_{max}\frac{x(d-x)}{2D_P}. \tag{57}$$

The densities of the current and of the corresponding stationary current are

$$i(t) = n_e F D_P \frac{4V_{max}}{d}\sum_{n=0}^{\infty}\frac{1-e^{-vt}}{v}, \tag{58}$$

$$I = \frac{n_e F V_{max} d}{2}. \tag{59}$$

Figure 3 shows the concentration profiles calculated numerically by implicit scheme (30) as well as analytically by (52)–(57).

When comparing the concentration profiles obtained at the zero-order reaction rate (Fig. 3) with those obtained at the first-order rate (Fig. 1), one can see a noticeable difference in the shape of the curves.

The evolution of the density $i(t)$ of the biosensor current is presented in Fig. 4. The biosensor response was modeled for the biosensors having the same values of the thickness d as in the case of the first-order reaction rate (see Fig. 2).

The biosensor current appears with some delay at relatively thick enzyme layers. This delay increases with the increase in the enzyme membrane thickness. When comparing the evolution of the biosensor current for different thicknesses of the enzyme membrane, one can see that the steady state response is always higher at the thicker membrane than at the thinner one. The direct proportionality of the steady state current on the membrane thickness was defined in (59). In the case of zero-order reaction rate (Fig. 4) the behaviour of the biosensor current importantly differs from that in the case of first-order reaction rate (Fig. 2).

3.4 Numerical Error Analysis

In the numerical simulation, the numerical error combines the calculation errors of two main kinds [110]. The first is caused by the finite precision of the computations involving the floating-point or the integer values. The second (usually called the truncation error) is defined as the difference (or relative difference) between the exact mathematical solution and the approximate solution obtained when simplifications are introduced into the mathematical equations.

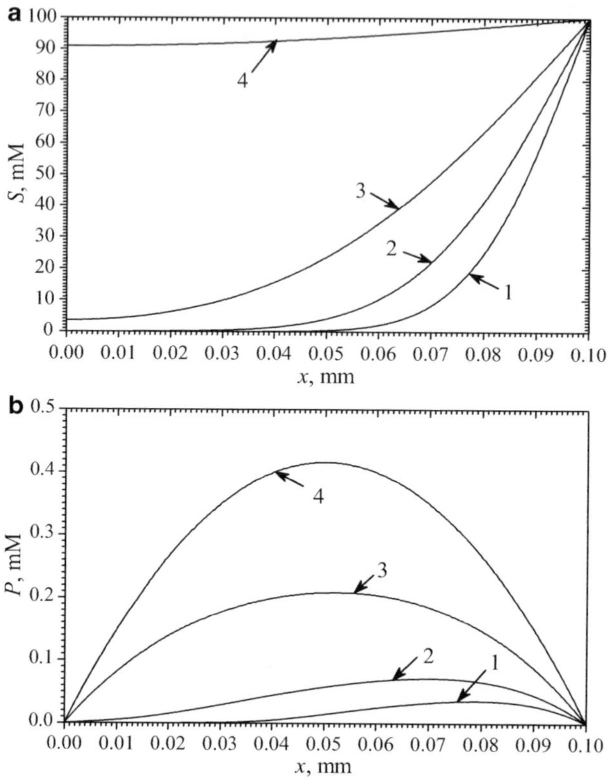

Fig. 3 Concentration profiles for the substrate (**a**) and product (**b**) at the following values of time t: 0.5 (1), 1 (2), 3 (3) s and steady state (4) in the case of zero-order reaction rate when $S_0 = 100\,\text{mM} = 1{,}000\ K_M$. The other parameters are the same as in Fig. 1

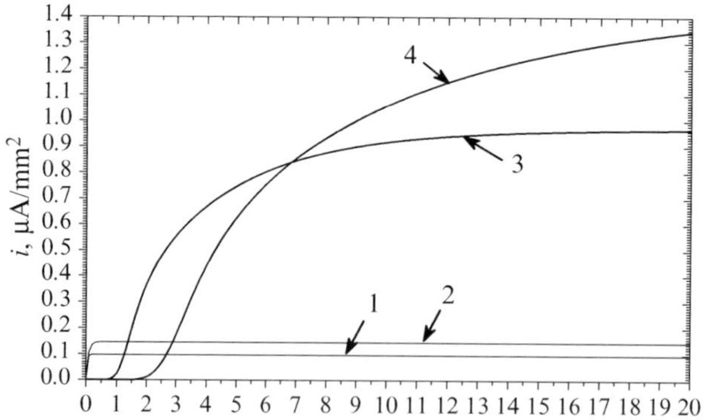

Fig. 4 Dynamics of the biosensor current at four values of the thickness d: 0.01 (1), 0.015 (2), 0.1 (3), 0.15 (4) mm in the case of zero-order reaction rate. The other parameters are the same as in Fig. 3

To decrease the numerical error of the first kind, the constants and the variables of the precision as high as possible, i.e. of the double or even the quadruple precision, have to be used.

Much more factors affect the numerical error of the second kind. The kind of the finite difference scheme to be used in the approximation of the mathematical model is among them. The next part of this book considers the truncation error of the finite difference schemes in detail.

The explicit finite difference schemes are much easier to program [209], however, the implicit ones are more precise and often more efficient. Although the processing speed of modern computers is high enough to ensure practical use of explicit schemes, the use of the faster implicit scheme is especially well justified when investigating the peculiarities of the biosensor response at wide ranges of the model parameters.

When building a computer simulation program, the validation (testing) of the software is of crucial importance. As has been mentioned in the previous section, the validation can be established by using the known analytical solution. Assuming the analytical solution as a true value of the biosensor response, the relative error E_C of the computer simulation is introduced,

$$E_C = \frac{|I_C - I|}{I}, \tag{60}$$

where I is the analytical solution (the density of the steady state current) of the problem, and I_C is the result (also the current density) of the computer simulation for that problem. If found value E_C is less than a small value then the program produces the right result at least at values of the parameters used in the analytical as well as the numerical calculations. The value of the relative error E_C decreases with increasing the number of points in the discrete grid (27) in both coordinates: space x and time t, i.e. with the increase of N and M introduced in (27).

For the initial boundary value problem (3)–(10) analytical solutions for two extreme cases of the substrate concentration S_0 are known, one for $S_0 \ll K_M$ and another for $S_0 \gg K_M$. At a relatively low concentration S_0 of the substrate, $S_0 \ll K_M$, the density I of the steady state current of the amperometric biosensor was defined in (49). At very high concentrations of the substrate, $S_0 \gg K_M$, the stationary current docs not depend on the substrate concentration and was defined in (59). It is more reliable to test the program at the substrate concentrations from both those extreme intervals.

Figure 5 shows some results of testing the simulation software programmed in JAVA language. The relative error E_C was calculated for two different values of S_0: $0.001K_M$ and $100K_M$. The values of the other parameters of the mathematical model were as follows:

$$\begin{aligned} &D_S = D_P = 300\,\mu\text{m}^2/\text{s}, \quad n_e = 2, \quad d = 0.1\,\text{mm} \\ &K_M = 100\,\mu\text{M}, \quad V_{max} = 100\,\mu\text{M/s}. \end{aligned} \tag{61}$$

The multiple numerical simulation was carried out changing the step size of the discrete grid (27) and using both finite difference schemes: implicit (30) and

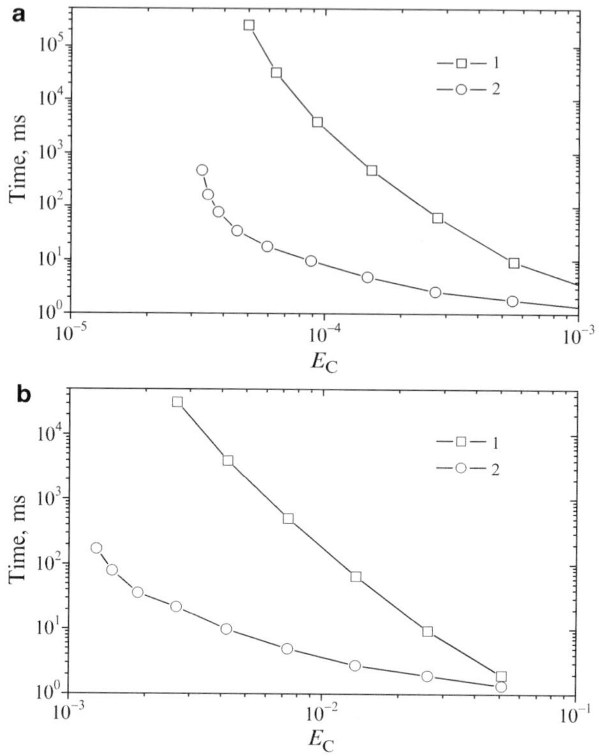

Fig. 5 Dependence of the calculation time on the relative error E_C for explicit (1) and implicit (2) schemes in the case of $S_0 \ll K_M$ (**a**) and $S_0 \gg K_M$ (**b**)

explicit (29). Figure 5 shows how the simulation time increases with the decrease in the relative error E_C. One can see in this figure that the implicit scheme is more efficient than the explicit one when very high accuracy of calculations is required.

4 Peculiarities of the Biosensor Response

Using numerical simulation the peculiarities of the biosensor response can be investigated.

4.1 *Effect of the Enzyme Membrane Thickness*

Figure 6 shows the density I of the steady state current while Fig. 7 presents the half time $T_{0.5}$ of the steady state versus the thickness d of the enzyme membrane. Figure 6 also presents the corresponding densities of the steady state current calculated analytically using (49).

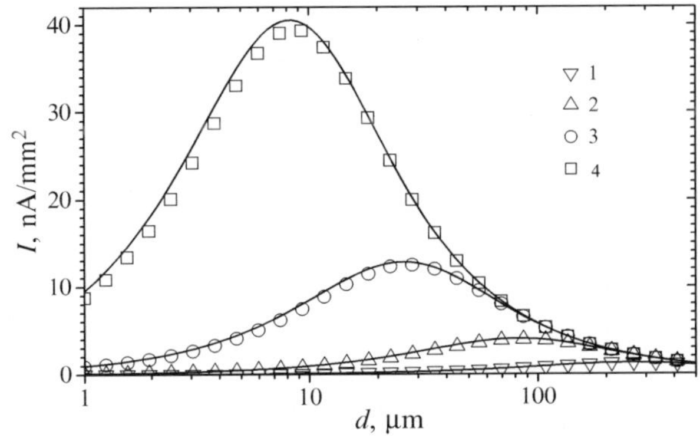

Fig. 6 The dependence of the steady state biosensor current I on the thickness d of the enzyme membrane at four maximal enzymatic rates V_{max}: 1 (1), 10 (2), 100 (3) and 1,000 (4) μM/s, $S_0 = 10\,\mu$M. Symbols are numerical solutions, while curves are analytical ones (formula (49))

One can see in Fig. 6 that the density I of the steady state biosensor current is a non-monotonous function of d at all values of the maximal enzymatic rate V_{max}. The higher maximal enzymatic rate V_{max} corresponds to the greater value of I. The results obtained by the numerical simulation shows that the maximum of I equals about 39.4 nA/mm^2 at $V_{max} = 1$ mM/s, while the maximal I is approximately equal to 12.5 nA/mm^2 at ten times smaller value of $V_{max} = 100\,\mu$M/s. The higher maximum of I corresponds to the thinner enzyme membrane. In the case of $V_{max} =$ 1 mM/s, the maximum of $I(d)$ gains at $d \approx 8.6\,\mu$m, while in the case of $V_{max} =$ 100 μM/s, the maximum of I gains at $d \approx 27\,\mu$m. Since $S_0 = 10\,\mu\text{M} = 0.1 K_M$, formula (49) can used to find analytically the membrane thickness d at which the steady state current gains the maximum at given values of n_e, D_S, S_0, V_{max}, K_M, where $S_0 \ll K_M$. At first, a derivative of $I(d)$ with respect to the thickness d is calculated [27],

$$\frac{\mathrm{d}I(d)}{\mathrm{d}d} = n_e F D_P S_0 \frac{-\cosh^2(\sigma) + \cosh(\sigma) + \sigma \sinh(\sigma)}{d^2 \cosh^2(\sigma)}. \tag{62}$$

Then the value of σ at which that derivative gets zero is found,

$$-\cosh^2(\sigma) + \cosh(\sigma) + \sigma \sinh(\sigma) = 0. \tag{63}$$

Equation 63 was solved numerically. The unique non-zero solution $\sigma = \sigma_{max} \approx$ 1.5055 was obtained. Sequentially, I gains the maximum at the membrane thickness d_{max}, where

$$d_{max} = \frac{1}{\sigma_{max}} \sqrt{\frac{D_S K_M}{V_{max}}}, \quad \sigma_{max} = 1.5055. \tag{64}$$

Accepting (61) and (64), one can calculate that $d_{max} \approx 8.25\,\mu$m and $I \approx 40\,$nA/mm^2 at $V_{max} = 1\,$mM/s; $d_{max} \approx 26.1\,\mu$m and $I \approx 12.8\,$nA/mm^2 at $V_{max} = 100\,\mu$M/s. These values compare sufficiently well with the corresponding values obtained by the numerical simulation of the biosensor operation. The corresponding values of the density I of the steady state current as well as thickness d varies by about 5 %. The difference in these values appears because of the substrate concentration $S_0 = 0.1\,K_M$. The analytical solution (49) is valid at $S_0 \ll K_M$ only, while the numerical one does not have such kind of restrictions at all. Because of this, values of d_{max} obtained by using the computer simulation at $S_0 = 10\,\mu$M and (61) are, in general, more accurate than the analytical ones.

Consequently, the steady state biosensor current as a function (49) of the membrane thickness d gains the maximum when the diffusion module σ equals $\sigma_{max} = 1.5055$. According to (13) and (64), the thickness d_{max} does not depend on the substrate concentration S_0. Nevertheless, using the numerical simulation, the following values of σ_{max} were obtained at four values of S_0: $\sigma_{max} \approx 1.51$ at $S_0 = 0.01\,\mu$M, $\sigma_{max} \approx 1.52$ at $S_0 = 1\,\mu$M, $\sigma_{max} \approx 1.57$ at $S_0 = 10\,\mu$M and $\sigma_{max} \approx 2.1$ at $S_0 = 100\,\mu$M. The diffusion module σ_{max} is approximately constant at $S_0 \ll K_M$, so that it is about coincident with the value obtained from the analytical solution (49). d_{max} increases with the increase in the substrate concentration S_0. That is especially notable at the substrate concentrations $S_0 \approx K_M$.

In the case of $S_0 \gg K_M$, the stationary current increases linearly with the increase in the thickness d (see (59)), i.e. $I \to \infty$ when $d_{max} \to \infty$ and $S_0 \gg K_M$. The dependence of d_{max} on V_{max} is fairly low, d_{max} varies by less than 3.5% when V_{max} changes from 1 to 1000 μM/s at any value of S_0 mentioned above.

Figure 7 shows, that the half time $T_{0.5}$ of the steady state increases notably with the increase in the thickness d of the enzyme membrane. That increase is slightly exponential. The half time $T_{0.5}$ reaches even one hundred seconds at $d = 400\,\mu$m

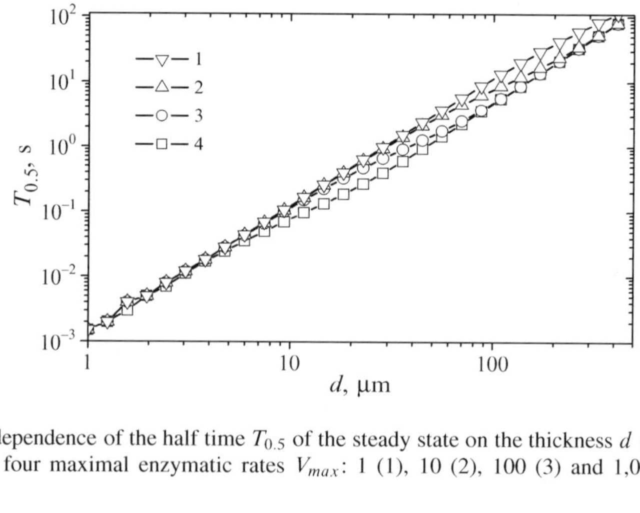

Fig. 7 The dependence of the half time $T_{0.5}$ of the steady state on the thickness d of the enzyme membrane at four maximal enzymatic rates V_{max}: 1 (1), 10 (2), 100 (3) and 1,000 (4) μM/s, $S_0 = 10\,\mu$M

while $T_{0.5}$ varies about only 0.1 s at $d = 10\,\mu m$. The effect of V_{max} on the response time is rather slight. A greater enzymatic activity V_{max}, i.e. a higher reaction rate, corresponds to a shorter response time. $T_{0.5}$ equals 11.4 s at $V_{max} = 1\,\mu M/s$, while $T_{0.5} = 4.7$ s at $V_{max} = 1{,}000\,\mu M/s$.

4.2 *Stability of the Response*

The stability of the response is one of the most critical characteristics of the biosensors [206]. It is very important to keep the analytical capability of the biosensors for as long as possible period. Usually, the maximal enzymatic rate V_{max} decreases permanently due to the enzyme inactivation. In general, the biosensor response is sensitive to changes in V_{max}. Figure 6 shows that the maximal biosensor current differs by orders of magnitude when changing V_{max}. The variation is especially notable in the cases of relatively thin enzyme membranes. In the case of relatively thick enzyme membrane, the steady state current does not practically vary by changing V_{max}. Consequently, the biosensor containing the thicker enzyme layer gives more stable response than the biosensor with the thinner layer. However, the thick membrane-based biosensors have very durable response time (see Fig. 7). Because of this the relatively thick biosensors are of limited applicability, e.g. in the flow injection systems, which are widely used for determination of various compounds [220].

Thus, the problem of the membrane thickness optimization arises. The task is to find the thickness d of the membrane as small as possible, ensuring the stability of the biosensor response at a range of V_{max} as wide as possible. Let V_1 and V_2 be two values of the maximal enzymatic rate, for which a stable biosensor response to the substrate of the concentration of S_0 is required. The minimal membrane thickness $d_\varepsilon(V_1, V_2, S_0)$ is introduced as the relative difference $R(d, V_1, V_2, S_0)$ between the biosensor response (the density I of the steady state biosensor current) at the thickness $d = d_\varepsilon$, $V_{max} = V_1$ and another response at $d = d_\varepsilon$, $V_{max} = V_2$, is less than the dimensionless decay rate ε,

$$R_d(d, V_1, V_2, S_0) = \frac{|I(d, V_1, S_0) - I(d, V_2, S_0)|}{I(d, V_1, S_0)}, \tag{65}$$

$$d_\varepsilon(V_1, V_2, S_0) = \min_{d>0}\{d : R_d(d, V_1, V_2, S_0) < \varepsilon\}, \tag{66}$$

where $I(d, V_{max}, S_0)$ is the density of the steady state biosensor current calculated at the membrane thickness d, the maximal enzymatic rate V_{max} and the substrate concentration S_0.

Let us assume $S_0 = 10\,\mu M$, $V_1 = 100\,\mu M/s$, $V_2 = 1{,}000\,\mu M/s$ and $\varepsilon = 0.01$. The numerical results presented in Fig. 6 and some additional calculations show that d_ε approximately equals $92\,\mu m$. To evaluate the biosensor stability at a wide

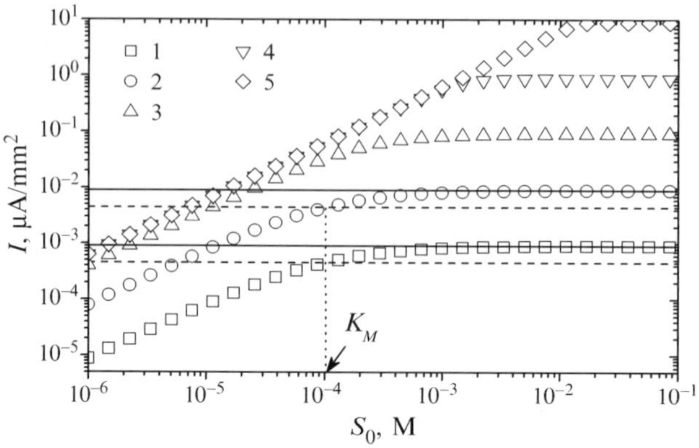

Fig. 8 The dependence of the density I of the steady state current on the concentration S_0 of the substrate at five maximal enzymatic rates V_{max}: 0.1 (1), 1 (2), 10 (3), 100 (4) and 1,000 (5) μM/s, $d = d_\varepsilon(100\,(\mu M/s), 1,000\,(\mu M/s), 10\,(\mu M) = 92\,\mu m$, calculated assuming $\varepsilon = 0.01$

range of the substrate concentration S_0, the response of the biosensor based on the membrane of the thickness $d = d_\varepsilon(V_1, V_2, S_0) = 92\,\mu m$ was calculated at different values of S_0 from this range. The following sections discuss this in detail.

4.3 The Response Versus the Substrate Concentration

Figure 8 shows the density I of the steady state current of the biosensor of the membrane thickness $d = 92\,\mu m$ versus the substrate concentration S_0 at the following five values of V_{max}: 0.1 (1), 1 (2), 10 (3), 100 (4) and 1000 (5) μM/s. One can see in this figure no noticeable difference between the values of I calculated at two values of V_{max}: 100 and 1,000 μM/s when the substrate concentration S_0 is less than about 10^{-3} M. Figure 9 explicitly shows the stable response of the biosensor based on the enzyme membrane of the thickness $d = 92\,\mu m$ when the maximal enzymatic rate reduces ten times: from 1,000 to 100 μM/s. Although the membrane thickness d_ε of 92 μm has been calculated at the substrate concentration $S_0 = 10^{-5}$ M, the biosensor response is sufficiently stable to the substrate of the concentration being up to about 10^{-3} M. The dependence of d_ε on the substrate concentration has already been noticed above. The biosensor response is especially sensitive to changes in V_{max} at high concentrations of the substrate. Figure 9 shows that the response of the biosensor of thickness $d = 92\,\mu m$ is approximately constant at the concentration higher than about 10^{-2} M. Because of this, such a biosensor is not practically useful to determine the substrate of the concentration larger than 10^{-2} M.

Figure 9 shows the effect of the substrate concentration S_0 on the half time $T_{0.5}$ of the steady state biosensor response. The thickness d of the enzyme membrane is the same as above, i.e. $d = d_\varepsilon = 92\,\mu m$. One can see in Fig. 9 that $T_{0.5}$ is

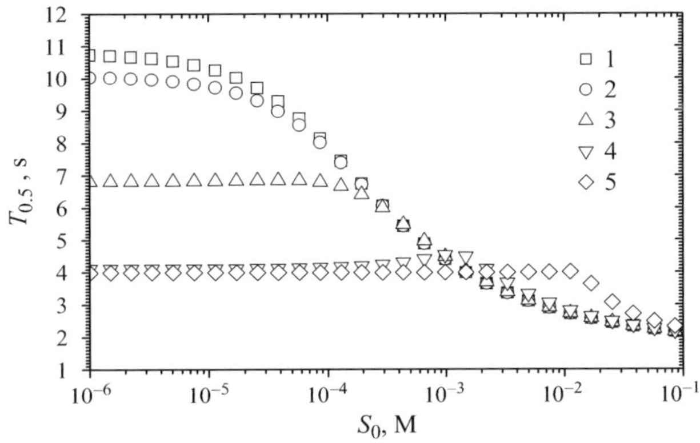

Fig. 9 The dependence of the half time $T_{0.5}$ of steady state biosensor response on the concentration S_0 of the substrate at five maximal enzymatic rates V_{max}: 0.1 (1), 1 (2), 10 (3), 100 (4) and 1000 (5) μM/s. The other parameters are the same as in Fig. 8

a monotonously decreasing function of S_0 at $V_{max} = 0.1, 1, 10\,\mu$M/s, and $T_{0.5}$ is a non-monotonic function of S_0 at two largest values of V_{max}: 100 and 1,000 μM/s. At S_0 being between 10^{-4} and 10^{-2} M, a shoulder on the curve appears for those two enzymatic rates. It seems possible that the shoulder on the curve arises because of the high V_{max}. At the substrate concentration $S_0 \gg K_M$ the reaction kinetics is of the zero order throughout the membrane whereas for $S_0 \ll K_M$ the kinetics is of the first order throughout. At intermediate values of the substrate concentration S_0 the kinetics changes from the zero order to the first order across the membrane [43, 46, 99, 129].

The biosensor based on the enzyme membrane of the thickness of 92 μm, gives a very stable response (Fig. 8) in a sufficiently short time (Fig. 9) when V_{max} is between 100 and 1,000 μM/s as well as the substrate concentration S_0 is less than about 10^{-3} M.

The Michaelis constant K_M is known to be the substrate concentration at which the reaction rate is half of its maximal value. Figure 8 shows the halving effect for two values of V_{max}: 0.1 and 1 μM/s. The half of the maximum of the steady state current is reached at substrate concentration of about $K_M = 0.1$ mM for both values of V_{max}. The relative difference between the half of the maximum of steady state current and the maximal current at K_M does not exceed 0.3% at $V_{max} = 0.1$ and 3% at $V_{max} = 1\,\mu$M/s. The maximum of the density I of the steady state current is equal to about 0.89 and to about 8.9 nA/mm^2 at the maximal enzymatic rate of 0.1 and 1 μM/s, respectively.

The effect of halving is not valid for the higher maximal enzymatic rates. Accepting the thickness d of 92 μm the diffusion module σ equals unity at $V_{max} = 3.5\,\mu$M/s. In the cases when $V_{max} > 3.5\,\mu$M/s the biosensor response is under the diffusion control. The effect of halving is not valid when the biosensor response is under the diffusion control. For example, in the case of $V_{max} = 10\,\mu$M/s ($\sigma \approx 1.7$) the half of the maximum of the steady state current is reached at

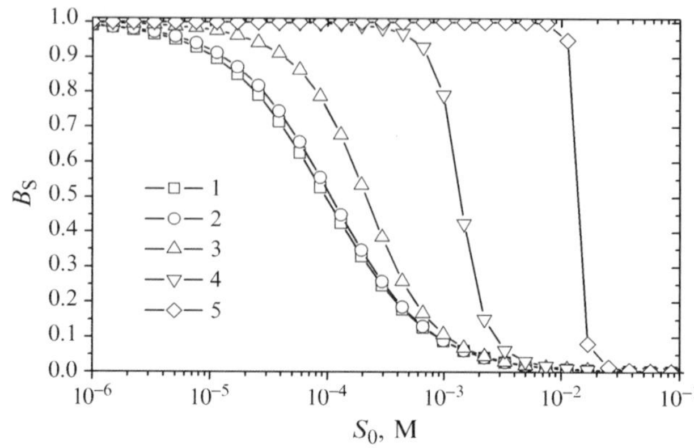

Fig. 10 The normalized biosensor sensitivity B_S versus the substrate concentration S_0. The parameters and notations are the same as in Fig. 8

$S_0 \approx 160\,\mu M = 1.6K_M$. If $V_{max} = 100\,\mu M/s$ ($\sigma \approx 5.3$) then the half of the maximum of steady state current is reached at $S_0 \approx 7.6\,K_M$.

Reaching the half of the maximum of steady state current at larger concentrations of the analyte usually means the longer calibration curve of the biosensor response. This is also expressed by the biosensor sensitivity presented in Fig. 10. One can see in this figure that the higher maximal enzymatic rate corresponds to the longer line close to the unity and consequently better selectivity. The zero biosensor sensitivity ($B_S = 0$) at the concentration S_0 of the substrate means that the change in the concentration does not effect the biosensor current, i.e. the current remains stagnant changing the concentration of the analyte. If the normalized biosensor sensitivity B_S is near to zero at the substrate concentration S_0 then the biosensor is not practically useful to recognize the analyte of such a concentration.

Figures 8 and 10 show that the biosensors based on the enzyme of the higher activity V_{max} are applicable to a wider range of substrate concentrations rather than those of the lower activity. This peculiarity of the amperometric biosensors can be reformulated in a more common way as follows: the biosensors characterized with the greater diffusion module are applicable to a wider range of the substrate concentrations rather than those of the smaller diffusion module.

4.4 The Response Versus the Maximal Enzymatic Rate

The effect of the maximal enzymatic rate V_{max} on the biosensor response at different concentrations (S_0) of the substrate was also investigated. The results of the computer simulation are depicted in Figs. 11 and 12. The calculations were done at the thickness $d = 100\,\mu m$ of the enzyme layer and the following five values of the concentration S_0: 10^{-6} (1), 10^{-5} (2), 10^{-4} (3), 10^{-3} (4) and 10^{-2} (5) M.

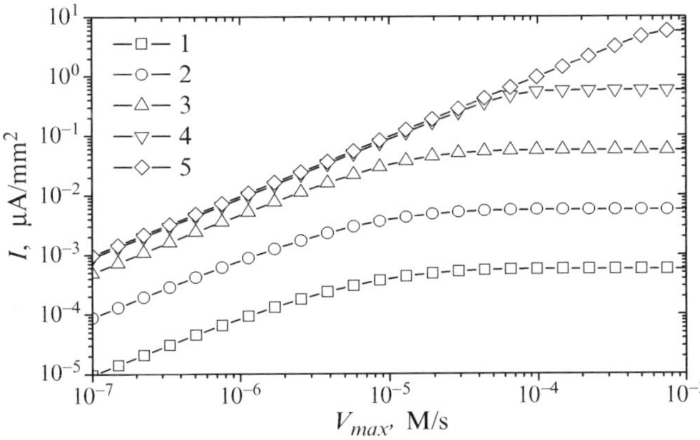

Fig. 11 The density I of steady state current versus the maximal enzymatic rate V_{max} at five concentrations S_0 of the substrate: 10^{-6} (1), 10^{-5} (2), 10^{-4} (3), 10^{-3} (4) and 10^{-2} (5) M, d = 100 μm

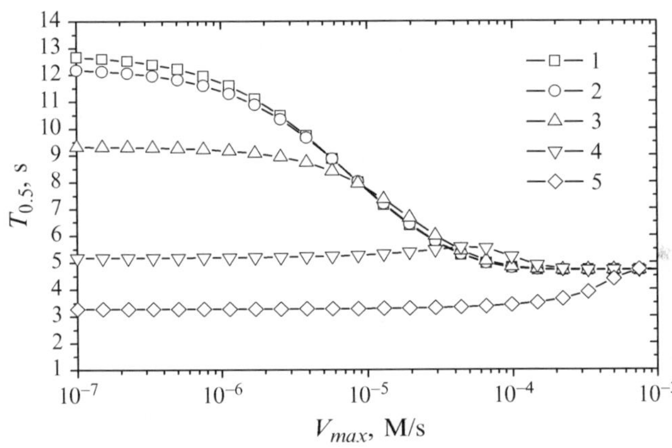

Fig. 12 The half time $T_{0.5}$ of the steady state versus the maximal enzymatic rate V_{max}. The other parameters and the notation are the same as in Fig. 11

As one can see in Fig. 11 the steady state current is approximately a linearly increasing function of V_{max} at a low activity of the enzyme. The range of the linearity increases with the increase in the substrate concentration S_0. In the case of the zero order reaction rate, i.e. at very high concentrations of the substrate ($S_0 \gg K_M$, curve 5), the current density I is practically the linear function of V_{max} at all values of the V_{max}. The linear dependence of I on the V_{max} also follows from the formulae (59). In the case of the first order reaction rate ($S_0 \ll K_M$, curves 1 and 2), the dependence is nonlinear as defined in (49). The density I of the steady state current is the linear function of V_{max} in all the cases when the diffusion module is less than unity ($\sigma < 1$). Let us notice that $\sigma = 1$ at $V_{max} = 3$ μM/s and values (61).

Figure 12 shows the dependence of the half time $T_{0.5}$ of the steady state. At low concentrations of the substrate ($S_0 \ll K_M$, curves 1 and 2), the half time $T_{0.5}$ is the monotonously decreasing function of V_{vax}. The response time changes slightly at moderate values of the substrate concentration S_0 (curves 3 and 4). At high concentrations of the substrate ($S_0 \gg K_M$, curve 5), the half time $T_{0.5}$ is the monotonously increasing function of the V_{vax}. However, at a very high enzyme activity ($V_{max} >\approx 0.8$ mM/s, $\sigma^2 >\approx 30$), $T_{0.5}$ depends neither on the S_0 nor on the V_{max}. Let us notice that at the substrate concentration S_0 of $10K_M$ and the enzyme activity V_{max} being between 10 and 100 μM/s, a shoulder on curve 4 appears.

4.5 Choosing the Enzyme Membrane Thickness

Formulas (65) and (66) introduce the concept of the minimal membrane thickness $d_\varepsilon(V_1, V_2, S_0)$, at which the relative difference $R_d(d, V_1, V_2, S_0)$ of the biosensor response is less than the decay rate ε. That concept can be considered as a framework to be used for the determination of the membrane thickness in the design of the biosensors producing a highly stable response to the substrate of concentration S_0 while the enzymatic rate changes from V_1 to V_2. In this case the minimal thickness d_ε needs to be calculated at concrete characteristics of the biosensor operation: the diffusion coefficients D_S, D_P, the number of electrons n_e, the Michaelis constant K_M and the substrate concentration S_0 approximate to the expected one.

Rather often the concentration of the analyte to be analysed varies within a known interval. Since the biosensor response is usually more stable at lower concentrations of the substrate (Fig. 8) than at higher ones, the larger value of the range of the expected concentrations should be used in the calculation of d_ε to ensure the stable response in the entire interval of the expected concentrations. In the case when $S_0 \ll K_M$, the density I of the steady state current may be calculated analytically from (18), otherwise the computer simulation is infinitely preferable for the calculation of $I(d, V_{max}, S_0)$, used in the framework and expressed by the formulas (65) and (66).

To be sure, that the framework, based on the definitions (65) and (66), really helps to find the membrane thickness at which the biosensor gives the relatively stable response, the biosensor response is also calculated in the case of the significantly thinner membrane. Figure 13 shows the density I of the steady state current versus the substrate concentration S_0 at the same values of the enzymatic rate V_{max} as in Fig. 8, however, the enzyme membrane is more than eight times thinner, $d = 10$ μm.

One can see in Fig. 13 that the biosensor response is very sensitive to changes of V_{max}. For example, in the case of $S_0 = 1$ μM, the current density I at $V_{max} = 1{,}000$ (μM/s) is about 4.7 times higher than I at $V_{max} = 100$ (μM/s) (Fig. 13), while the corresponding values of I are approximately the same in the case of the membrane thickness $d = d_\varepsilon(100(\mu M/s), 1{,}000(\mu M/s), 10(\mu M)) = 92$ μm (Fig. 8).

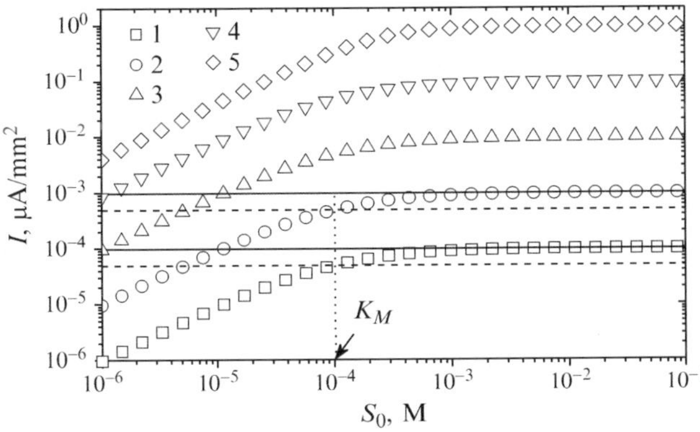

Fig. 13 The dependence of the density I of the steady state current on the substrate concentration S_0 at the enzyme membrane thickness $d = 10\,\mu\text{m}$ and five maximal enzymatic rates V_{max}: 0.1 (1), 1 (2), 10 (3), 100 (4) and 1,000 (5) µM/s

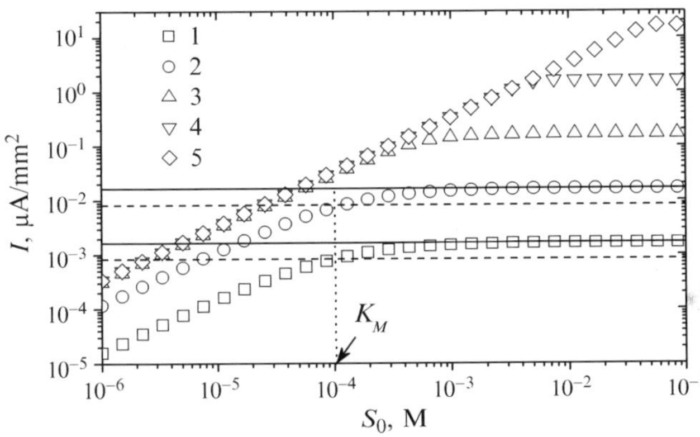

Fig. 14 The dependence of the density I of the steady state current on the concentration S_0 of the substrate at $d = d_\varepsilon(10\,\mu\text{M/s}, 1{,}000\,\mu\text{M/s}, 10\,\mu\text{M}) = 0.17\,\text{mm}$, $\varepsilon = 0.1$. The other parameters and the notations are the same as in Fig. 8

Let us notice (Fig. 8) that at $d = 10\,\mu\text{m}$, the relative difference R_d (formula (65)) between I at $V_{max} = 10\,\mu\text{M/s}$ and another value of I at $V_{max} = 1{,}000\,\mu\text{M/s}$ equals approximately 0.59 when $S_0 = 10\,\mu\text{M}$. This difference keeps approximately unchanged at all S_0 less than about 100 µM. As one more example of the framework application, let us choose the enzyme membrane thickness d to reduce that difference. Accepting $V_1 = 10\,\mu\text{M/s}$, $V_2 = 100V_1 = 1{,}000\,\mu\text{M/s}$, $\varepsilon = 0.1$ and using the definition (66) as well as the results presented in Fig. 6, the value of $d_\varepsilon(V_1, V_2, 10\,(\mu\text{M})) \approx 0.17\,\text{mm}$ was found. Figure 14 plots the density I of the steady state current versus S_0 at $d = 0.17\,\text{mm}$, at the same values of V_{max} as in Figs. 8 and 13.

No notable difference is observed between values of I, calculated at the three values of V_{max}, 10, 100 and 1,000 μM/s, when the substrate concentration S_0 is less than about $5K_M = 500$ μM. Figure 14 presents the stable response of the biosensor based on the enzyme membrane of the thickness $d = 0.17$ mm, when the maximal enzymatic rate reduces 100 times, from 1000 to 10 μM/s, while analysing the substrate of the concentration less than 100 μM.

4.6 Biosensor Resistance

Figure 6 shows the significant influence of the membrane thickness on the biosensor response. However, the significance of the influence is different at different membrane thickness. The resistance of the membrane-based biosensors to changes of the membrane thickness is introduced. The normalized dimensionless resistance B_R of the biosensor is expressed as the gradient of the steady state biosensor current with respect to the membrane thickness d,

$$B_R = \frac{d}{I(d)} \times \frac{\mathrm{d}I(d)}{\mathrm{d}d}. \tag{67}$$

where $I(d)$ is the density of the steady state biosensor current calculated at the membrane thickness of d. Since I is a non-monotonous function of d (Fig. 6), the B_R varies between -1 and 1. The cases when B_R is close to 1 or -1 correspond to the biosensors the response of which is very sensitive to changes in the thickness d of the enzyme membrane. If the resistance B_R is about to zero, then the corresponding biosensors are very resistant to relatively small changes in d.

In the case of the first order reaction rate when the substrate concentration S_0 is significantly less than the Michaelis constant K_M ($S_0 \ll K_M$), the normalized biosensor resistance can be calculated analytically from Eqs. (49) and (62) as follows:

$$B_R = \frac{\sigma \tanh(\sigma)}{\cosh(\sigma) - 1} - 1. \tag{68}$$

One can see in the formula that the biosensor resistance does not depend on the substrate concentration. However, (68) is valid for only small substrate concentrations, $S_0 \ll K_M$.

Figure 15 plots the biosensor resistance B_R versus the membrane thickness d. The substrate concentration S_0 and the other parameters are the same as in Fig. 6. In Fig. 15, the symbols are the numerically calculated solutions of the model, while the lines are the analytical ones (formula (68)). One can see in Fig. 15 that the shape of all the curves of the normalized resistance is very similar. The higher maximal enzymatic rate V_{max} corresponds to the thinner enzyme membrane at which $B_R = 0$. When the enzyme activity V_{max} equals 1 μM/s the biosensor based on the enzyme membrane of the thickness $d \approx 260$ μm is mostly resistant to changes in d. For the

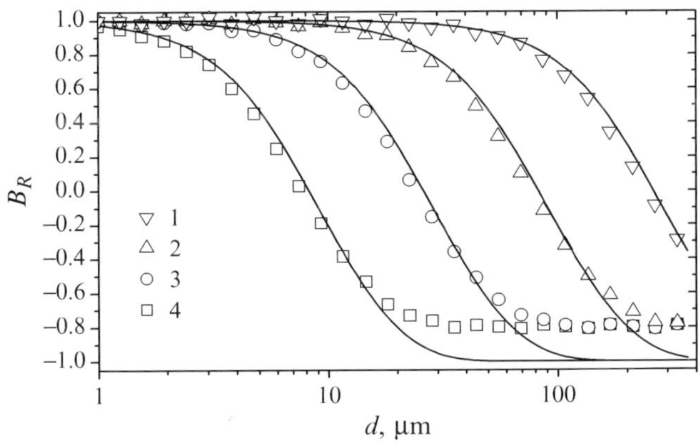

Fig. 15 The dependence of the biosensor resistance B_R on the thickness d of the enzyme membrane. The parameters and the notations are the same as in Fig. 6. Symbols are the numerically calculated solutions, while curves are analytical solutions (formula (68))

tenfold greater V_{max}, B_R falls about zero at $d \approx 82\,\mu$m. In general, the hundredfold increase in V_{max} corresponds to the tenfold thinner enzyme membrane at which $B_R = 0$.

From (68), the value of σ at which $B_R = 0$ can be calculated as a solution of the following equation:

$$\frac{\sigma \tanh(\sigma)}{\cosh(\sigma) - 1} - 1 = 0. \tag{69}$$

Rewriting terms shows that Eqn. (69) is equivalent to (63). The biosensors satisfying $\sigma \approx 1.51$ are mostly resistant to changes in the enzyme membrane thickness.

In Fig. 15, the relative difference between the numerical solutions and the analytical ones reaches about 20%. The largest difference is notable at the thickest enzyme membranes. This difference can be explained by the substrate concentration $S_0 = 0.1K_M$ used in the numerical simulation. The analytical expression (68) is valid for $S_0 \ll K_M$.

4.7 *Maximal Gradient of the Current*

When changing the substrate concentration S_0 as well as the maximal enzymatic rate V_{max}, the normalized maximal gradient B_G of the biosensor current stays practically unchanged. In all the numerical experiments the results of which are depicted in Fig. 8, the maximal gradient B_G varies only between 44.6 and 44.8. The time T_G of the maximal gradient equals about 0.08 s in all these calculations.

The thickness d of the enzyme layer has a notable effect on the maximal gradient B_G. The dependence of B_G on the thickness d is presented in Fig. 16. Figure 17 shows the corresponding times of the maximal gradients.

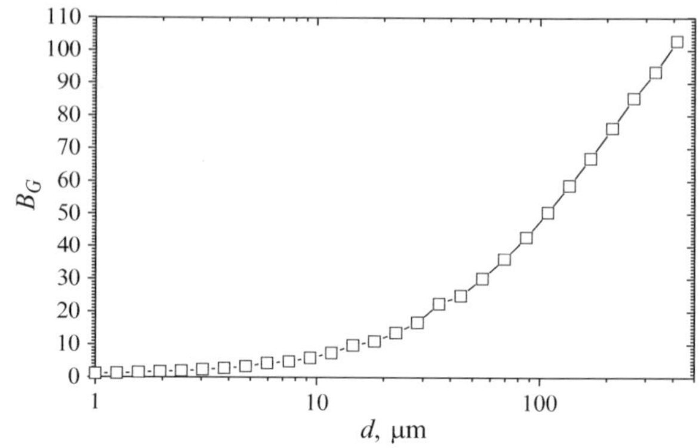

Fig. 16 The normalized maximal gradient B_G versus the thickness d of the enzyme membrane at $V_{max} = 100\,\mu M/s$. The other parameters are the same as in Fig. 6

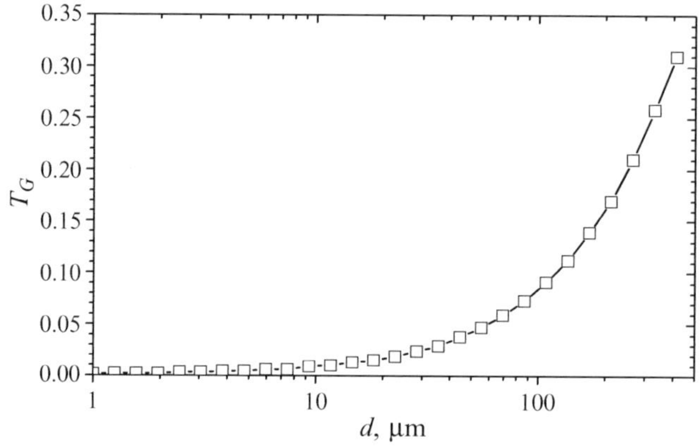

Fig. 17 The time T_G of the maximal gradient of the biosensor current versus the thickness d of the enzyme membrane. All the parameters are the same as in Fig. 16

Since no noticeable difference was observed when changing V_{max} and S_0, both Figs. 16 and 17, present the results only for the fixed values of V_{max} and S_0. As one can see in these figures, the maximal gradient G_D and the time T_G vary non-linearly in a few orders of magnitude when changing the thickness d. However, the time T_G stays very short even for thick enzyme membranes, e.g. $T_G = 0.37$ s in the case of $d = 0.5$ mm.

5 Flow Injection Analysis

The mathematical model (3)–(10) describes the operation of the biosensors acting in the batch mode where the concentration of the substrate as well as the product over the enzyme surface (bulk solution/membrane interface) remains constant while the biosensor comes into contact with the substrate. In the flow injection mode the biosensor contacts the substrate only for a short term (seconds to tens of seconds) [151, 220]. When the analyte disappears, the buffer solution swills the enzyme surface reducing the substrate concentration at this surface to zero.

5.1 Mathematical Model

When applying Eqs. (3)–(10) to the biosensors acting in the injection mode, the boundary conditions (9) and (10) have to be adjusted for the bulk solution/membrane interface. In the case of the flow injection analysis, these two conditions are only valid in the beginning of the biosensor action. Later, the concentrations of the substrate as well as the product are reduced to zero,

$$S(d,t) = \begin{cases} S_0, & t \le T_F, \\ 0, & t > T_F, \end{cases} \tag{70}$$

$$P(d,t) = \begin{cases} P_0, & t \le T_F, \\ 0, & t > T_F, \end{cases} \tag{71}$$

where T_F is the injection time, i.e. the time when the substrate disappears from the bulk solution/membrane interface. Usually, the zero concentration of the reaction product is also accepted for the initial phase of the operation, i.e. $P_0 = 0$.

The governing equations (3) together with the initial conditions (4), (5) and the boundary conditions (6)–(8), (70), (71) form the mathematical model of the amperometric biosensor acting in the injection mode.

Let us notice that the boundary conditions (70) and (71) for the injection analysis generalize the conditions (9) and (10) used in the batch analysis. Assuming $T_F = T$, the boundary conditions (70), (71) reduce to (9) and (10), respectively.

Accepting $T_F = T$ the model of the biosensor acting in the injection mode can also be used for the biosensors acting in the batch mode.

5.2 Numerical Solution

When solving the initial boundary value problem of the injection analysis, the boundary conditions (70), (71) can be approximated as follows:

$$S_N^j = S_0, \quad P_N^j = P_0, \quad j = 1, \ldots, M_F, \tag{72}$$

$$S_N^j = 0, \quad P_N^j = 0, \quad j = M_F + 1, \ldots, M, \tag{73}$$

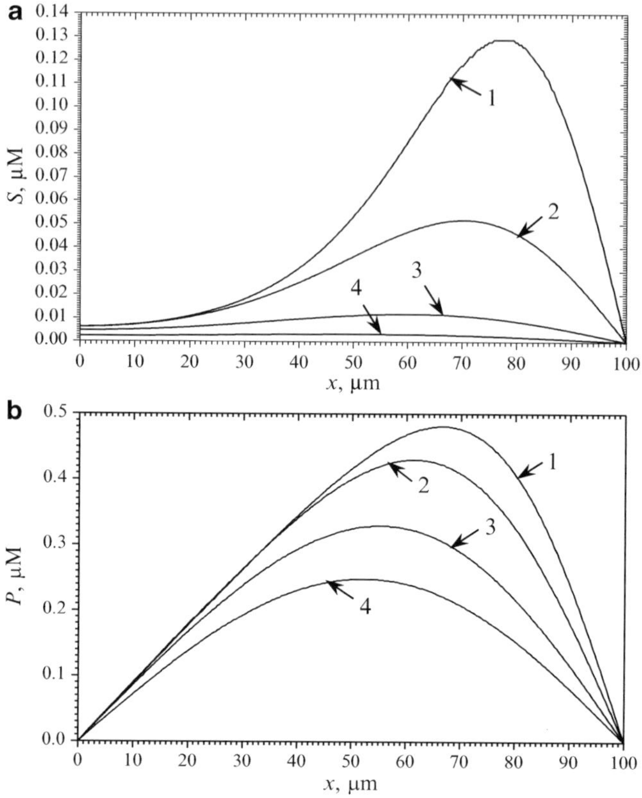

Fig. 18 The concentration profiles of the substrate (**a**) and the product (**b**) at the following values of time t: 10.5 (1), 11 (2), 12 (3) and 13 s (4), $T_F = 10$ s, $S_0 = 1\,\mu$M. The other parameters are the same as in Fig. 1

where

$$M_F = \max_{j=1,\ldots,M} \{j : t_j \leq T_F\}. \tag{74}$$

Figure 18 shows the concentration profiles calculated by using the numerical simulation. When comparing this figure with Fig. 1, one can see notably different shapes of the corresponding curves. However, until the moment T_F, i.e. while the enzyme surface contacts the analyte, the concentration profiles are practically identical for both modes of the analysis, i.e. the batch and the injection.

The boundary conditions (70) and (71) are discontinuous. Discontinuity in the boundary conditions increases the error of the numerical solution [5, 222]. In order to circumvent the discontinuity, the following boundary conditions were also used instead of (70) and (71):

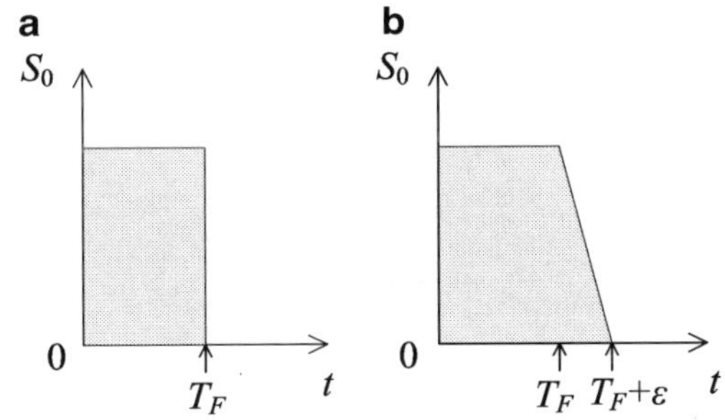

Fig. 19 The profiles of the substrate concentration S_0 in the case of the discontinuous (**a**) boundary conditions (70) and (71) as well as the continuous (**b**) conditions (75) and (76)

$$S(d,t) = \begin{cases} S_0, & t \le T_F, \\ \varphi(S_0, t, \varepsilon), & T_F < t \le T_F + \varepsilon, \\ 0, & t > T_F + \varepsilon, \end{cases} \tag{75}$$

$$P(d,t) = \begin{cases} P_0, & t \le T_F, \\ \varphi(P_0, t, \varepsilon), & T_F < t \le T_F + \varepsilon, \\ 0, & t > T_F + \varepsilon, \end{cases} \tag{76}$$

where $\varphi(C, t, \varepsilon)$ is a continuous function such as $\varphi(C, T_F, \varepsilon) = C$, $\varphi(C, T_F + \varepsilon, \varepsilon) = 0$ and $\varphi(C, t, \varepsilon)$ monotonously decreases when t changes form T_F to $T_F + \varepsilon$.

Several different expressions of φ as well as small values of ε were used in the numerical simulation. The following linear function is one of the simplest functions to be used in (75) and (76):

$$\varphi(C, t, \varepsilon) = \frac{C}{\varepsilon}(T_F - t + \varepsilon), \quad T_F \le t \le T_F + \varepsilon. \tag{77}$$

Figure 19 shows the profiles of the substrate concentration S_0 in the case of the discontinuous boundary conditions (70) and (71) as well as the continuous boundary conditions (75) and (76) assuming the φ definition (77).

Accepting a relatively small value of ε, e.g. $\varepsilon = 0.01T_F$, no notable difference between the numerical solutions was observed.

5.3 *Biosensor Response*

Figure 20 shows the dynamics of the current of the biosensor acting in the injection mode.

One can see that the current of the biosensor acting in the injection mode is a non-monotonous function of time t. Because of the substrate remaining in the enzyme

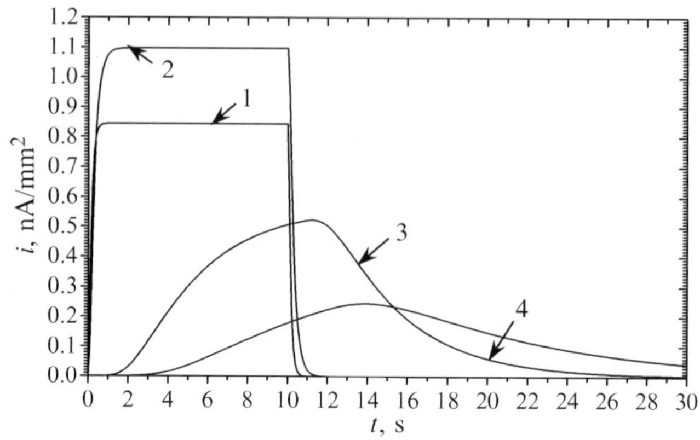

Fig. 20 The dynamics of the density i of the biosensor current at four values of the membrane thickness d: 0.01 (1), 0.015 (2), 0.1 (3), 0.15 (4) mm and the duration $T_F = 10$ s of the injection. The other parameters are the same as in Fig. 18

membrane, the mass diffusion as well as the enzyme reaction remains for some time even after the disconnection of the biosensor and the substrate.

Figure 20 shows that the shapes of two curves (1 and 2) corresponding to the thin enzyme membranes (0.01 and 0.015 mm) noticeably differ from the other two curves (3 and 4). The enzyme kinetics determines mainly the response of the two thin biosensors while the response of the two thicker biosensors is controlled by the diffusion. In the case of the thin membranes, the diffusion module σ^2 equals 0.33 and 0.75, respectively, i.e. both of them are less than unity. σ^2 for the other two biosensors is ten times greater, i.e. greater than unity. The small diffusion module shows a very fast mass transport by diffusion. Because of this the reaction–diffusion process in the thin biosensors approaches the steady state before the time T_F, and the current falls to zero very shortly. In the case of the thick enzyme layer ($\sigma^2 > 1$), the diffusion is relatively slow and it determines the smooth curve of the current.

The system (3)–(8), (70), (71) describing the biosensor action in the injection analysis approaches to a steady state as $t \to \infty$. Because of the zero concentration of the surrounding substrate at $t > T_F$, the density I of the steady state current falls to zero,

$$I = \lim_{t \to \infty} i(t) = 0. \tag{78}$$

Because of this, the steady state current is not practically useful in the analytical systems. Since the current density $i(t)$ of the biosensor acting in the injection mode is a non-monotonous function, the maximal current is one of the mostly used characteristics for this kind of the biosensors,

$$I_{max} = \max_{t>0} \{i(t)\}, \tag{79}$$

where I_{max} is the maximal density of the biosensor current.

The corresponding time T_{max} of the maximal biosensor current is used instead of the steady state one,

$$T_{max} = \{t : i(t) = I_{max}\}. \tag{80}$$

In the numerical simulation, the calculation can be stopped when the current reaches the maximum. The minimal time at which the current starts to decrease may be accepted as the time of the maximal current,

$$T_{max} \approx \min_{j=1,\dots} \left\{t_j : i_j > i_{j+1}\right\}, \tag{81}$$

where $t_j = j\tau$, $i_j = i(t_j)$, τ is the step of the discrete grid in direction of time.

Using the maximal current as the main characteristics of the biosensor acting in the injection mode leads to the following definition of the dimensionless sensitivity B_S:

$$B_S(S_0) = \frac{S_0}{I_{max}(S_0)} \times \frac{\mathrm{d}I_{max}(S_0)}{\mathrm{d}S_0}, \tag{82}$$

where $I_{max}(S_0)$ is the density of the maximal biosensor current calculated at the substrate concentration S_0.

5.4 *Peculiarities of the Biosensor Response*

Using the computer simulation the dependence of the biosensor response on the substrate concentration and the injection duration was investigated.

5.4.1 Maximal Current Versus Substrate Concentration

Figure 21 shows the density I_{max} of the maximal current versus the substrate concentration S_0 at different values of the maximal enzymatic rate V_{max}. As it is possible to notice, the shape of curves in the injection analysis is very similar to that in the batch analysis (Figs. 8 and 13).

As has been mentioned in Sect. 4.3, the Michaelis constant K_M is the substrate concentration at which the reaction rate is the half its maximal value. The effect of halving was verified if it is valid in the injection analysis as in the batch one.

The maximum of the density of the maximal current is approximately equal to $1.44\,\mathrm{nA/mm^2}$ at the maximal enzymatic rate of $0.1\,\mu\mathrm{M/s}$ (see Fig. 21). The half of the maximum $0.72\,\mathrm{nA/mm^2}$ of the maximal current is reached at $S_0 \approx 0.5\,\mathrm{mM} = 5\,K_M$ which is noticeably farther from K_M.

For the other two values of V_{max}, 1 and $10\,\mu\mathrm{M/s}$, the half of the maximum of the maximal current is also reached at $S_0 \approx 5\,\mathrm{K_M}$. This is not valid for greater values of V_{max}. So, the effect of halving is valid in the injection analysis in cases when the biosensor response is under the enzyme kinetics control. However, in the case of the enzyme membrane thickness $d = 0.15\,\mathrm{mm}$ and the injection duration $T_F = 10\,\mathrm{s}$,

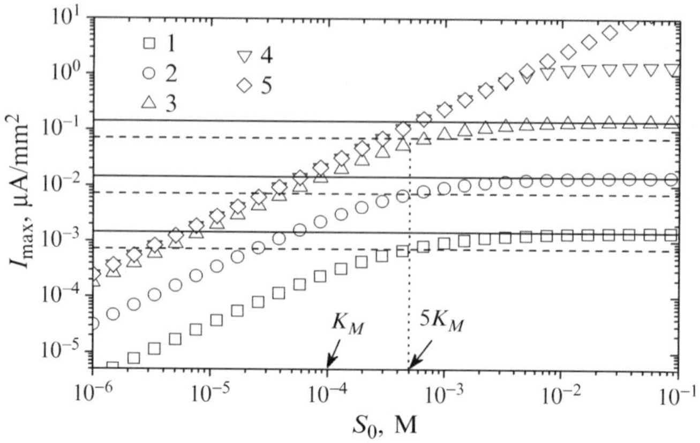

Fig. 21 The density I_{max} of the maximal biosensor current versus the substrate concentration S_0 at five maximal enzymatic rates V_{max}: 0.1 (1), 1 (2), 10 (3), 100 (4) and 1,000 (5) μM/s, the injection duration $T_F = 10$ s and the membrane thickness $d = 0.15$ mm

the half of maximum of the maximal current is reached at the substrate concentration $S_0 = 5K_M$ which considerably differs from the Michaelis constant K_M.

In the case of the thicker enzyme membrane keeping the values of the other parameters unchanged, the half of the maximum of the maximal current is reached even at the higher substrate concentration S_0 [24]. The higher half of the maximum of the maximal current corresponds to a longer linearly increasing part of the maximal biosensor current as a function of S_0. Consequently, the biosensors with a higher half of the maximum of the maximal current are sensitive to a wider range of the substrate concentrations than those with a lower half of the maximum of the maximal current. Below this is discussed in detail.

5.4.2 Response Time Versus Substrate Concentration

Figure 22 shows the evolution of the time T_{max} of the maximal current vs. the substrate concentration S_0 in the injection analysis. When comparing curves in Fig. 22 with the corresponding curves of the half of the steady state time obtained in the batch analysis (Fig. 9), it is possible to notice a considerable difference in the shape. In the injection analysis (Fig. 22) T_{max} is a monotonous increasing function of S_0 at all values of V_{max}, while in the batch analysis (Fig. 9), $T_{0.5}$ is a monotonous decreasing or even a non-monotone function of S_0.

Figure 22 shows the considerable increase of the time T_{max} of the maximal response with the increase of the substrate concentration S_0 only at high values of S_0. This can be explained by a sufficient supply of the substrate after the time T_F when the substrate disappears from the surrounding analyte. The more substrate penetrates

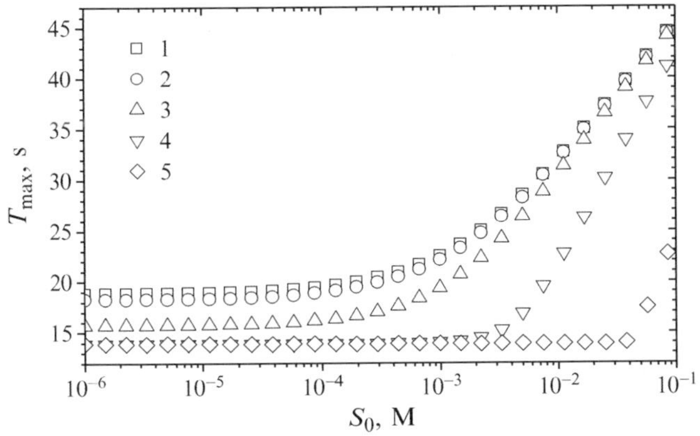

Fig. 22 The time T_{max} of the maximal biosensor current versus the substrate concentration S_0. The parameters and the notations are the same as in Fig. 21

into the enzyme layer until the time T_F, the longer enzyme reaction occurs. At low concentrations of the substrate, the time of the maximal current increases very weakly.

5.4.3 Sensitivity Versus Substrate Concentration

As has been mentioned in Sect. 5.4.1 the biosensors acting in the injection mode have the higher half of the maximum of the maximal current than the corresponding biosensors acting in the batch mode. The half of the maximum of the maximal current influences directly the biosensor sensitivity. Biosensors with the higher half of the maximum of the maximal current are sensitive to a wider range of the substrate concentration than those with a lower half of the maximum of the maximal current.

Figure 23 shows the dependence of the biosensor sensitivity for a wide range of the substrate concentrations calculated for several values of the enzyme activity V_{max}. When comparing this figure with Fig. 10 one can notice that at low substrate concentrations, the amperometric biosensors are very sensitive in both modes of the analysis: the batch and the injection. At higher concentrations of S_0 the sensitivity of the biosensor acting in the injection analysis is noticeably higher than the sensitivity of the corresponding biosensor acting in the batch mode.

To find out if the injection duration T_F really influences the sensitivity of biosensors, the normalized sensitivity B_S is calculated for different values of the duration T_F and of the enzymatic activity V_{max}. The results of calculation are depicted in Fig. 24.

Figure 24 shows a notable effect of the injection duration T_F on the biosensor sensitivity for low maximal enzymatic activities V_{max}. As has been discussed above (see Fig. 20), the thickness of the enzyme membrane also effects the behaviour of

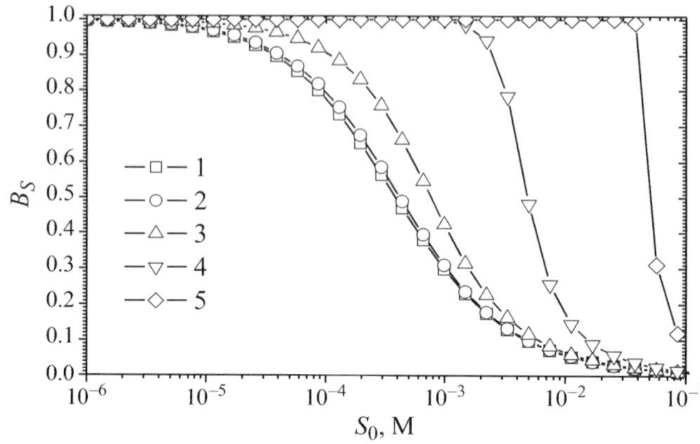

Fig. 23 The normalized biosensor sensitivity B_S versus the substrate concentration S_0. The parameters and the notations are the same as in Fig. 21

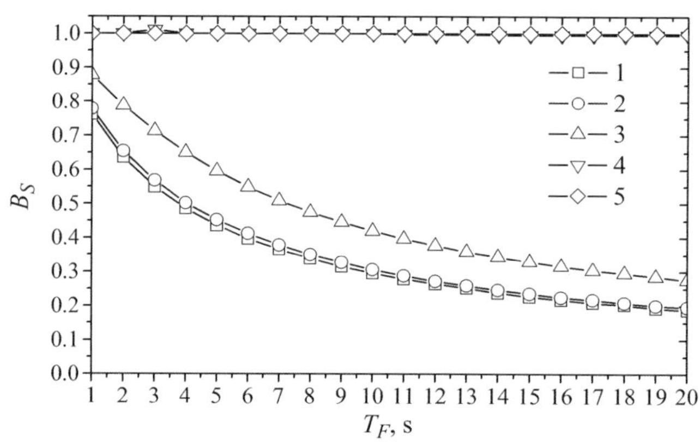

Fig. 24 The normalized biosensor sensitivity B_S versus the injection duration T_F at concentration $S_0 = 1\,\text{mM} = 10K_M$ of the substrate. The other parameters and the notations are the same as in Fig. 21

the response of the biosensor acting in the injection mode. Additional calculations showed that the injection duration T_F effects the biosensor sensitivity when the diffusion module σ^2 is less than unity, i.e. when the biosensor response is under the enzyme kinetics control.

5.5 Sequential Injection Analysis

The injection analysis can be generalized to the sequential injection analysis [220], where two surrounding environments of the biosensor action, i.e. the substrate

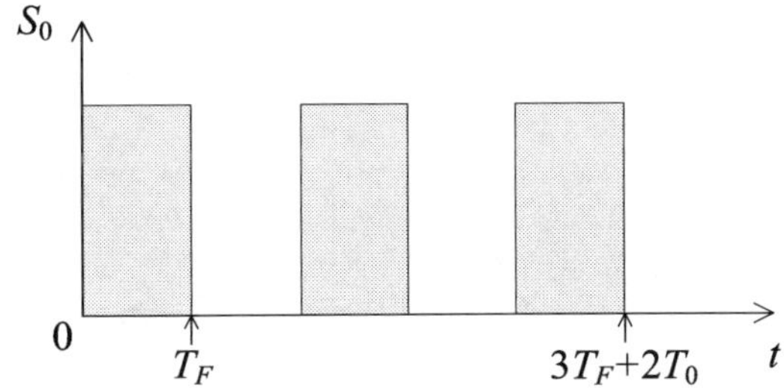

Fig. 25 The profile of the substrate concentration S_0 in the case of the sequential injection

solution of the concentration S_0 and the buffer solution containing no substrate, alternate sequentially. Figure 25 shows the profile of the concentration S_0 of the substrate surrounding the biosensor in the sequential injection analysis. The product concentration out of the enzyme layer equals zero at all value of $t > 0$.

Assuming the time intervals of the injection and the swilling of the uniform duration, the boundary conditions corresponding the profile of the substrate concentration depicted in Fig. 25 are defined as follows:

$$S(d,t) = \begin{cases} S_0, & (T_F + T_0)k \le t \le (T_F + T_0)k + T_F, \\ 0, & (T_F + T_0)k + T_F < t < (T_F + T_0)(k+1), \end{cases} \quad k = 1, \dots, K, \tag{83}$$

$$P(d,t) = \begin{cases} P_0, & (T_F + T_0)k \le t \le (T_F + T_0)k + T_F, \\ 0, & (T_F + T_0)k + T_F < t < (T_F + T_0)(k+1), \end{cases} \quad k = 1, \dots, K, \tag{84}$$

where T_F is the duration of the injection, T_0 is the duration of the swilling, and K is the number of injections.

The steady state current is practically meaningless for the sequential injection analysis. One can see in Fig. 26 that the biosensor current oscillates. The number of peaks corresponds to the number of injections. If the next injection starts when the current is above the zero then the next peak may be higher than the previous one. For example, the first local maximum of curve 4 in Fig. 26 equals 0.244 nA/mm^2 (at $t = 13.9$ s), the second one equals 0.269 ($t = 33.6$ s) and the third – 0.271 (at $t = 53.6$ s).

Starting from the second peak, the local maximum of the current as well as the time period between two peaks practically unchanges. As one can see in Fig. 26, $i(t)$ is practically the periodical function with the period of $T_F + T_0$.

6 Biosensors with Chemical Amplification

The detection limit of the enzyme electrodes depends on the sensitivity of amperometric systems [52,92]. The sensitivity of the enzyme electrodes can be increased by a cyclic conversion of the substrate [76, 129, 136, 227, 231]. The cyclic conversion

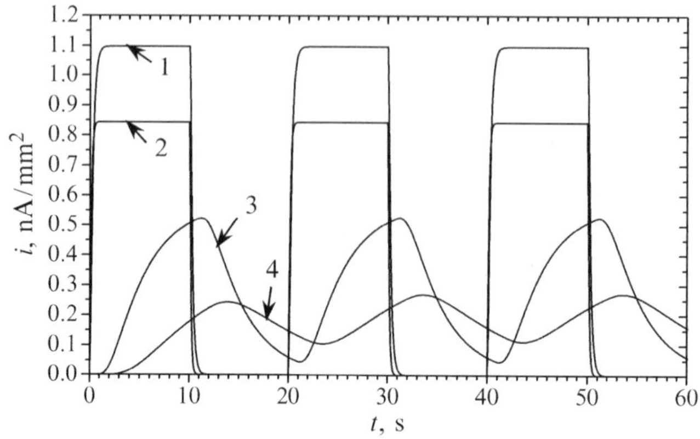

Fig. 26 The dynamics of the biosensor current at four values of the thickness d, 0.01 (1), 0.015 (2), 0.1 (3), 0.15 (4) mm, the injection duration $T_F = 10$ s and the swilling duration $T_F = 10$ s. The other parameters are the same as in Fig. 18

of the substrate and the regeneration of the analyte are typically performed by using a membrane containing two enzymes. The biosensors containing two enzymes with the regeneration of the analyte will be discussed later.

Another type of highly sensitive amperometric electrodes were developed by utilizing the substrate cyclic conversion in single enzyme membranes [70, 135, 144, 173, 231]. In these electrodes the cyclic conversion of the substrate was carried out by a conjugation of the enzymatic reaction with chemical or electrochemical process.

This section presents a model allowing the computer simulation of the response of the biosensor, utilizing the amplification by the conjugated electrochemical and the enzymatic substrates conversions.

The scheme of the substrate (S) electrochemical conversion to the product (P) following catalyzed with the enzyme (E) the product conversion to the substrate is considered [144],

$$S \longrightarrow P \stackrel{E}{\longrightarrow} S \tag{85}$$

In the case of phenol sensitive biosensors, for example, 1,2-benzoquinone (S) is electrochemically reduced and pyrocatechol (P) formed is oxidised in the membrane with immobilized laccase [144].

6.1 Mathematical Model

Assuming the symmetrical geometry of the electrode and the homogeneous distribution of the immobilized enzyme in the membrane, the dynamics of the biosensor can be described by the reaction–diffusion system,

$$\frac{\partial S}{\partial t} = D_S \frac{\partial^2 S}{\partial x^2} + \frac{V_{max} P}{K_M + P},$$
$$\frac{\partial P}{\partial t} = D_P \frac{\partial^2 P}{\partial x^2} - \frac{V_{max} P}{K_M + P}, \quad x \in (0, d), \quad t > 0, \tag{86}$$

where x and t stand for space and time, respectively, $S(x,t)$ and $P(x,t)$ are the substrate and the reaction product concentrations, respectively, d is the thickness of the enzyme membrane, D_S, D_P are the diffusion coefficients, V_{max} is the maximal enzymatic rate and K_M is the Michaelis constant [29].

Let $x = 0$ represent the electrode surface, and $x = d$ represent the boundary layer between the analyzed solution and the enzyme membrane. Assuming the zero concentration of the reaction product in the bulk solution, the initial conditions are the same as in the case of amperometric biosensors with no cyclic conversion (2),

$$S(x,0) = 0, \quad x \in [0, d),$$
$$S(d,0) = S_0, \tag{87}$$

$$P(x,0) = 0, \quad x \in [0, d),$$
$$P(d,0) = 0, \tag{88}$$

where S_0 is the concentration of the substrate in the bulk solution.

In the scheme (85) the substrate is an electro-active substance. Due to the amperometry the electrode potential is chosen to keep the zero concentration of the substrate at the electrode surface. During the electrochemical conversion the product is generated, S $\longrightarrow$ P. The rate of the product generation at the electrode is proportional to the rate of conversion of the substrate ($t > 0$),

$$S(0,t) = 0, \tag{89}$$

$$D_P \frac{\partial P}{\partial x}\bigg|_{x=0} = -D_S \frac{\partial S}{\partial x}\bigg|_{x=0}. \tag{90}$$

When the buffer solution is well-stirred and in a powerful motion, then the diffusion layer ($0 < x < d$) remains at a constant thickness. Consequently, the concentration of the substrate as well as the product over the enzyme surface (bulk solution/membrane interface) does not change while the biosensor contacts the substrate solution.

Assuming the zero concentration of the reaction product and S_0 as the concentration of the substrate in the bulk solution, the boundary conditions at the solution/membrane interface are identical to those of the amperometric biosensor without the cyclic conversion (2) ($t > 0$),

$$S(d,t) = S_0, \tag{91}$$

$$P(d,t) = 0. \tag{92}$$

The biosensor current depends upon the flux of the electro-active substance (substrate) at the electrode surface, i.e. at the border $x = 0$. Consequently, the density

$i(t)$ of the biosensor current at time t is explicitly obtained from Faraday's and Fick's laws

$$i(t) = n_e F D_S \frac{\partial S}{\partial x}\bigg|_{x=0} = -n_e F D_P \frac{\partial P}{\partial x}\bigg|_{x=0}, \tag{93}$$

where n_e is the number of electrons involved in a charge transfer, F is the Faraday constant.

All other characteristics of the biosensor response defined for the amperometric biosensors in Sect. 2 can also be applied to the biosensor with the substrate cyclic conversion.

6.2 Finite Difference Solution

Accepting the uniform discrete grid for space and time (27) and the notation (28), the partial differential equations (86) can be replaced with the following difference equations:

$$\begin{aligned}
&\frac{S_i^{j+1} - S_i^j}{\tau} = D_{S_e} \frac{S_{i+1}^{j+1} - 2S_i^{j+1} + S_{i-1}^{j+1}}{h^2} + \frac{V_{max} P_i^j}{K_M + P_i^j},\\
&\frac{P_i^{j+1} - P_i^j}{\tau} = D_{P_e} \frac{P_{i+1}^{j+1} - 2P_i^{j+1} + P_{i-1}^{j+1}}{h^2} - \frac{V_{max} P_i^j}{K_M + P_i^j},\\
&i = 1, \ldots, N-1, \quad j = 1, \ldots, M.
\end{aligned} \tag{94}$$

The initial conditions (87) and (88) are approximated by

$$\begin{aligned}
S_i^0 &= 0, \qquad i = 0, \ldots, N-1;\\
S_N^0 &= S_0,\\
P_i^0 &= 0, \qquad i = 0, \ldots, N.
\end{aligned} \tag{95}$$

The boundary conditions (89)–(92) are approximated as follows:

$$\begin{aligned}
&S_0^j = 0,\\
&D_P(P_1^j - P_0^j) = -D_S(S_1^j - S_0^j),\\
&S_N^j = S_0,\\
&P_N^j = 0,\\
&j = 1, \ldots, M.
\end{aligned} \tag{96}$$

Having the numerical solution of the problem, the density of the biosensor current at time $t = t_j$ can be calculated by

$$i(t_j) = n_e F D_S (S_1^j - S_0^j)/h, \quad j = 0, \ldots, M. \tag{97}$$

or by

$$i(t_j) = -n_e F D_P (P_1^j - P_0^j)/h, \quad j = 0, \ldots, M. \tag{98}$$

6.3 Concentration Profiles

Figure 27 shows the distribution of the substrate and the product for the catalytically active ($V_{max} > 0$) as well as for the inactive ($V_{max} = 0$) electrodes at the steady state condition.

In the case of the catalytically inactive electrode ($V_{max} = 0$) the mathematical model (86)–(92) describes the action of the following scheme:

$$\mathrm{S} \longrightarrow \mathrm{P} \tag{99}$$

In scheme (99) no substrate regeneration is observed. It is only the product that is produced as the result of the electrochemical reaction. Comparing the response of the catalytically active electrode ($V_{max} > 0$) with the inactive one ($V_{max} = 0$) allows us to evaluate the effect of the catalytical activity on the response.

It is obvious that the gradient of both species is constant in the entire membrane in the case of the absence of the catalytic activity. The non-linear distribution of the concentration of both compounds, however, is observed at $V_{max} > 0$. The gradient of the substrate concentration significantly increases with the increase in the maximal enzymatic rate V_{max}. Consequentially, the biosensor current also increases. The response is greater at larger maximal enzymatic rate. Because of the boundary conditions (89)–(92), the symmetry with respect to the axis $S = P = 0.5S_0$ can be noticed at all values of V_{max} in Fig. 27. In the case of the symmetry, $P(x, t) = S_0 - S(x, t)$ at $x \in [0, d]$ and $t \to \infty$. This can be observed only at the steady state conditions, i.e. $\partial S/\partial t = \partial P/\partial t = 0$.

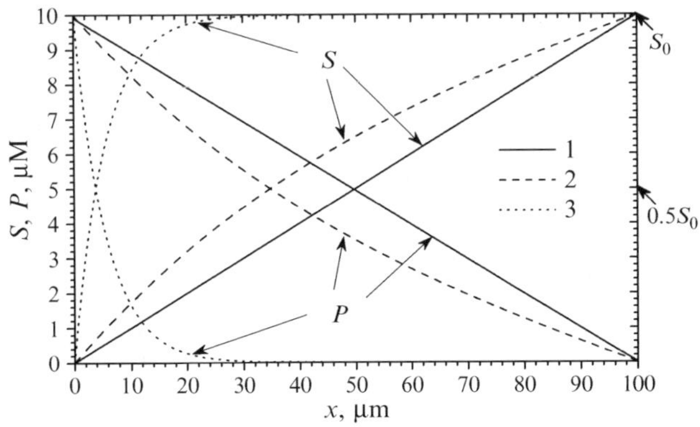

Fig. 27 The concentration profiles of the substrate (S) and the product (P) in the enzyme layer at three maximal enzymatic rates V_{max}, 0 (1), 10 (2) and 1,000 (3) µM/s, $S_0 = 10$ µM, $d = 100$ µm

6.4 *Peculiarities of the Biosensor Response*

6.4.1 Dependence of Response on the Substrate Concentration

Using computer simulation the dependence of the steady state biosensor current as well as the biosensor response time on the substrate concentration S_0 has been investigated [29]. To get the results for a wide range of the values of the maximal enzymatic rate the investigation was carried out at the following values of V_{max}: 0, 1, 10, 100 and 1,000 μM/s. The results of the calculations are depicted in Figs. 28, 29 and 30. In these calculations the biosensor was modeled accepting the relatively thick enzyme layer, $d = 150\,\mu$m.

Figure 28 presents the dependence of the biosensor steady state current on the substrate concentration S_0. As it is possible to notice, the density I of the steady state biosensor current is the monotonous increasing function of S_0 at all values of V_{max}. In the case of the catalytically inactive electrode ($V_{max} = 0$) the current density I is the linear function of S_0 in the entire domain of the investigation. The non linear biosensor current versus S_0 is observed with the catalytically active membrane, i.e. $V_{max} > 0$. The calculations show that at low concentrations of the substrate, $S_0 < K_M$, the density I of the steady state current is approximately the linear function of S_0 even at $V_{max} > 0$. That is because of the linearity of the Michaelis Menten equation at $S_0 < K_M$. The higher V_{max} (Fig. 28) evokes the higher biosensor response. However, at the high concentration of the substrate, when $S_0 > K_M$, the amplification of the biosensor response decreases. At all values of $V_{max} > 0$ the current converges to the maximal value, calculated at $V_{max} = 0$, while the substrate concentration S_0 increases.

The ratio of the steady current increase measured with the enzyme active electrode ($V_{max} > 0$) to the steady state current measured with the catalytically inactive

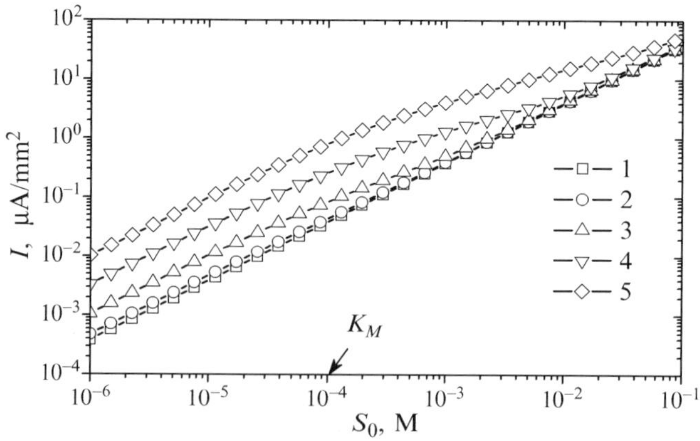

Fig. 28 The dependence of the density of the steady state current on the substrate concentration S_0 at five maximal enzymatic rates V_{max}: 0 (1), 1 (2), 10 (3), 100 (4) and 1,000 (5) μM/s, $d = 150\,\mu$m

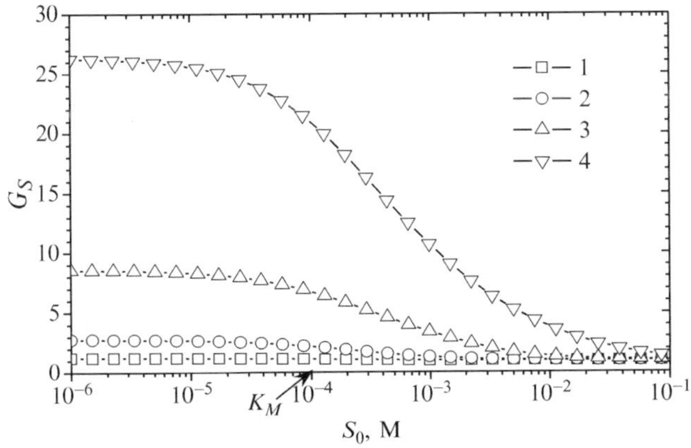

Fig. 29 The dependence of the signal gain G_S on the substrate concentration S_0 at four maximal enzymatic rates V_{max}: 1 (1), 10 (2), 100 (3) and 1,000 (4) μM/s, $d = 150\,\mu$m

electrode ($V_{max} = 0$) can be considered as the gain of the sensitivity, G_S, of the sensor with chemically amplified response,

$$G_S(V_{max}) = \frac{I(V_{max})}{I(0)}. \tag{100}$$

where $I(V_{max})$ is the density of the steady state biosensor current calculated at the maximal enzymatic activity of V_{max}.

Figure 29 shows the dependence of the gain G_S of the steady state current on the substrate concentration S_0. As it is possible to notice in Fig. 29, the biosensor response increases up to 26 times ($G_S \approx 26$) at $S_0 = 1\,\mu$M and $V_{max} = 1$ mM/s. In comparison with the response of the catalytically inactive electrode, this increase of the biosensor sensitivity compares fairly with the increase of the response of the biosensor with the immobilized laccase [144].

As has been mentioned above, the biosensor response is slightly amplified at high concentrations of the substrate. For all $V_{max} \leq 1$ mM/s, the gain G_S of the steady state current is less than 4 at $S_0 > 100K_M$. According to the definition (100), $G_S \to 1$ when $V_{max} \to 0$, $G(0) = 1$, for all values of S_0. Figure 29 distinctly shows that the amplification is especially significant at low concentrations of the substrate.

Figure 30 shows the effect of the substrate concentration S_0 on the half time $T_{0.5}$ of the steady state current. As it is possible to notice in Fig. 30, $T_{0.5}$ is the monotonous decreasing function of S_0 at $V_{max} > 0$. $T_{0.5}$ is approximately the constant function, $T_{0.5} = 10.5$ s, at $V_{max} = 0$. The most significant difference in $T_{0.5}$ is noticed at low substrate concentrations when the largest amplification appears (see Figs. 29 and 30). For example, at $V_{max} = 1$mM/s and $S_0 = 1\,\mu$M, $T_{0.5}$ is about 2.7 (28.6/10.5) times greater than at $V_{max} = 0$ and the same S_0. The variation of $T_{0.5}$ vs. S_0 decreases with the increase in S_0.

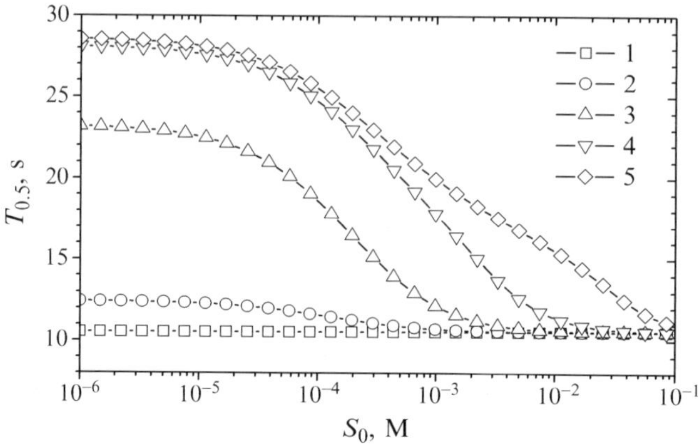

Fig. 30 The half time $T_{0.5}$ of the steady state biosensor response versus the substrate concentration S_0 at five maximal enzymatic rates V_{max}: 0 (1), 1 (2), 10 (3), 100 (4) and 1,000 (5) μM/s, $d = 150\,\mu\text{m}$

In Fig. 30 the shapes of curves of $T_{0.5}$ vs. S_0 are very similar for all values of V_{max} being between $0.01K_M$ and K_M. At S_0 being greater than $10K_M$ a shoulder on the curve appears for $V_{max} = 1$ mM/s. This shoulder is rather similar to the other one observed in Fig. 9 when simulating the response of the amperometric biosensor with no cyclic conversion. It seems possible that the reason of the appearance of the shoulders on the curves is the same.

6.4.2 Sensitivity Versus Substrate Concentration

As it is possible to notice in Fig. 28 the steady state current as a function of the substrate concentration S_0 does not become a constant as it was in the case of biosensors without amplification (Figs. 8, 13, 14 and 21). This means that the enzyme electrodes with the substrate cyclic conversion have a long calibration curve. This also means that the enzyme electrodes with the substrate cyclic conversion are sensitive enough even at relatively high concentrations of the substrate.

Figure 31 shows the dependence of the biosensor sensitivity for a wide range of the substrate concentrations calculated for several values of the enzyme activity V_{max}.

One can see in Fig. 31 that at a wide range of the enzymatic activities V_{max}, the biosensor sensitivity never approaches zero when the substrate concentration changes from 1 μM to 0.1 M. The highest sensitivity is proper to the catalytically inactive electrode. The sensitivity of the catalytically active enzyme electrodes is the non-monotonous function of substrate concentration S_0. The cave sharpness on the curve depends on the enzymatic activity V_{max}. The concentration of the minimal sensitivity differs. In the case of a very low enzymatic activity ($V_{max} = 1\,\mu$M/s),

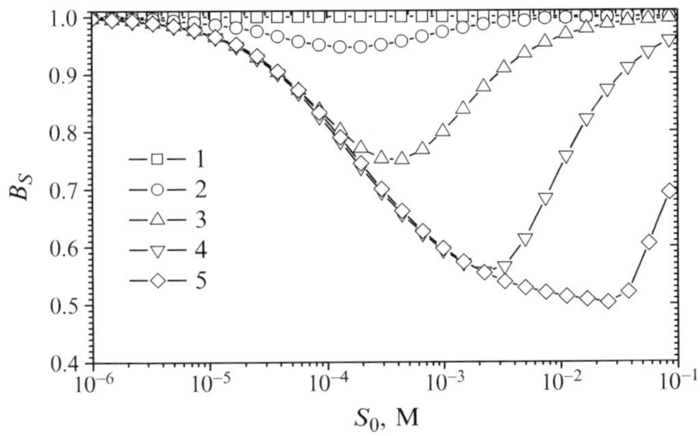

Fig. 31 The normalized biosensor sensitivity B_S versus the substrate concentration S_0. The parameters and the notations are the same as in Fig. 28

the minimum appears at $S_0 \approx K_M = 100\,\mu\text{M}$. For a higher value of V_{max} the minimum of the sensitivity gains at a higher concentration S_0.

6.4.3 Effect of the Enzyme Membrane Thickness

To investigate the influence of the thickness d of the enzyme layer on the dynamics of the biosensor response the density I of the steady state current are calculated at a wide range of the membrane thickness d: from 1 to 500 μm and different values of the maximal enzymatic rate V_{max}. The results of the calculations are depicted in Fig. 32.

No notable signal amplification occurs in the case of a very thin enzyme layer, $d < 3\,\mu\text{m}$, at all used values of the maximal enzymatic rate V_{max} (Fig. 32). Let us notice that a noticeable amplification starts at the thickness d at which the diffusion module σ^2 (see (13)) is approximately equal to the unity. The diffusion coefficient D_S and the Michaelis–Menten constant K_M were constant in all the numerical experiments,

$$D_S = 300\,\mu\text{m}^2/\text{s}, \quad K_M = 100\,\mu\text{M}. \tag{101}$$

The enzyme layer thickness d_σ, at which $\sigma^2 = 1$, can be defined as a function of V_{max},

$$d_\sigma(V_{max}) = \sqrt{\frac{D_S K_M}{V_{max}}}. \tag{102}$$

In Fig. 32 the values of the thickness $d_\sigma(V_{max})$ are presented for all non-zero enzymatic rates. The figure shows a fair correlation between the values of d_σ and the enzyme layer thickness, at which the amplification starts. From the thickness d considerably greater than d_σ, the steady state I becomes approximately

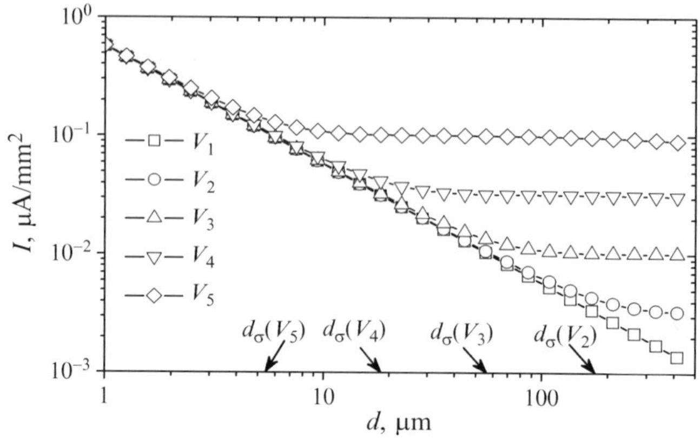

Fig. 32 The dependence of the density I of the steady state current on the membrane thickness d at five enzymatic rates V_{max}: $V_1 = 0$, $V_2 = 1$, $V_3 = 10$, $V_4 = 100$ and $V_5 = 1,000$ (μM/s), $S_0 = 10$ μM/s

constant. For example, in the case of $V_{max} = V_5 = 1$ mM/s, the current density $I = 106\,\text{nA/mm}^2$ at $d = 11.6\,\mu\text{m} \approx 2d_\sigma(d_\sigma = \sqrt{30}\,\mu\text{m}$ at $d = 1\,\mu\text{m})$ differs from the other one $I = 88.7\,\text{nA/mm}^2$ at about 45 times ticker membrane, $d = 500\,\mu\text{m} \approx 91d_\sigma$, only by about 16.3%. The steady state biosensor current decreases linearly with the increase in the membrane thickness d when the enzyme reaction rate distinctly ($\sigma < 0.5$) controls the biosensor response, and it changes slightly only when the biosensor response is significantly under the diffusion control ($\sigma > 2$) (Fig. 32). The approximately linear decreasing function I of d turns to the approximately constant one for all values of V_{max}.

6.4.4 Effect of the Reaction Rate

The results of the investigation of the effect of the maximal enzymatic rate V_{max} on the signal gain G_S are presented in Fig. 33 at two values of the substrate concentration S_0: 10, 100 μM/s and two values of the thickness d: 15 and 150 μm.

As it is possible to notice in Fig. 33, the increase of the gain G_S is approximately linear at relatively high maximal enzymatic rates. The significant increase of G_S starts at V_{max} at which the diffusion module approximately equals one.

Similarly to the function d_σ, the maximal enzymatic activity V_σ is introduced as the function of the thickness d, at which $\sigma^2 = 1$,

$$V_\sigma(d) = \frac{D_S K_M}{d^2}. \tag{103}$$

The increase of G_S starts at $V_{max} \approx V_\sigma$. The gain G_S becomes approximately the linear increasing function of V_{max} at $V_{max} >\approx 4V_\sigma$, i.e. when $\sigma >\approx 2$. This

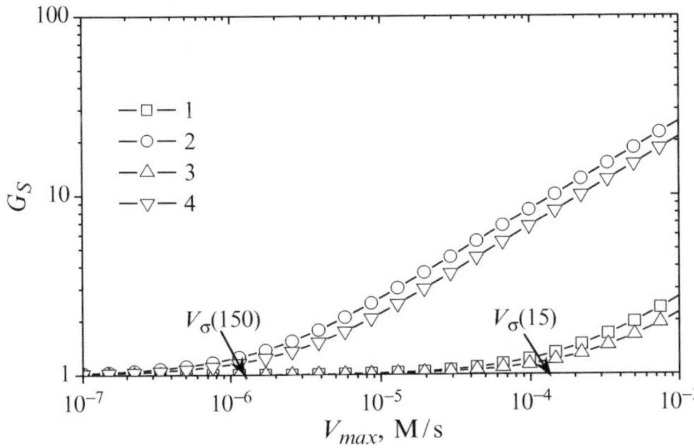

Fig. 33 The response gain G_S versus the maximal enzymatic rate V_{max} at two values of the substrate concentration S_0: 10 (1,2), 100 (3,4) μM/s, and two values of the thickness d: 15 (1,3) and 150 (2,4) μm

property is valid for both concentrations of the substrate and both thicknesses of the enzyme layer. The change of the substrate concentration S_0 by 10 times slightly influences the behaviour of G_S versus V_{max}.

7 Potentiometric Biosensors

Potentiometric sensors are based on the measurement of the potential under no current flow. The analytical information is obtained by converting the recognition process into the potential, which is proportional (in a logarithmic fashion) to the concentration of the reaction product. These devices have been widely used in environmental, medical and industrial applications [84, 106, 218, 245, 246, 275].

For the potentiometric biosensors, the same schemes as for the amperometric ones can be considered. In one of the simplest cases of the potentiometric biosensors, the enzyme (E) catalyzes the substrate (S) conversion to the product (P) [229, 258],

$$S \xrightarrow{E} P \tag{104}$$

7.1 Mathematical Model

The mathematical model of the potentiometric biosensor can be expressed very similarly to the model of the amperometric biosensor [39, 103]. Coupling the enzyme-catalyzed reaction in the enzyme with the diffusion described by Fick's law leads to the following equations of the reaction–diffusion type ($t > 0$):

$$\begin{aligned}\frac{\partial S}{\partial t} &= D_S \frac{\partial^2 S}{\partial x^2} - \frac{V_{max} S}{K_M + S},\\ \frac{\partial P}{\partial t} &= D_P \frac{\partial^2 P}{\partial x^2} + \frac{V_{max} S}{K_M + S}, \quad 0 < x < d,\end{aligned} \tag{105}$$

where x stands for space, t stands for time, $S(x,t)$ is the concentration of the substrate, $P(x,t)$ is the concentration of the reaction product, d is the thickness of the enzyme membrane, D_S and D_P are the diffusion coefficients, V_{max} is the maximal enzymatic rate and K_M is the Michaelis constant.

The biosensor operation starts when the substrate appears over the surface of the enzyme region. This is expressed in the initial conditions ($t = 0$)

$$S(x,0) = 0, \quad 0 \le x < d, \tag{106}$$

$$S(d,0) = S_0, \tag{107}$$

$$P(x,0) = 0, \quad 0 \le x \le d, \tag{108}$$

where S_0 is the concentration of the substrate to be analyzed, $x = 0$ corresponds to the electrode surface, while $x = d$ represents the bulk solution/membrane interface.

Assuming the well-stirred buffer solution, the concentration of the substrate as well as the product at the enzyme membrane surface remains constant [56, 234, 267] ($t > 0$),

$$S(d,t) = S_0, \tag{109}$$

$$P(d,t) = 0. \tag{110}$$

Following the scheme (104) the substrate (S) is electro-inactive,

$$D_S \left.\frac{\partial S}{\partial x}\right|_{x=0} = 0, \quad t > 0. \tag{111}$$

Although, the reaction product (P) is the electro-active substance, in the case of potentiometry, the non-leakage boundary condition is also defined for the product,

$$D_P \left.\frac{\partial P}{\partial x}\right|_{x=0} = 0, \quad t > 0. \tag{112}$$

7.2 *Biosensor Response*

Typically, the change of the potential caused by the reaction product concentration change is measured. The biosensor potential is proportional in the logarithmic fashion to the concentration of the reaction product at the electrode surface ($x = 0$),

$$E(t) = E_0 + \frac{R_c T_K}{zF} \ln P(0,t), \tag{113}$$

where E is the measured potential (in volts), E_0 is the characteristic constant for the ion-selective electrode, R_c is the universal gas constant, $R_c = 8.314$ J/mol K, T_K is the absolute temperature (K), z is the signed ionic charge, F is the Faraday constant [51, 128].

The system (105)–(112) approaches a steady state as $t \to \infty$,

$$E_\infty = \lim_{t\to\infty} E(t), \tag{114}$$

where E_∞ is assumed to be the steady state biosensor potential.

7.3 Finite Difference Solution

The exact analytical solution of the reaction–diffusion problem (105)–(112) only exist in the case of the linear reaction term. In the common case the numerical methods are usually applied.

The governing equations (105) describing the action of the potentiometric biosensor are of the same type as of the amperometric biosensor (3). The numerical approximation of the system (105) has earlier been discussed in detail.

To find the numerical solution of the problem in the domain $[0, d] \times [0, T]$ a discrete grid has to be introduced. Let $\omega_h \times \omega_\tau$ be a uniform discrete grid defined as follows:

$$\begin{aligned} \omega_h &= \{x_i : x_i = ih,\ i = 0, 1, \ldots, N,\ hN = d\}, \\ \omega_\tau &= \{t_j : t_j = j\tau,\ j = 0, 1, \ldots, M;\ \tau M = T\}. \end{aligned} \tag{115}$$

The following notation is used in the finite difference approximations:

$$\begin{aligned} &S_i^j = S(x_i, t_j), \quad P_i^j = P(x_i, t_j), \quad E^j = E(t_j), \\ &i = 0, \ldots, N,\ j = 0, \ldots, M. \end{aligned} \tag{116}$$

The initial conditions (106)–(108) are approximate as follows:

$$S_i^0 = 0, \quad i = 0, \ldots, N-1, \tag{117}$$

$$S_N^0 = S_0, \tag{118}$$

$$P_i^0 = 0, \quad i = 0, \ldots, N. \tag{119}$$

The differential equations (105) can be approximated by the scheme

$$\begin{aligned} \frac{S_i^{j+1}-S_i^j}{\tau} &= D_S\frac{S_{i+1}^{j+1}-2S_i^{j+1}+S_{i-1}^{j+1}}{h^2}-\frac{V_{max}S_i^j}{K_M+S_i^j}, \\ \frac{P_i^{j+1}-P_i^j}{\tau} &= D_P\frac{P_{i+1}^{j+1}-2P_i^{j+1}+P_{i-1}^{j+1}}{h^2}+\frac{V_{max}S_i^j}{K_M+S_i^j}, \\ i &= 1,\dots,N-1,\ j=0,\dots,M-1. \end{aligned} \tag{120}$$

The boundary conditions (109)–(112) are approximated as follows:

$$S_0^j = S_1^j, \quad S_N^j = S_0, \quad j = 1,\dots,M, \tag{121}$$

$$P_0^j = P_1^j, \quad P_N^j = 0, \quad j = 1,\dots,M. \tag{122}$$

The systems of linear finite difference equations can be solved efficiently because of the tridiagonality of the matrices of the systems. Having the numerical solution of the problem, the biosensor potential at time $t = t_j$ is calculated by

$$E^j = E_0 + \frac{R_c T_K}{zF}\ln\left(P_0^j\right), \quad j = 1,\dots,M. \tag{123}$$

7.4 Validation of Numerical Solution

The existing exact analytical solution of the reaction–diffusion problem (105)–(112) can be applied to validate the numerical solution.

When the substrate concentration S_0 to be measured is very small in comparison with the Michaelis constant K_M, the biochemical reaction takes place under the first-order reaction rate. Accepting the substrate and the product equal diffusivities and a relatively low concentration of the substrate, $S_0 \ll K_M$, the product concentration at the electrode surface was found by Carr using the Fourier transformation [61],

$$\begin{aligned} P(0,t) &= S_0\left(1-\frac{1}{\cosh(\sigma)}\right) \\ &\quad -\frac{4}{\pi}\exp\left(-\frac{\pi^2 Dt}{4d^2}\right)\left(\frac{4+\alpha\pi^2\left(1-\exp\left(-\frac{V_{max}}{K_M}t\right)\right)}{4+\alpha\pi^2}\right), \end{aligned} \tag{124}$$

where $D = D_S = D_P$, σ^2 is the dimensionless diffusion module (the Damköhler number) defined by (13).

Assuming $t \to \infty$ the corresponding product concentration at the steady state conditions can be obtained from formulae (124),

$$P(0,t) = S_0 \left(1 - \frac{1}{\cosh(\sigma)}\right). \tag{125}$$

Having (125) the steady state potential E_∞ is easily calculated,

$$E_\infty = E_0 + \frac{R_c T_K}{zF} \ln\left(S_0 \left(1 - \frac{1}{\cosh(\sigma)}\right)\right). \tag{126}$$

In the case of the high enzymatic activity V_{max}, Eq. (124) may be significantly reduced [128],

$$P(0,t) = S_0 \left(1 - \frac{1}{\cosh(\sigma)}\right) - \frac{4}{\pi} \exp\left(-\frac{\pi^2 Dt}{4d^2}\right). \tag{127}$$

In the case of the especially high substrate concentration (the zero-order reaction rate), $S_0 \gg K_M$, the product concentration at the electrode surface was also found analytically [61, 128],

$$P(0,t) = \frac{V_{max} d^2}{D} \left(\frac{1}{2} - \frac{16}{\pi^3}\right) \exp\left(-\frac{\pi^2 Dt}{4d^2}\right). \tag{128}$$

Assuming $t \to \infty$, the corresponding stationary potential E_∞ is calculated as follows ($S_0 \gg K_M$):

$$E_\infty = E_0 + \frac{R_c T_K}{zF} \ln\left(\frac{V_{max} d^2}{2D}\right) = E_0 + \frac{R_c T_K}{zF} \ln\left(\frac{\sigma^2 K_M}{2}\right). \tag{129}$$

7.5 Simulated Biosensor Response

Figure 34 shows both the substrate and the product concentration profiles simulated at the following values of the parameters:

$$\begin{aligned} &D_S = D_P = 300\,\mu\text{m}^2/\text{s}, \quad d = 100\,\mu\text{m}, \\ &K_M = 100\,\mu\text{M}, \quad V_{max} = 100\,\mu\text{M/s}, \quad S_0 = 0.1 K_M = 10\,\mu\text{M}, \\ &E_0 = 0\,\text{V}, \quad z = 1, \quad T_K = 298\,K. \end{aligned} \tag{130}$$

Figure 34 presents the concentrations at the approximate steady state conditions ($t = 133$ s) and the following intermediate values of time t: 1, 5, 10, 20 s. Since the profiles of the substrate concentration S at $t \geq 5$ s are practically identical to the profile at the steady state conditions, the intermediate concentration profiles are not marked individually in this figure.

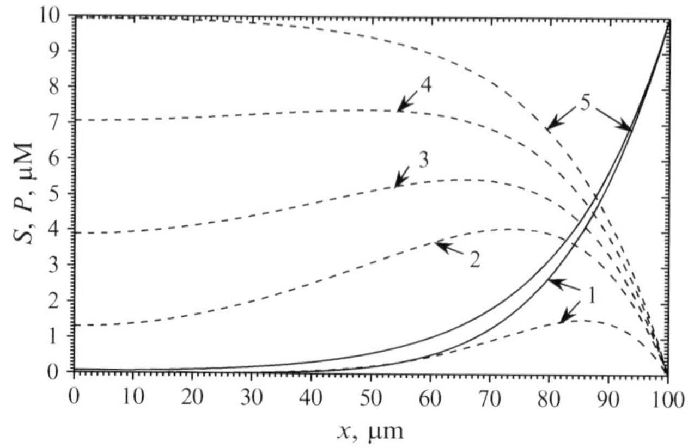

Fig. 34 The concentration profiles of the substrate (*solid lines*) and the product (*dashed lines*) at the following values of time t: 1 (1), 5 (2), 10 (3), 20 (4) and 133 (5) s. The values of the parameters are defined in (130)

At the steady state conditions ($\partial S/\partial t = \partial P/\partial t = 0$), the governing equations (105) and the boundary conditions (109)–(112) lead to

$$P(x,t) + S(x,t) = S_0, \quad t \to \infty,\ x \in [0,\ d]\,. \tag{131}$$

Because of this, the symmetry with respect to the axis $S = P = 0.5S_0$ can be noticed at $t = 133$ s in Fig. 34.

In addition, one can see in Fig. 34 that $P(0,133) \approx S_0$ and $S(0,133) \approx 0$, i.e. under the steady state conditions, the product concentration at the electrode surface equals the concentration of the substrate in the buffer solution. However, this feature of the potentiometric biosensors is only valid in the case of high enzymatic rate. Figure 35 shows the substrate and the product concentration profiles for different values of the maximal enzymatic rate V_{max} at the steady state conditions.

In the numerical simulation, the steady state time T_R was assumed as the time when the absolute potential slope value falls below a given small dimensionless decay rate ε normalized with the potential value,

$$T_R \approx \min_{P(0,t)>0} \left\{ t : \left| \frac{t}{E(t)} \times \frac{\mathrm{d}E(t)}{\mathrm{d}t} \right| < \varepsilon \right\}. \tag{132}$$

The potential $E_R = E(T_R)$ at the biosensor response time T_R was assumed as the steady state biosensor potential E_∞, $E_R \approx E_\infty$. In calculations, $\varepsilon = 10^{-2}$ was used.

The response time T_R as an approximate steady state time is very sensitive to the decay rate ε, i.e. $T_R \to \infty$, when $\varepsilon \to 0$. To investigate the behaviour of the response time, the time $T_{0.001}$ of the potential differing from the steady state potential by 0.001 V is usually used,

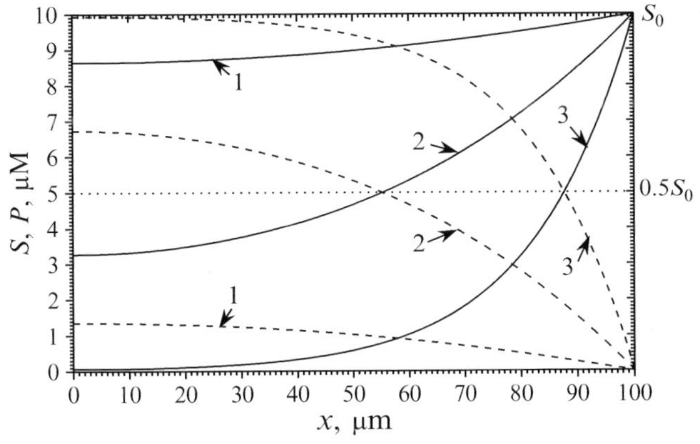

Fig. 35 The steady state concentration profiles of the substrate (*solid lines*) and the product (*dashed lines*) at three maximal enzymatic rates V_{max}: 1 (1), 10 (2), 100 (3) μM/s. The other parameters are defined in (130)

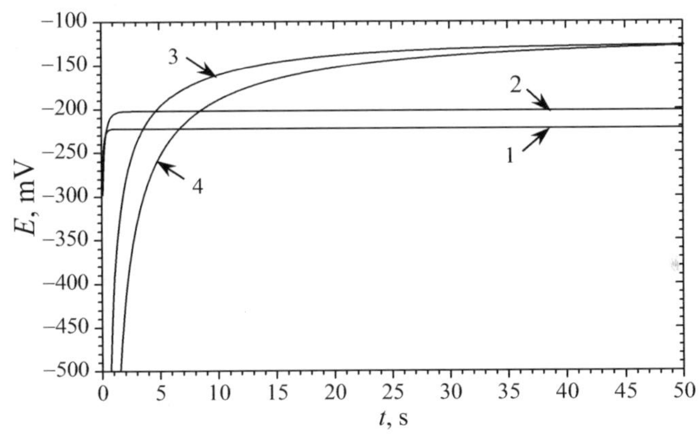

Fig. 36 The dynamics of the biosensor potential E at four values of the membrane thickness d: 10 (1), 15 (2), 100 (3), 150 (4) μm. The other parameters are the same as in Fig. 34

$$T_{0.001} = \min\{t : E(t) > E_R - 0.001\}, \quad E_R = E(T_R), \quad E_\infty \approx E_R. \quad (133)$$

At the conditions defined in (130), the steady state time $T_{0.001}$ is approximately equal to 78 s.

The evolution of the biosensor potential $E(t)$ obtained by using computer simulation is presented in Fig. 36. The biosensor response was modeled for the enzyme membranes of different thickness d: 10, 15, 10 and 150 μm.

When comparing the evolution of the biosensor current in two cases of relatively thin ($d = 10$ and 15 μm) membranes, one can see that the biosensor response is noticeably higher at the thicker membrane ($d = 15$ μm) than at the thinner one

($d = 10\,\mu m$). However, when comparing the biosensor responses in the other two cases of ten times thicker ($d = 100$ and $150\,\mu m$) membranes, one can see the opposite tendency, i.e. the biosensor with the thicker ($d = 150\,\mu m$) membrane generates the lower response than with the thinner one ($d = 100\,\mu m$). However, this difference decreases with the increase in time. When simulating response for a longer time ($t > 50\,s$), the curves intersects so, that the biosensor with the thicker ($d = 150\,\mu m$) membrane generates the higher potential than with the thinner one ($d = 100\,\mu m$) as it has been noticed for the tenfold thinner membranes. So, the effect of the membrane thickness on the biosensor response in the case of the potentiometry differs from that in the case of the amperometry (Fig. 2).

7.6 *Peculiarities of the Biosensor Response*

7.6.1 Effect of the Enzyme Membrane Thickness

The dependence of the biosensor response on the thickness d of the enzyme membrane has been investigated. Figure 37 shows the steady state potential while Fig. 38 presents the time $T_{0.001}$ versus the thickness d of the enzyme membrane.

As it is possible notice in Fig. 37 the steady state potential E_R is a monotonously increasing function of the thickness d of the enzyme layer at all values of the maximal enzymatic rate V_{max}. However, at high values of d the potential E_R is practically a constant function of d. When increasing the thickness d, the function $E_R(d)$ reaches approximately the same maximum for all values of V_{max}. At moderate thickness d the higher maximal enzymatic rate V_{max} corresponds to the greater value of E_R.

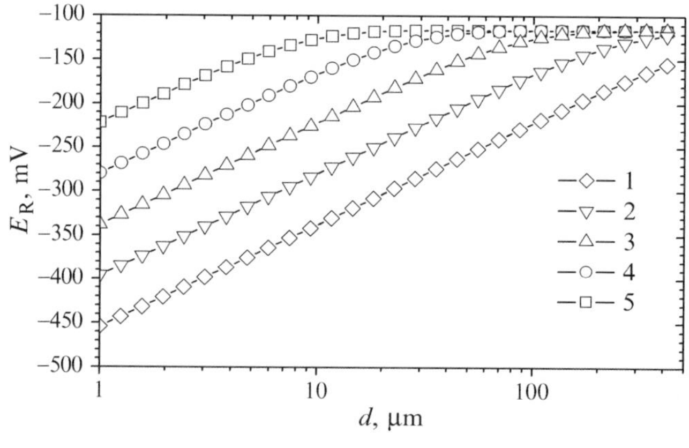

Fig. 37 The dependence of the steady state potential E_R on the thickness d of the enzyme membrane at five values of the maximal enzymatic rate V_{max}: 0.1 (1), 1 (2), 10 (3), 100 (4) and 1,000 (5) $\mu M/s$. The values of the other parameters are defined in (130)

Since in the calculations above the substrate concentration was relatively low, $S_0 = 10\,\mu\text{M} = 0.1K_M$, formula (126) can be applied to find analytically the membrane thickness d at which the steady state potential E gains the maximum for given values of z, T_K, D_S, D_P, S_0, V_{max}, K_M. At first, the derivative of $E(d)$ with respect to the thickness d is calculated,

$$\frac{\mathrm{d}E_\infty(d)}{\mathrm{d}d} = \frac{RT_K}{zF} \times \frac{\tanh(\sigma)}{\cosh(\sigma) - 1} \times \frac{\sigma}{d}. \tag{134}$$

It is rather easy to see that the derivative $\mathrm{d}E_\infty(d)/\mathrm{d}d$ never gets zero and is positive at all positive values of σ as well as d,

$$\forall \sigma > 0 : \quad \frac{\mathrm{d}E_\infty(d)}{\mathrm{d}d} > 0 \quad \text{if} \quad S_0 \ll K_M. \tag{135}$$

Consequently, $E_\infty(d)$ is a monotonously increasing function of the enzyme membrane thickness d. Since

$$\frac{\mathrm{d}E_\infty(\sigma)}{\mathrm{d}\sigma} = \frac{RT_K}{zF} \times \frac{\sigma \tanh(\sigma)}{\cosh(\sigma) - 1} > 0 \quad \text{at } \sigma > 0, \tag{136}$$

then E_∞ is a monotonously increasing function also of the diffusion module σ.

As has been mentioned above, in the case of $S_0 \ll K_M$ when increasing the enzyme membrane thickness d the steady state potential E_∞ approaches practically the same values for all enzymatic activities V_{max} (see Fig. 37). That property of potentiometric biosensors can be proved as follows:

$$\begin{aligned} \lim_{\sigma\to\infty} E_\infty &= E_0 + \frac{R_c T_K}{zF} \lim_{\sigma\to\infty} \ln\left(S_0\left(1 - \frac{1}{\cosh(\sigma)}\right)\right) \\ &= E_0 + \frac{R_c T_K}{zF} \ln S_0 \,. \end{aligned} \tag{137}$$

In the case of $S_0 \gg K_M$ the steady state potential also increases monotonously with the increase in the diffusion module σ as well as in the thickness d (see (129)). However, there is no finite limit in this case.

From (126) and (129) follows that the steady state potential extremely decreases when the membrane thickness d reduces to zero, i.e. $E_\infty \to -\infty$ when $d \to 0$. Since $E_\infty \to -\infty$ also at $\sigma \to 0$, this property is valid for all values of the enzymatic activity.

Figure 38 shows that the response time $T_{0.001}$ of the potentiometric biosensor noticeably increases with the increase in the thickness d of the enzyme membrane. This increase is approximately linear. The effect of V_{max} on the response time is slight.

When comparing the effect of the membrane thickness on the response time of the potentiometric biosensor (Fig. 38) with that of the corresponding amperometric

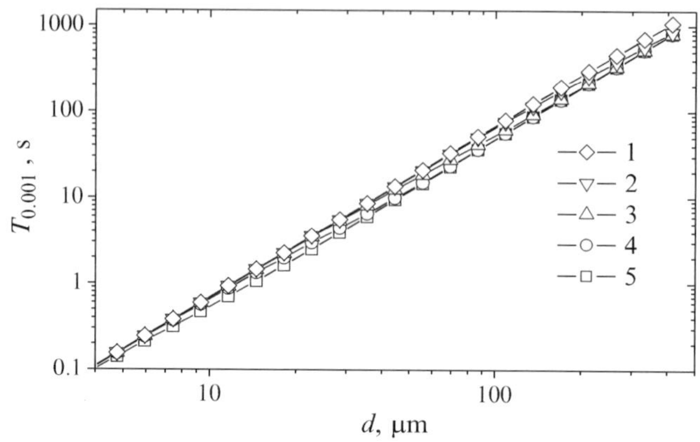

Fig. 38 The dependence of the response time $T_{0.001}$ on the thickness d of the enzyme layer. The notation and values of the parameters are the same as in Fig. 37

biosensor (Fig. 7), one can see that the response time $T_{0.001}$ of a potentiometric biosensor is about tenfold greater than the time $T_{0.5}$ of the corresponding amperometric biosensor.

7.6.2 Stability of the Response

As has been mentioned above, it is very important to have the biosensors which would keep their analytical capability for a long period. Usually, the maximal enzymatic rate V_{max} decreases permanently due to the enzyme inactivation. Figure 37 shows how sensitive is the response of the potentiometric biosensor to changes of the enzymatic activity V_{max}. The variation of the response is especially notable in the cases of relatively thin enzyme membranes. In the cases of the relatively thick enzyme membrane, E_R practically does not vary by changing V_{max}. Consequently, the biosensor containing the thicker enzyme layer gives more stable response than the biosensor with the thinner layer. However, the thick membrane-based biosensors have very durable response time (see Fig. 38). For example, the response time $T_{0.001}$ varies from 47 to 68 s when the membrane thickness d equals 100 μm. The time $T_{0.001}$ is even more durable at thicker enzyme membrane. The time $T_{0.001}$ is approximately an exponentially increasing function of the thickness d. The influence of the maximal enzymatic rate V_{max} on the response time $T_{0.001}$ is rather slight.

Similar to the amperometric biosensors, the problem of the membrane thickness optimization can also be formulated for the potentiometric biosensors. The goal is to find the thickness of the membrane as small as possible, ensuring the stability of the biosensor response at a range of V_{max} as wide as possible. Let V_1 and V_2 be two values of the maximal enzymatic rate V_{max} ($V_1 < V_2$) for which the biosensor response has to be stable. The minimal membrane thickness $d_\varepsilon(V_1, V_2, S_0)$ ensuring the stable response is introduced as the relative difference $R(d, V_1, V_2, S_0)$ between the biosensor response (the steady state biosensor potential E_R) at $d = d_\varepsilon$, $V_{max} = V_1$ and the other response at $d = d_\varepsilon$, $V_{max} = V_2$ is less than the dimensionless decay rate ε,

$$R(d, V_1, V_2, S_0) = \left| \frac{E_R(d, V_1, S_0) - E_R(d, V_2, S_0)}{E_R(d, V_1, S_0)} \right|, \tag{138}$$

$$d_\varepsilon(V_1, V_2, S_0) = \min_{d>0} \{d : R(d, V_1, V_2, S_0) < \varepsilon\}, \tag{139}$$

where $E_R(d, V_{max}, S_0)$ is the biosensor potential calculated at the membrane thickness d, the maximal enzymatic rate V_{max} and the substrate concentration S_0.

Let us accept $S_0 = 10\,\mu$M, $V_1 = 100\,\mu$M/s, $V_2 = 1{,}000\,\mu$M/s and $\varepsilon = 0.01$. From the numerical results presented in Fig. 37 follows that $d_\varepsilon \approx 66\,\mu$m. To evaluate the biosensor stability, the response of the biosensor based on the membrane of the thickness $d = d_\varepsilon(V_1, V_2, S_0) = 66\,\mu$m has to be calculated at wide ranges of the substrate concentration S_0 and the maximal enzymatic rate V_{max}.

Figure 39 shows the potential E_R versus the substrate concentration S_0 at five values of V_{max}: 0.1, 1, 10, 100 and 1,000 μM/s. No notable difference (Fig. 39) is observed between the values of E_R, calculated at two largest values of V_{max}, 100 and 1,000 μM/s, when the substrate concentration S_0 is less than about 1 mM. Figure 39 expressively shows the stable response of the biosensor based on the enzyme membrane of the thickness $d = 66\,\mu$m, when the maximal enzymatic rate reduces ten times, from 1,000 to 100 μM/s. Although the membrane thickness d_ε was calculated for the substrate concentration $S_0 = 10\,\mu$M, the biosensor response to the substrate of the concentration S_0 is also sufficiently stable up to about 500 μM. Figure 39 also shows that the response of the biosensor of the thickness

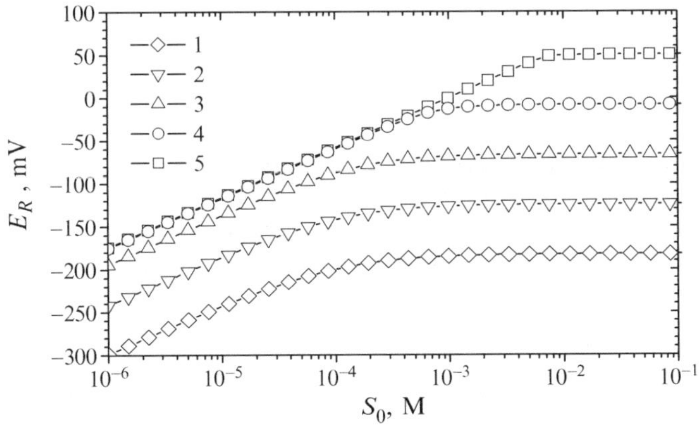

Fig. 39 The dependence of the steady state potential E_R on the substrate concentration S_0 at five maximal enzymatic rates (V_{max}): 0.1 (1), 1 (2), 10 (3), 100 (4) and 1,000 (5) μM/s, $d = 66\,\mu$m

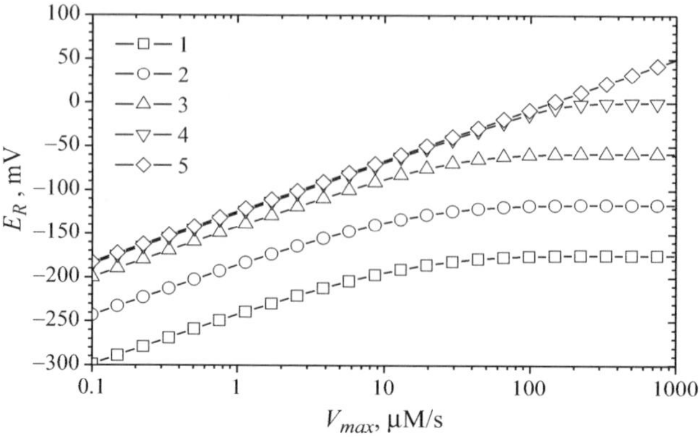

Fig. 40 The dependence of the steady state potential E_R on the maximal enzymatic rate V_{max} at five substrate concentrations (S_0), 1 (1), 10 (2), 100 (3), 1,000 (4) and 10,000 (5) μM, $d = 66\,\mu$m

of 66 μm is approximately constant at the concentrations higher than about 10 mM changing V_{max} from 0.1 to 1,000 μM/s. Because of this, such a biosensor is not practically useful to determine the substrate of larger concentrations.

7.6.3 Effect of the Reaction Rate

Figure 40 shows the steady state potential E_R versus the maximal reaction rate V_{max} at five values of the substrate concentration S_0, 1, 10, 100, 1,000 and 10,000 μM, and the same membrane thickness d as above, i.e. $d = 66\,\mu$m. One can see in Fig. 40 that the response of the biosensor is stable to changes in the maximal enzymatic rate V_{max} from 100 to 1,000 μM/s when the concentration S_0 is equal to or less than 100 μM. In the case of $V_{max} < 100\,\mu$M/s the biosensor response is very sensitive to changes in V_{max}.

7.6.4 Dependence of the Response on the Diffusion Module

The influence of the membrane thickness d and the maximal enzymatic rate V_{max} on the biosensor response has discussed. The dimensionless diffusion module σ^2 directly combines both these parameters of the mathematical model (105)–(112). The diffusion module σ^2 increases with the increase of both parameters V_{max} and d.

The numerical experiments presented in Fig. 37 were repeated at three additional substrate concentrations (S_0): 10^{-7}, 10^{-4}, 10^{-1} M. Figure 41 shows the calculated values of the potential E_R as a function of the diffusion module σ^2. One can see in this figure that the values of E_R calculated at a concrete enzymatic rate V_{max} form a continuous curve. In both extreme cases of S_0, $S_0 \ll K_M$ and $S_0 \gg K_M$, the steady state potential is expressed through the module σ with no additional entries

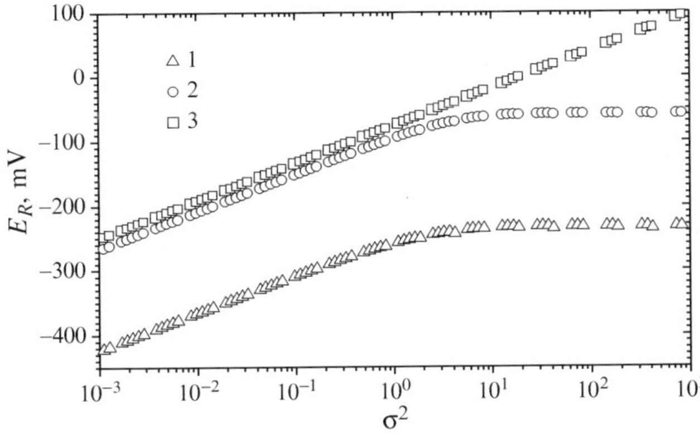

Fig. 41 The dependence of the steady state potential E_R on the dimensionless diffusion module σ^2 at three substrate concentrations (S_0), 10^{-7} (1), 10^{-4} (2), 10^{-1} (3) M, and five maximal enzymatic rates (V_{max}), 0.1, 1, 10, 100 and 1,000 μM/s, changing the membrane thickness d

of V_{max} and d (see formulas (126) and (129)). Figure 41 shows that this property is also valid for intermediate values of S_0 when $S_0 \approx K_M$. So, the decrease in the steady state potential, which appears due to the decrease of V_{max}, can really be compensated by the increase in the membrane thickness d. This feature of the response is used in the framework defined by (138) and (139).

8 Enzyme Inhibition

Very often the kinetics of enzyme action is much more complicated than the Michaelis–Menten kinetics. Inhibition, activation, allostery and other types of non-Michaelis–Menten kinetics are known for the diversity of enzymes [72, 81, 98].

A number of substances may cause a reduction in the rate of an enzyme catalyzed reaction. Enzyme inhibitors are molecules that interact with enzymes and decrease their activity [80, 81]. Substrate and product inhibition is where either the substrate or the product of an enzyme reaction inhibit the enzyme's activity.

8.1 *Substrate Inhibition*

Along with its normal effect of enhancing the rate of reaction, the substrate also acts as an inhibitor. In the substrate inhibition there is a progressive decrease in the activity at high substrate concentrations.

The scheme of the enzyme (E) catalyzed conversion of the substrate (S) to the product (P)

$$\mathrm{S} \longrightarrow \mathrm{P} \tag{140}$$

is a simplified version of the following scheme

$$E + S \longleftrightarrow ES \longrightarrow E + P \tag{141}$$

According to scheme (141), the substrate (S) combines reversibly with the enzyme (E) to form the complex (ES). The complex then dissociates into product (P) and the enzyme is regenerated.

The simplest scheme of non-Michaelis–Menten kinetics may be, for example, described by adding the relationship of the interaction of the enzyme substrate complex (ES) with another substrate molecule (S) followed by the generation of the non-active complex (ESS) to the Michaelis–Menten scheme (141),

$$ES + S \longleftrightarrow ESS \tag{142}$$

8.1.1 Mathematical Model

In the case of the substrate (S) conversion to the product (P) catalyzed by the enzyme (E) and proceeded according to the scheme (141), (142), the reaction rate decreases with the increase in the substrate concentration which then acts as a negative effect. The kinetic equations for the substrate inhibition are

$$\begin{aligned} \frac{\partial S}{\partial t} &= D_S \frac{\partial^2 S}{\partial x^2} - \frac{V_{max} S}{K_M + S(1 + S/K_s)}, \\ \frac{\partial P}{\partial t} &= D_P \frac{\partial^2 P}{\partial x^2} + \frac{V_{max} S}{K_M + S(1 + S/K_s)}, \quad x \in (0,\ d), \quad t > 0, \end{aligned} \tag{143}$$

where x and t stand for space and time, respectively, $S(x,t)$ is the concentration of the substrate, $P(x,t)$ is the concentration of the reaction product, d is the thickness of the enzyme layer, D_S, D_P are the diffusion coefficients, V_{max} is the maximal enzymatic rate and K_M and K_s are the Michaelis and the substrate inhibition constants, respectively [148].

The governing equations (143) together with the following initial and the boundary conditions form the initial boundary value problem ($t > 0$):

$$S(x,0) = 0, \quad x \in [0,\ d), \tag{144}$$

$$S(d,0) = S_0, \tag{145}$$

$$P(x,0) = 0, \quad x \in [0,\ d], \tag{146}$$

$$P(0,t) = P(d,t) = 0, \tag{147}$$

$$D_S \frac{\partial S}{\partial x}\bigg|_{x=0} = 0, \tag{148}$$

$$S(d,t) = S_0. \tag{149}$$

Mathematical model (143)–(149) describes the behaviour of the amperometric biosensor with the substrate inhibition.

Let us notice, that the governing equations (143) are reduced to the equations describing the Michaelis–Menten kinetics (3) in the limit as K_s approaches infinity, $K_s \to \infty$.

The density $i(t)$ of the biosensor current at time t is expressed as usual,

$$i(t) = n_e F D_P \left.\frac{\partial P}{\partial x}\right|_{x=0} . \tag{150}$$

The system (143)–(149) approaches a steady state as $t \to \infty$. However, the biosensor current can be a non-monotonous function. Because of this, the density I of steady state and the density I_{max} of the maximal current are considered,

$$I = \lim_{t\to\infty} i(t), \quad I_{max} = \max_{t>0} \{i(t)\} . \tag{151}$$

8.1.2 Solution of the Problem

Due to highly nonlinear reaction term in (143), the problem (143)–(149) was solved numerically.

The discretization of the initial (144)–(146) and the boundary (147)–(149) conditions has been discussed in Sect. 3. The governing equation (143) can be approximated with the following implicit finite difference scheme:

$$\begin{aligned}
\frac{S_i^{j+1} - S_i^j}{\tau} &= D_S \frac{S_{i+1}^{j+1} - 2S_i^{j+1} + S_{i-1}^{j+1}}{h^2} - \frac{V_{max} S_i^j}{K_M + S_i^j (1 + S_i^j / K_s)}, \\
\frac{P_i^{j+1} - P_i^j}{\tau} &= D_P \frac{P_{i+1}^{j+1} - 2P_i^{j+1} + P_{i-1}^{j+1}}{h^2} + \frac{V_{max} S_i^j}{K_M + S_i^j (1 + S_i^j / K_s)}, \\
i &= 1, \ldots, N-1, \; j = 0, \ldots, M-1 .
\end{aligned} \tag{152}$$

8.2 *Effect of Substrate Inhibition*

Figure 42 shows the results of the numerical modeling of the biosensor at a relatively low enzyme activity of $V_{max} = 1\,\mu\text{M/s}$ and the thickness of the enzyme layer of $d = 100\,\mu\text{m}$ [148]. The values of the other parameters of the mathematical model are as follows:

$$\begin{aligned}
&D_S = D_P = 300\,\mu\text{m}^2/\text{s}, \quad n_e = 2, \quad d = 100\,\mu\text{m} \\
&K_M = 100\,\mu\text{M}, \quad K_s = 0.1 K_M .
\end{aligned} \tag{153}$$

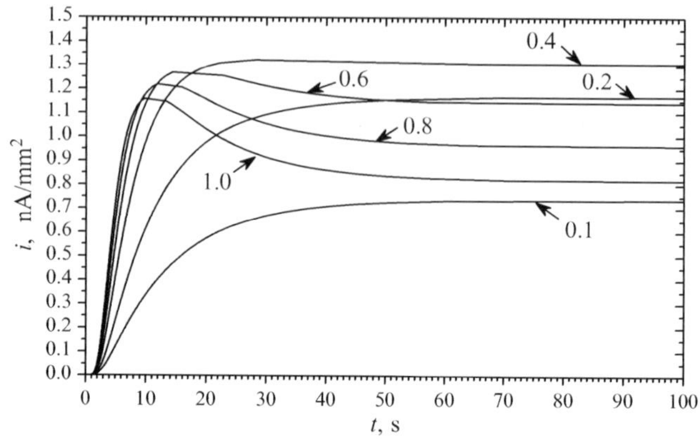

Fig. 42 The dynamics of the density i of the biosensor current at different values of the normalized substrate concentration. The numbers on the curves show S_0/K_M values, $V_{max} = 1\,\mu M/s$, $d = 100\,\mu m$

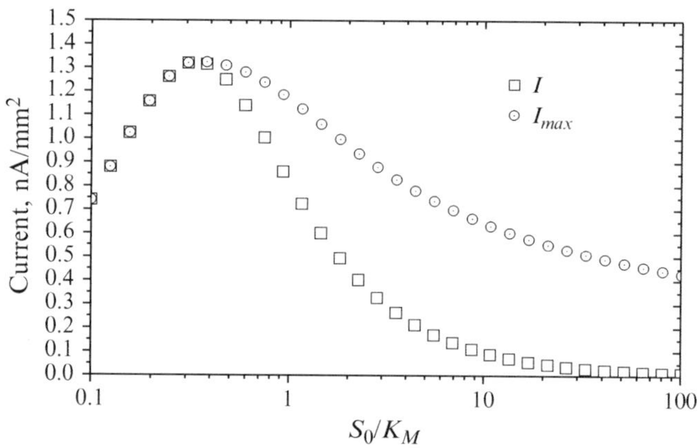

Fig. 43 The dependence of the density of the steady state (I) as well as the maximal (I_{max}) current on the normalized substrate concentration. The calculation parameters are the same as in Fig. 42

The results indicate a complex character of the transition state. At low substrate concentration, the kinetic curves resembled the diffusion shape, whereas at S_0/K_M above 0.4 they showed the maximal values.

The dependence of the maximal and the steady state responses on the normalized substrate concentration is shown in Fig. 43.

The appearance of the maximal response value is associated with the enzyme kinetics. The differentiation of the enzyme reaction rate on the bulk substrate concentration gives the maximal value at $S_0/K_M = 0.34$ (Fig. 43).

The appearance of the maximal response at the non-steady conditions (Fig. 42) is associated with the change of the substrate concentration in the enzyme layer that agrees with the dependence of the enzyme rate on the substrate concentration. The steady state response is determined by the stationary substrate concentration and the enzymatic rate. This response is less than the maximal value due to the inhibition of the enzyme activity with the substrate.

The modeling of the amperometric biosensor was also performed at a significantly greater value of the maximal enzymatic activity V_{max} of 1 mM/s (Figs. 44 and 45). At this value of V_{max} the diffusion module σ^2 is approximately equal to 333 while in the case of a low activity of 1 μM/s, $\sigma^2 = 1/3$. This means that at higher enzymatic activity the biosensor response is strongly limited by the diffusion while at lower activity the biosensor acts in the kinetic regime.

When the enzyme activity is high, the response change is quite different at different substrate concentrations (Figs. 44 and 45). At S_0/K_M less than 11, the

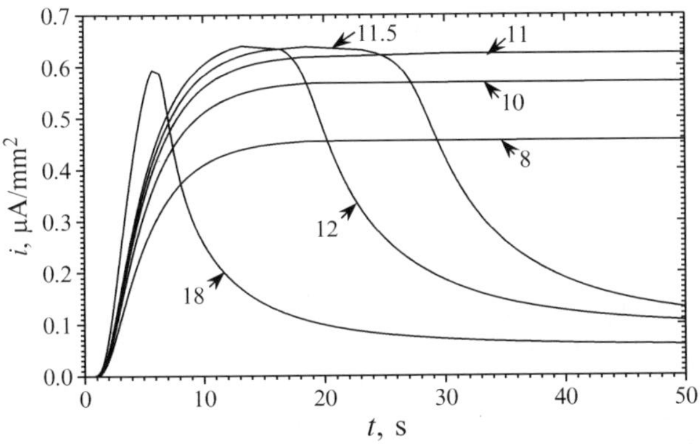

Fig. 44 The density i of the biosensor current at different values of the normalized substrate concentration. The numbers on the curves show values of S_0/K_M, $V_{max} = 1$ mM/s, $d = 100$ μm

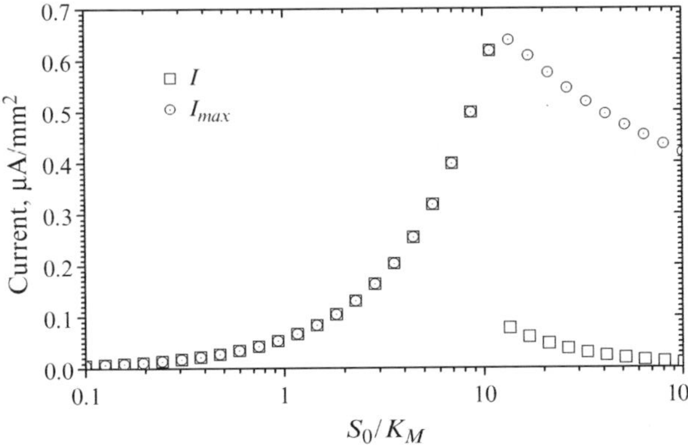

Fig. 45 The dependence of the density of the steady state (I) as well as the maximal (I_{max}) current on the normalized substrate concentration at $V_{max} = 1$ mM/s. The other model parameters are the same as in Fig. 43

response monotonous increases. However, a farther small increase in the substrate concentration causes a sharp decrease in the response for a time of about 36 s. This time shortens to 17 s at $S_0/K_M = 12$, but the character of the response change is similar. The farther increase in S_0 changes the character of the response kinetics as shown for S_0/K_M value 18 (Fig. 44). These sharp changes in response more precisely demonstrate the data presented in Fig. 45.

The modeling of the amperometric biosensor with the substrate inhibition reveals the complex kinetics of the biosensor response. At low substrate concentration, the kinetics looks like a simple substrate diffusion. At the substrate concentration comparable to the Michaelis–Menten constant K_M, the response change shows the maximal value. At high enzyme activity V_{max} and high (in comparison with K_M) substrate concentration S_0 a sharp response change was indicated (Figs. 44 and 45). The numerical simulations show that the substrate inhibition produces different calibration graphs for the biosensor response at the transition and the steady state.

8.3 Product Inhibition

The products of many enzyme-catalyzed reactions behave as inhibitors when they are presented in the reaction mixture. From this point of view they are called product inhibitors and the phenomenon is known as the product inhibition.

The product inhibition rather often causes a loss in the productivity of the enzyme process at high degrees of conversion. The loss results from the inhibition of the enzyme by the product. This is due to the structural similarity of the product to the substrate.

The product inhibition can be described by adding the relationship of the interaction of the product (P) with the enzyme (E) to the Michaelis–Menten scheme. Due to the interaction the product enzyme complex (EP) is produced. The complex then dissociates into the product (P), and the enzyme (E) is regenerated,

$$\mathrm{E} + \mathrm{S} \longleftrightarrow \mathrm{ES} \longrightarrow \mathrm{E} + \mathrm{P} \longleftrightarrow \mathrm{EP} \tag{154}$$

The reversible interaction of the product with the enzyme is often neglected, although it is an essential component of the entire process.

8.3.1 Mathematical Model

The kinetic equations for the product inhibition are

$$\begin{aligned} \frac{\partial S}{\partial t} &= D_S \frac{\partial^2 S}{\partial x^2} - \frac{V_{max} S}{K_M(1 + P/K_p) + S}, \\ \frac{\partial P}{\partial t} &= D_P \frac{\partial^2 P}{\partial x^2} + \frac{V_{max} S}{K_M(1 + P/K_p) + S}, \quad x \in (0, d), \quad t > 0, \end{aligned} \tag{155}$$

where K_p is the product inhibition constant. The meaning of all other parameters is the same as in the case of the substrate inhibition.

The governing equations (155) with the initial conditions (144), (146) and the boundary conditions (147), (149) form together the initial boundary value problem describing the behaviour of an amperometric biosensor with the product inhibition.

The governing equations (155) are reduced to the Michaelis–Menten equations (3) in the limit as K_p approaches infinity, $K_p \to \infty$.

The density $i(t)$ of the biosensor current and the density I of the steady state current are expressed as in the case of the substrate inhibition.

8.3.2 Solution of the Problem

The governing equation (155) can be approximated with the following implicit finite difference scheme:

$$\begin{aligned}
\frac{S_i^{j+1} - S_i^j}{\tau} &= D_S \frac{S_{i+1}^{j+1} - 2S_i^{j+1} + S_{i-1}^{j+1}}{h^2} - \frac{V_{max} S_i^j}{K_M(1 + P_i^j/K_p) + S_i^j}, \\
\frac{P_i^{j+1} - P_i^j}{\tau} &= D_P \frac{P_{i+1}^{j+1} - 2P_i^{j+1} + P_{i-1}^{j+1}}{h^2} + \frac{V_{max} S_i^j}{K_M(1 + P_i^j/K_p) + S_i^j}, \\
i &= 1, \dots, N-1, \; j = 0, \dots, M-1.
\end{aligned} \tag{156}$$

Since all other equations of the mathematical model are the same as for a biosensor with the substrate inhibition, the same numerical approximation can be applied.

8.4 *Effect of Product Inhibition*

Figure 46 shows the results of the numerical simulation of the action of the biosensor with the product inhibition at low ($V_{max} = 1\,\mu\text{M/s}$) as well as at high ($V_{max} = 1\,\text{mM/s}$) enzyme activities. The simulated values of the steady state current of the corresponding amperometric biosensor without the product inhibition are also presented in Fig. 46. To see the effect of the product inhibition on the biosensor response, the inhibition constant K_p should be significantly less than K_M, e.g. $K_p = 0.001\,K_M$. The values of the other parameters used in the simulation can be the same as in the case of the substrate inhibition (153).

As one can see in Fig. 46 the effect of the product inhibition is rather different from that of the substrate inhibition (Figs. 44 and 45).

In both opposite cases of the enzyme activity, low and high, no notable effect of the product inhibition on the steady state current was observed at low and

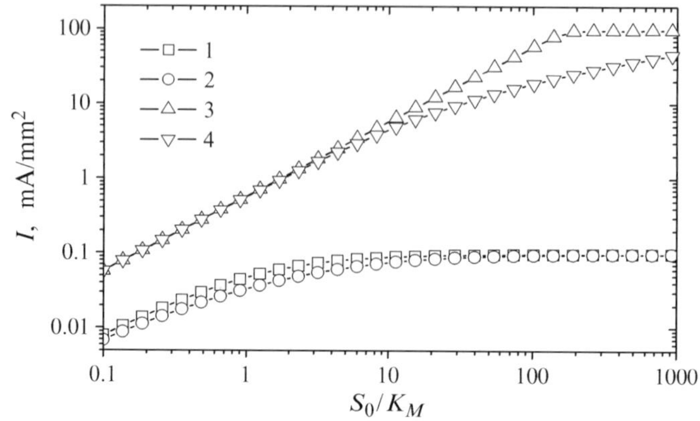

Fig. 46 The dependence of the steady state current on the normalized substrate concentration for the biosensors with (2, 4) and without (1, 3) product inhibition at two values of V_{max}: 1 (1, 2) and 1,000 (3, 4) μM/s

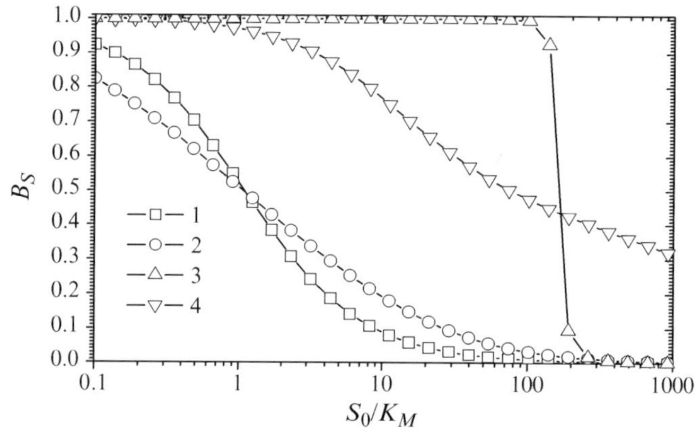

Fig. 47 The sensitivity of the biosensors with (2,4) and without (1,3) the product inhibition versus the normalized substrate concentration. The parameters and the notations are the same as in Fig. 46

high substrate concentrations. Due to the product inhibition the steady state current decreases notably only at intermediate concentrations of the substrate.

At low enzymatic activity ($V_{max} = 1\,\mu M/s$), due to the product inhibition the steady state current differs from that of the corresponding biosensor without the product inhibition when the normalized substrate concentration S_0/K_M varies between 0.1 and 10. The peak in difference is observed at $S_0/K_M \approx 1.5$ when the density of the steady state current decreases about 1.4 times due to the product inhibition.

At the high enzymatic activity ($V_{max} = 1\,mM/s$), the peak in difference between the responses of the biosensors with and without the product inhibition appears at significantly greater substrate concentration of $S_0/K_M \approx 300$. In this case the product inhibition causes the decrease of the steady state current in about 3.9 times.

As it is possible to notice in Fig. 46 the shape of curves is rather different. The curves of the steady state current vs. the substrate concentration S_0 for the biosensors with the product inhibition are slightly smoother than those of the biosensors without the inhibition. Since the shape of these curves has a direct relation to the sensitivity of the biosensors the normalized sensitivity B_S was calculated as defined by (20). The calculation results are depicted in Fig. 47.

Figure 47 shows noticeable effect of the product inhibition on the biosensor sensitivity. That effect depends on the enzyme activity as well as the substrate concentration. At high enzymatic activity ($V_{max} = 1$ mM/s) and high substrate concentration ($S_0/K_M > 200$) the product inhibition causes the substantial increase in the biosensor sensitivity. However, at this activity and S_0/K_M being between 1 and 100, the product inhibition decreases the sensitivity. Similar effect can be noticed in the case of low enzymatic activity ($V_{max} = 1$ μM/s). Depending on the enzyme activity and the substrate concentration the product inhibition causes the increase or the decrease in the biosensor sensitivity (see Fig. 47).

One-Layer Multi-Enzyme Models of Biosensors

Contents

1 Biosensors Response to Mixture of Compounds 114
1.1 Mathematical Model 114
1.2 Solution of the Problem 116
1.3 Generation of Data Sets 117
1.4 Concluding Remarks 120
2 Biosensors Acting in Trigger Mode 120
2.1 Mathematical Models 121
2.2 Finite Difference Solution 126
2.3 Simulated Response 128
2.4 Peculiarities of the Response 129
2.5 Concluding Remarks 136

The amperometric biosensors have proved to be reliable and low-cost in various analytical systems with applications in biotechnology, medicine and environmental monitoring [106, 218, 229, 246, 275]. However, amperometric biosensors possess a number of serious drawbacks. One of the main reasons restricting wider use of the biosensors is a relatively short linear range of the calibration curve. Increasing the concentration range of detectable analyte, the sensitivity and specificity of the detection event improves the prospects for commercialising biosensors [176, 196, 217, 228, 246].

One way of overcoming those problems is to couple different enzymes either in sequence, in competition or in recycle pathways. Due to the appropriate combination of enzymes, the range of analyte species accessible to measurement, the selectivity and the sensitivity of the biosensor may be enhanced [63, 94, 137, 150, 274].

Mathematical modeling of two-enzyme biosensors started in 1980 [129, 130] with the modeling of an amperometric mono-layer enzyme electrode with two co-immobilized enzymes. It was only the first-order reaction rate that was considered in the first mathematical models [129, 130, 133, 188, 233]. The dynamic response of these electrodes was analysed solving diffusion equations and using Green's function [133]. Further analysis of dual enzyme biosensors response was performed by Scheller, Schulmeister and others [227, 234]. A comprehensive review of the models developed during the first decade was given in [235].

R. Baronas et al., *Mathematical Modeling of Biosensors*, Springer Series on Chemical Sensors and Biosensors 9, DOI 10.1007/978-90-481-3243-0_7,

Later, non-linear mathematical models have been developed for amperometric two-enzyme biosensors [171,237,242,243]. It should be noted that the basic methods of the problems solution are the same as in the one-layer one-enzyme case.

In this chapter mathematical models of two types of amperometric multi-enzyme biosensors are considered. One type of the biosensors utilizes enzymatic reactions assuming no interaction between the analyzed substrates and reaction products [25, 31, 38]. The mathematical model of such biosensors allows to simulate the biosensor response to a mixture of compounds (substrates). The second type of biosensors utilizes an enzymatic reaction followed by a cyclic product conversion. The two kinds of the product regeneration in those two-enzyme biosensors are analyzed: enzymatic and electrochemical [32,74,200,211].

1 Biosensors Response to Mixture of Compounds

Let us consider the response of an amperometric biosensor to a mixture of substrates. Assuming no interaction between the substrates (compounds) of the mixture, enzyme-catalyzed reactions are considered,

$$\mathrm{S}_k \xrightarrow{\mathrm{E}} \mathrm{P}_k, \quad k = 1, \ldots, K, \tag{1}$$

where the substrate S_k binds to the enzyme (E) to form an enzyme–substrate complex, following the substrate S_k conversion to the product P_k, $k = 1, \ldots, K$, K is the number of the substrates present in the mixture [25,31].

A biosensor is assumed to be an enzyme electrode, having a layer of an enzyme immobilized onto the surface of the probe.

1.1 Mathematical Model

Assuming no interaction between the analyzed substrates (compounds) of the mixture, the symmetrical geometry of the electrode, the homogeneous distribution of immobilized enzyme in the enzyme membrane, and considering one-dimensional-in-space mass transport by the diffusion, coupling the enzyme reaction with the diffusion as described by Fick's law leads to the following equations:

$$\begin{aligned}
\frac{\partial S^{(k)}}{\partial t} &= D_S^{(k)} \frac{\partial^2 S^{(k)}}{\partial x^2} - \frac{V_{max}^{(k)} S^{(k)}}{K_M^{(k)} + S^{(k)}}, \\
\frac{\partial P^{(k)}}{\partial t} &= D_P^{(k)} \frac{\partial^2 P^{(k)}}{\partial x^2} + \frac{V_{max}^{(k)} S^{(k)}}{K_M^{(k)} + S^{(k)}}, \\
x \in (0, d), \quad & t \in (0, T), \quad k = 1, \ldots, K,
\end{aligned} \tag{2}$$

where x and t stand for space and time, respectively, $S^{(k)}(x,t)$ is the concentration of the substrate S_k, $P^{(k)}(x,t)$ is the concentration of the reaction product P_k, d is the thickness of the enzyme layer, T is the full time of the biosensor operation, $D_S^{(k)}$, $D_P^{(k)}$ are the diffusion coefficients of the substrate S_k and the product P_k, respectively, $V_{max}^{(k)}$ is the maximal enzymatic rate of the biosensor attainable with that amount of enzyme, when the enzyme is fully saturated with the substrate (compound) S_k, and $K_M^{(k)}$ is the Michaelis constant.

The biosensor operation starts when some substrate appears over the surface of the enzyme layer. This is defined by the initial conditions ($t = 0$)

$$
\begin{aligned}
&S^{(k)}(x,0) = \begin{cases} 0, & 0 \le x < d, \\ S_0^{(k)}, & x = d, \end{cases}, \\
&P^{(k)}(x,0) = 0, \quad 0 \le x \le d, \\
&k = 1,\ldots,K,
\end{aligned}
\tag{3}
$$

where $S_0^{(k)}$ is the concentration of substrate S_k in the bulk solution.

At the electrode surface ($x = 0$), the boundary conditions depend on the electric activity of the substance. Following the scheme (1) the substrates are electro-inactive substances,

$$
D_S^{(k)} \frac{\partial S^{(k)}}{\partial x}\bigg|_{x=0} = 0, \quad 0 < t \le T. \tag{4}
$$

The reaction products are electro-active substances. In the case of amperometry, the potential at the electrode is chosen to keep zero concentration of the product,

$$
P^{(k)}(0,t) = 0, \quad 0 < t \le T. \tag{5}
$$

If the substrate is well-stirred and in a powerful motion, then the diffusion layer ($0 < x < d$) remains at a constant thickness. Consequently, the concentration of all the substrates as well as the products over the enzyme surface (bulk solution/membrane interface) remains constant during the biosensor action. In the case of the batch analysis, the biosensor contacts the mixture during the whole time of the biosensor operation. However, in the case of the injection analysis, the biosensor contacts the mixture only for a short time (the injection time) [29, 220] (Sect. 5 of first chapter of Part II). When the analyte disappears, a buffer solution swills the enzyme surface reducing the concentrations of the substrate and the product at the biosensor surface to zero,

$$
\begin{aligned}
&S^{(k)}(d,t) = \begin{cases} S_0^{(k)}, & 0 < t \le T_F, \\ 0, & T_F < t \le T, \end{cases} \\
&P^{(k)}(d,t) = 0, \quad 0 < t \le T, \\
&k = 1,\ldots,K,
\end{aligned}
\tag{6}
$$

where T_F is the injection time.

Assuming T_F as the full time of the biosensor operation, $T = T_F$, the mathematical model (2)–(6) may also be applied to the batch analysis. In the case of the batch analysis, the boundary condition (6) reduces to $S^{(k)}(d,t) = S_0^{(k)}$ and $P^{(k)}(d,t) = 0$. So, a model of a biosensor used in the batch analysis can be considered as a special case of the model (2)–(6) of the corresponding biosensor operating in the flow injection mode.

The anodic current depends upon the flux of the electro-active analyte (product) at the electrode surface ($x = 0$). The density $i^{(k)}(t)$ of the biosensor current , as a result of the reaction of the substrate S_k with the enzyme at time t, is proportional to the concentration gradient of the product P_k at the surface of the electrode as described by the Faraday and the Fick laws [128, 235],

$$i^{(k)}(t) = n_e F D_P^{(k)} \frac{\partial P^{(k)}}{\partial x}\bigg|_{x=0}, \tag{7}$$

where n_e is a number of electrons involved in a charge transfer at the electrode surface and F is the Faraday constant.

Assuming that the overall biosensor response to a mixture represents the total sum of individual responses to each constituent substrate and having values of the current density $i^{(k)}(t)$ for all compounds, $k = 1, \dots, K$, the total density $i(t)$ of the biosensor current can be calculated additively,

$$i(t) = \sum_{k=1}^{K} i^{(k)}(t). \tag{8}$$

1.2 Solution of the Problem

When analyzing the problem (2)–(6), one can notice that there is no direct relationship between pairs of the unknown variables $S^{(k_1)}$, $P^{(k_1)}$ and $S^{(k_2)}$, $P^{(k_2)}$, at $k_1 \neq k_2$, $k_1, k_2 = 1, \dots, K$. Because of this, the initial boundary value problem (2)–(6) consisting of $8K$ equations can be split into K problems containing only eight equations (2)–(6) at a given k, $k = 1, \dots, K$. The problem (2)–(6) formulated for a given compound S_{k_1} of the mixture, can be solved individually and independently from the problem formulated for another compound S_{k_2}, $k_1, k_2 = 1, \dots, K$, $k_1 \neq k_2$.

Let us assume the formulation of the problem (2)–(6) for a single substrate $S = S_k$ and the reaction product $P = P_k$, $k = K = 1$. Let V_{max} be the maximal enzymatic rate of the modeled biosensor, K_M is the corresponding Michaelis constant, S is the concentration of the substrate S, and P is concentration of the reaction product P. The problem (2)–(6), formulated for a single substrate S and the reaction product P, can be solved numerically as discussed in Sect. 3 of first chapter of Part II.

In the common case of K compounds, having responses of the biosensor to each constituent compound, Eq. 8 allows to calculate the common biosensor response to the mixture of K compounds [25]. Consequently, to obtain values $i(t_j)$, $j = 0, \ldots, N$, $t_0 = 0$, $t_N = T$, of the density of the total biosensor current, it is required:

1. To run computer simulation K times to obtain values $i^{(k)}(t_j)$ of the density of the biosensor current for each compound of the mixture, $k = 1, \ldots, K$, $j = 1, \ldots, N$
2. To calculate the density of the total biosensor current as defined by (8)

In step (1) only values of the following parameters: $D_S^{(k)}$, $D_P^{(k)}$, $V_{max}^{(k)}$, $K_M^{(k)}$ and $S_0^{(k)}$ vary when one computer simulation changes to the next one. This procedure of computation is valid for both modes of analysis: the batch and the injection.

1.3 Generation of Data Sets

An accurate and reliable calibration of an analytical system as well as the proper test of the methods of chemometrics requires a lot of experimental data [11, 57, 174]. Assuming a good enough adequacy of the mathematical model to the physical phenomena, the data synthesised using a computer simulation can be employed instead of the experimental ones [31, 38].

The computational experiments are usually much cheaper and faster than the physical ones. The computer simulation is perfectly reasonable when the biosensors to be used in practice are still in a stage of development. Then the design of smart biosensors to be used in analytical systems and the development of effective methods of data analysis may take place in parallel.

Let us discuss an application of the computer simulation to generate data for a calibration of an amperometric biosensor [25, 31, 38]. A system of four ($K = 4$) compounds is considered. Let each compound of eight ($M = 8$) different concentrations be used in the calibration to have the biosensor response to a wide range of substrate concentrations.

It is necessary to solve the problem (2)–(6) for a given component S_k numerically $K \times M = 4 \times 8 = 32$ times at four different values of the maximal enzymatic rate and eight values of the substrate concentration. The following values of the parameters of the mathematical model (2)–(6) were assumed constant in all numerical experiments:

$$
\begin{gathered}
D_S^{(k)} = D_P^{(k)} = 300\,\mu\text{m}^2/\text{s},\ K_M^{(k)} = 100\,\mu\text{M}, \quad k = 1, \ldots, K, \\
n_e = 2, \quad d = 200\,\mu\text{m}.
\end{gathered} \tag{9}
$$

The enzymatic reaction for each component of the mixture was characterized by the individual maximal enzymatic rate:

$$
V_{max} = 10^{-k}\ \text{mM}/s, \quad k = 1, \ldots, K. \tag{10}
$$

To have the biosensor responses for a wide range of the substrate concentrations, the following values of the $S_0^{(k)}$ for every of the substrates $S_1, \ldots, S_K$ of the mixture were taken:

$$\begin{aligned} S_0^{(k)} \in \{\varsigma_m \times S_0,\ m = 1, \ldots, M\}, \quad k = 1, \ldots, K, \\ K = 4, \quad M = 8, \quad S_0 = 10\,\mu\text{M}, \\ \varsigma_1 = 1,\ \varsigma_2 = 2,\ \varsigma_3 = 4,\ \varsigma_4 = 8, \\ \varsigma_5 = 12,\ \varsigma_6 = 16,\ \varsigma_7 = 32,\ \varsigma_8 = 64. \end{aligned} \tag{11}$$

In the simulation of the biosensor response for all the values defined in (9)–(11), only the values of three parameters, $V_{max}^{(k)}$, $K_M^{(k)}$ and $S_0^{(k)}$, of the model vary when one computer simulation changes to the next one. In addition, every computation was repeated twice to simulate the biosensor response in the batch as well as the injection mode of the analysis. Hence, the biosensor response was simulated $K \times M \times 2$ times.

Let $i_m^{(k)}(t_j)$ be a value of the density $i^{(k)}(t_j)$ of the biosensor current calculated at concentration $S_0^{(k)} = S_{0,m}^{(k)}$ of the substrate S_k, $m = 1, \ldots, M$, $k = 1, \ldots, K$, $j = 0, \ldots, N$, $t_N = T$. Having M numerical solutions (M sets of values of the biosensor current) $i_m^{(k)}(t_j)$, $j = 0, \ldots, N$, for each $k = 1, \ldots, K$ (in total $K \times M$ solutions), the full factorial $i_{m_1,\ldots,m_K}^{(k)}(t_j)$ of $M^K = 8^4 = 4{,}096$ solutions can be produced additively,

$$i_{m_1,\ldots,m_K}(t_j) = \sum_{k=1}^{K} i_{m_k}^{(k)}(t), \quad m_1, \ldots, m_K = 1, \ldots, M,\ j = 0, \ldots, N. \tag{12}$$

During an ordinary computer simulation, the values of the biosensor current were stored in a file every second of the simulation. Thus, $N = T$ values of $i_m^{(k)}(t_j)$, $t_j = j$ (seconds), $j = 1, \ldots, N$, for each $k = 1, \ldots, K$ and $m = 1, \ldots, M$ were produced as a result of the computer simulation of the biosensor response (in total $K \times M \times N$ values). The results of $K \times M$ simulations were stored into $K \times M$ files for both modes of the analysis.

Later, using an additional simple utility of summation, a matrix $M^K \times N$ of the biosensor response data was produced following (8) and stored in a new file. This procedure was repeated for the batch as well as the injection mode of the analysis.

The calculation results are depicted in Figs. 1 and 2. Figure 1 shows every 64th only of M^K (the full factorial of $M \times K$) simulated biosensor responses for $K = 4$ values of the maximal enzymatic rate and $M = 8$ substrate concentrations in the case of the batch analysis. Figure 2 presents the generated biosensor responses in the case of the injection analysis. An evolution of biosensor currents are depicted for the first 80 s of the biosensor action only because of a marginal change of the biosensor current at greater values of time t.

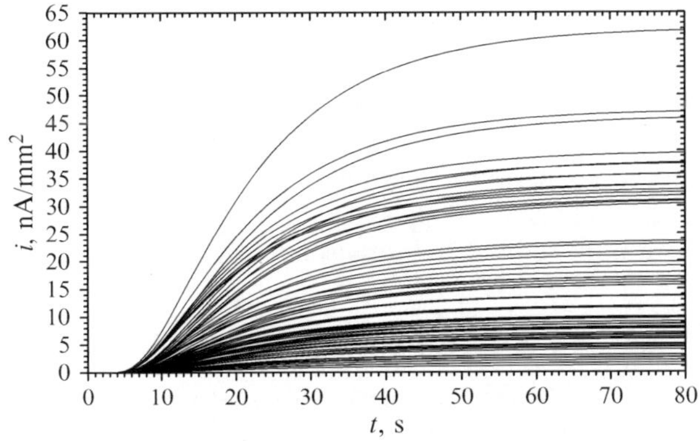

Fig. 1 Every 64th biosensor response curve of M^K responses at $K = 4$ values of the maximal enzymatic rate and $M = 8$ substrate concentrations in the batch analysis

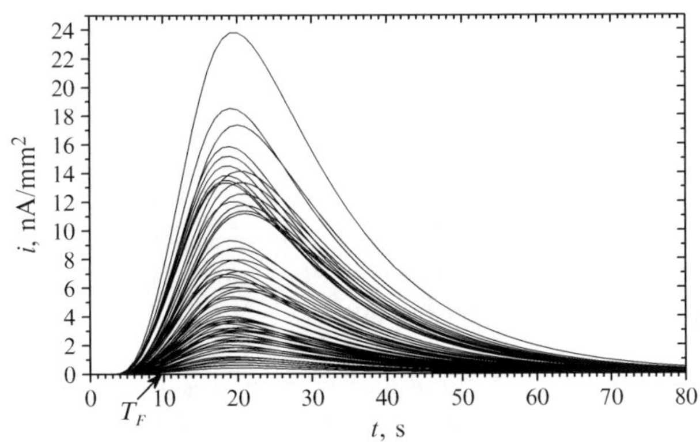

Fig. 2 Every 64th biosensor response curve of M^K responses at $K = 4$ values of the maximal enzymatic rate and $M = 8$ substrate concentrations in the injection analysis, $T_F = 10$ s

The data synthesized using the computer simulation was successfully applied to calibrate and validate the sensor system based on an amperometric biosensor and an artificial neural network [31, 38]. Coupling the biosensors with artificial neural networks is growing in importance as a tool for multi-component analysis [101, 169, 197, 212, 282].

Data collected in complex processes contains a wealth of redundant information, since the variables are collinear. The data pre-processing methods can be applied in such situations to enhance the relevant information, to make the resulting models simpler and easier to interpret. The correlation coefficients analysis and the principal component analysis are among the approaches widely used to reduce the dimensionality of the vectors $(i_{m_1,\ldots,m_K}(t_0), \ldots, i_{m_1,\ldots,m_K}(t_N))$ of input data [163, 177].

1.4 Concluding Remarks

Assuming no interaction between the components of a mixture, the mathematical model (2)–(6) describes an operation of amperometric biosensors in the batch as well as in the injection analysis. The initial boundary value problem (2)–(6) can be solved for each component independently. The total biosensor current is calculated additively form the individual biosensor responses to each component of the mixture.

Computer simulation of the biosensor response can be used to generate pseudo experimental biosensor responses to mixtures of compounds. If K is a number of mixture component and M is a number of different concentrations of each component, then the biosensor responses for full factorial of mixtures (M^K samples) can be synthesized by a simple routine of summation from the results of $K \times M$ computer simulations of the response.

2 Biosensors Acting in Trigger Mode

Chemical amplification is known as a powerful approach for increasing the sensitivity of biosensors [72, 78, 229, 258]. The amplification in the biosensors response has been achieved by cyclic conversion of substrates [92, 129, 136, 227, 231, 250, 275]. A mono-layer single-enzyme biosensor utilizing the amplification by conjugated electrochemical and enzymatic substrates conversion has also been discussed in Sect. 6 of first chapter of Part II.

The cyclic conversion of the substrate and the regeneration of the analyte are usually carried out by using a membrane containing two enzymes [129, 208]. If a biosensor contains an enzyme that starts analyte conversion following the cyclic product conversion the scheme of the biosensor action can be called as "triggering". An amperometric detection of alkaline phosphatase based on hydroquinone recycling might be an example of this type of conversion [74].

The substrate of the alkaline phosphatase, i.e. p-hydroxyphenyl phosphate, is usually hydrolyzed by alkaline phosphatase to hydroquinone. The hydroquinone, instead of being detected directly, enters into an amplification cycle where it is oxidized to quinone at the electrode surface and then reduced back to hydroquinone by glucose oxidase in the presence of glucose. The consumption–regeneration cycle of the hydroquinone resulted in an amplification factor about eight. Another example of utilizing trigger scheme might be a high sensitive determination of β-galactosidase used as a label in heterogeneous immunoassay [200]. As substrate p-aminophenyl-β-galactopyranoside was used. Produced p-aminophenol that is electrochemically active compound can be detected directly [211]. To increase the sensitivity of the determination, the p-aminophenol is entered into a bioelectrocatalytic amplification cycle using glucose dehydrogenase (GDH). Both presented schemes include enzymatic trigger reactions together with the electrochemical and enzymatic amplification steps. Therefore, by analogy with the electrochemical nomenclature they may be abbreviated as acting following the CEC mechanism.

The triggering of the consecutive substrate conversion can also be realized by an enzymatic conversion of substrate (trigger reaction) following the second enzymatic reaction and electrochemical conversion. This scheme can be abbreviated as CCE. The scheme may be realized, for example, by using peroxidase and glucose dehydrogenase. The peroxidase produces an oxidized product that is reduced by the GDH so realizing the cyclic conversion of the product.

This section presents mathematical models of biosensors acting in a trigger mode. One mathematical model describes the dynamics of the response of biosensors utilizing a trigger enzymatic reaction following the electrochemical and enzymatic product cyclic conversion (CEC scheme), while the other model describes the behaviour of biosensors utilizing a trigger enzymatic reaction following enzymatic and electrochemical conversion of the product (CCE scheme) [32]. The modeled response of the trigger biosensors is compared with the response of a single-enzyme mono-layer biosensor.

2.1 *Mathematical Models*

The biosensor is considered as an electrode, containing a mono-layer membrane with immobilized two enzymes applied onto the surface of the electrochemical transducer. The symmetrical geometry of the electrode and homogeneous distribution of immobilized enzymes in the enzyme membrane of a uniform thickness are assumed.

2.1.1 Modeling Biosensor Acting in CEC Mode

Consider the CEC scheme

$$\mathrm{S} \xrightarrow{\mathrm{E_1}} \mathrm{P_1}, \tag{13}$$

$$\mathrm{P_1} \longrightarrow \mathrm{P_2}, \tag{14}$$

$$\mathrm{P_2} \xrightarrow{\mathrm{E_2}} \mathrm{P_1}, \tag{15}$$

where the substrate (S) is enzymatically (E_1) converted to the product (P_1) followed by the electrochemical conversion of the product (P_1) to another product (P_2) that in turn is enzymatically (E_2) converted back to P_1.

Coupling the enzyme-catalyzed reactions (13) and (15) with the one-dimensional-in-space diffusion described by Fick's law leads to the following system of equations:

$$
\begin{aligned}
\frac{\partial S}{\partial t} &= D_S \frac{\partial^2 S}{\partial x^2} - \frac{V_{max}^{(1)} S}{K_M^{(1)} + S}, \\
\frac{\partial P^{(1)}}{\partial t} &= D_P^{(1)} \frac{\partial^2 P^{(1)}}{\partial x^2} + \frac{V_{max}^{(1)} S}{K_M^{(1)} + S} + \frac{V_{max}^{(2)} P^{(2)}}{K_M^{(2)} + P^{(2)}}, \\
\frac{\partial P^{(2)}}{\partial t} &= D_P^{(2)} \frac{\partial^2 P^{(2)}}{\partial x^2} - \frac{V_{max}^{(2)} P^{(2)}}{K_M^{(2)} + P^{(2)}}, \quad x \in (0,\, d), \quad t > 0,
\end{aligned}
\tag{16}
$$

where x and t stand for space and time, respectively, $S(x,t)$ and $P^{(i)}(x,t)$ denote the concentrations of the substrate S and product P_i, respectively, $V_{max}^{(i)}$ is the maximal enzymatic rate, $K_M^{(i)}$ is the Michaelis constant, d is the thickness of the enzyme membrane, D_S and $D_P^{(i)}$ are the diffusion coefficients, $i = 1, 2$. The model parameters $V_{max}^{(1)}$ and $K_M^{(1)}$ stand for the enzyme-catalyzed reaction (13), while $V_{max}^{(2)}$ and $K_M^{(2)}$ – for the next enzymatic reaction (15).

Let $x = 0$ represent the electrode surface and $x = d$ – the bulk solution/enzyme membrane interface. Assuming the substrate concentration of S_0 and zero concentration of the reaction product in the bulk solution, the initial conditions are defined as follows ($t = 0$):

$$
\begin{aligned}
&S(x,0) = 0, \quad x \in [0,\, d), \\
&S(d,0) = S_0, \\
&P^{(i)}(x,0) = 0, \quad x \in [0,\, d],\ i = 1,2.
\end{aligned}
\tag{17}
$$

The electrode potential is chosen to keep zero concentration of the product P_1 at the electrode surface,

$$P^{(1)}(0,t) = 0, \quad t > 0. \tag{18}$$

The substrate is electro-inactive substance,

$$D_S \left.\frac{\partial S}{\partial x}\right|_{x=0} = 0, \quad t > 0. \tag{19}$$

Due to the electrochemical reaction (14), the generation rate of the product P_2 at the electrode surface is proportional to the generation rate of the product P_1,

$$D_P^{(2)} \left.\frac{\partial P^{(2)}}{\partial x}\right|_{x=0} = -D_P^{(1)} \left.\frac{\partial P^{(1)}}{\partial x}\right|_{x=0}, \quad t > 0. \tag{20}$$

When the bulk solution is well-stirred outside and in a powerful motion, the diffusion layer remains at a constant thickness ($0 < x < d$). Thus, the concentrations of the substrate and of both products over the enzyme surface (bulk solution/membrane interface) remain constant while the biosensor contacts the solution of the substrate ($t > 0$),

$$\begin{aligned} S(d,t) &= S_0, \\ P^{(1)}(d,t) &= 0, \\ P^{(2)}(d,t) &= 0. \end{aligned} \tag{21}$$

The biosensor current depends upon the flux of the product P_1 at the electrode surface, i.e. at the border $x = 0$. The density $i_{CEC}(t)$ of the current at time t is obtained explicitly from Faraday's and Fick's laws using the flux of the product P_1 at the surface of the electrode,

$$i_{CEC}(t) = n_e F D_P^{(1)} \left.\frac{\partial P^{(1)}}{\partial x}\right|_{x=0}, \tag{22}$$

where n_e is the number of electrons involved in a charge transfer at the electrode surface, and F is the Faraday constant.

Taking into account (20), $i_{CEC}(t)$ can also be calculated as follows:

$$i_{CEC}(t) = -n_e F D_P^{(2)} \left.\frac{\partial P^{(2)}}{\partial x}\right|_{x=0}. \tag{23}$$

The system (16)–(21) approaches a steady state as $t \to \infty$:

$$I_{CEC} = \lim_{t\to\infty} i_{CEC}(t). \tag{24}$$

where I_{CEC} is taken as the density of the steady state current of the biosensor acting in CEC mode.

2.1.2 Modeling Biosensor Acting in CCE Mode

In the CCE scheme, the substrate S is enzymatically (E_1) transformed into the product P_1 followed by the enzymatic (E_2) conversion of P_1 into another product P_2 that is electrochemically converted back to the product P_1,

$$\mathrm{S} \xrightarrow{\mathrm{E}_1} \mathrm{P}_1, \tag{25}$$

$$\mathrm{P}_1 \xrightarrow{\mathrm{E}_2} \mathrm{P}_2, \tag{26}$$

$$\mathrm{P}_2 \longrightarrow \mathrm{P}_1. \tag{27}$$

Assuming the same geometry of the enzyme membrane as in the CEC mode, the coupling the reactions (25)–(26) with the diffusion leads to the following equations:

$$\begin{aligned}
\frac{\partial S}{\partial t} &= D_S \frac{\partial^2 S}{\partial x^2} - \frac{V_{max}^{(1)} S}{K_M^{(1)} + S}, \\
\frac{\partial P^{(1)}}{\partial t} &= D_P^{(1)} \frac{\partial^2 P^{(1)}}{\partial x^2} + \frac{V_{max}^{(1)} S}{K_M^{(1)} + S} - \frac{V_{max}^{(2)} P^{(1)}}{K_M^{(2)} + P^{(1)}}, \\
\frac{\partial P^{(2)}}{\partial t} &= D_P^{(2)} \frac{\partial^2 P^{(2)}}{\partial x^2} + \frac{V_{max}^{(2)} P^{(1)}}{K_M^{(2)} + P^{(1)}}, \quad x \in (0, d), \quad t > 0.
\end{aligned} \tag{28}$$

Here and below, the notations are the same as in the model of a biosensor acting in CEC mode (Sect. 2.1.1).

The initial conditions are the same as in the case of CEC scheme (17).

When the biosensor acts in the CCE mode, the electrode potential is chosen to keep zero concentration of the product P_2 at the electrode surface,

$$P^{(2)}(0, t) = 0, \quad t > 0. \tag{29}$$

The rate of the product P_1 generation at the electrode surface is proportional to the rate of conversion of the product P_2. Thus, the boundary conditions (19)–(21) are also applicable to the system (25)–(27).

The density $i_{CCE}(t)$ of the biosensor current is proportional to the concentration gradient of the product P_2 at the surface of the electrode:

$$i_{CCE}(t) = n_e F D_P^{(2)} \frac{\partial P^{(2)}}{\partial x}\bigg|_{x=0} = -n_e F D_P^{(1)} \frac{\partial P^{(1)}}{\partial x}\bigg|_{x=0}. \tag{30}$$

The density I_{CCE} of the steady state current of the biosensor acting in the CCE mode is calculated as follows:

$$I_{CCE} = \lim_{t \to \infty} i_{CCE}(t). \tag{31}$$

2.1.3 Modeling Biosensor Acting in CE Mode

To compare the responses of trigger and ordinary amperometric biosensors, the action of the CE biosensor was analyzed. A CE biosensor contains a mono-layer membrane with only one immobilized enzyme applied onto the surface of the electrochemical transducer. The number of immobilized enzymes is the only fundamental difference between the CE biosensor and the trigger one.

In accordance with the CE scheme, the substrate S is enzymatically (E_1) converted to the product P_1 followed by the electrochemical product (P_1) conversion to another product (P_2):

$$S \xrightarrow{E_1} P_1, \tag{32}$$

$$P_1 \longrightarrow P_2. \tag{33}$$

In this scheme only the product P_1 is the electroactive substance.

The mathematical model of the biosensor acting in the CE mode can be derived from the model (16)–(21) of the biosensor acting in the CEC mode by accepting the inactive enzyme E_2, i.e. $V_{max}^{(2)} = 0$. Assuming $i_{CE}(t)$ to be the density of the anodic current, the transient current $i_{CE}(t)$ and the corresponding stationary current I_{CE} of the biosensor acting in the CE mode, can be calculated as of the CEC biosensor using (22) and (34), respectively,

$$i_{CE}(t) = n_e F D_P^{(1)} \frac{\partial P^{(1)}}{\partial x}\bigg|_{x=0}, \qquad I_{CE} = \lim_{t\to\infty} i_{CE}(t). \tag{34}$$

Let us notice, that following the schemes (32) and (33), the product P_2 has no effect on the response of the CE biosensor. Consequently, the response of that biosensor is identical to the response of a biosensor utilizing only the scheme (32). This type of the biosensors has been in detail analyzed, particularly, in the previous Sects. 1 to 4 of first chapter of Part II.

2.1.4 Enzymatic Amplification

Both of presented trigger schemes (CEC and CCE) include enzymatic reactions together with electrochemical and enzymatic amplification steps [32, 74, 200]. The response amplification is one of the most important characteristics of the trigger biosensors. To compare the amplified biosensor response with the response without amplification, the gain of the sensitivity is defined as the ratio of the steady state current of the trigger biosensor to the steady state current of the corresponding CE biosensor,

$$G_{CEC} = \frac{I_{CEC}}{I_{CE}}, \qquad G_{CCE} = \frac{I_{CCE}}{I_{CE}}. \tag{35}$$

When comparing the CEC and CCE schemes with the CE scheme and the corresponding mathematical models, one can see that the enzymatic activity of the enzyme E_2 is the determinant characteristics. Because of this, it is reasonable to analyze the signal gains G_{CEC} and G_{CCE} as functions of the enzymatic activity $V_{max}^{(2)}$ assuming all other parameters as being identical,

$$\begin{aligned} G_{CEC}\left(V_{max}^{(2)}\right) &= \frac{I_{CEC}\left(V_{max}^{(2)}\right)}{I_{CE}} = \frac{I_{CEC}\left(V_{max}^{(2)}\right)}{I_{CEC}(0)}, \\ G_{CCE}\left(V_{max}^{(2)}\right) &= \frac{I_{CCE}\left(V_{max}^{(2)}\right)}{I_{CE}} = \frac{I_{CCE}\left(V_{max}^{(2)}\right)}{I_{CEC}(0)}, \end{aligned} \tag{36}$$

where $I_{CEC}(V_{max}^{(2)})$ and $I_{CCE}(V_{max}^{(2)})$ are the densities of the steady state currents of the trigger biosensors acting in the CEC and the CCE mode, respectively, at the maximal activity $V_{max}^{(2)}$ of the enzyme E_2. Since at zero activity of the enzyme E_2, the response of the CEC biosensor becomes identical to the response of CE biosensor, $I_{CEC}(0)$ equals I_{CE}.

2.2 Finite Difference Solution

Since the type of equations of the mathematical models of trigger biosensors are the same as in the one-layer one-enzyme case, the corresponding initial boundary value problems can be solved similarly.

To find a numerical solution of the problem in the domain $[0, d] \times [0, T]$ a discrete grid has to be introduced. Let $\omega_h \times \omega_\tau$ be a uniform discrete grid defined as follows:

$$\begin{aligned} \omega_h &= \{x_i : x_i = ih,\ i = 0, 1, \ldots, N,\ hN = d\}, \\ \omega_\tau &= \{t_j : t_j = j\tau,\ j = 0, 1, \ldots, M;\ \tau M = T\}. \end{aligned} \tag{37}$$

The following notation is used in the finite difference approximations presented below:

$$\begin{aligned} &S_i^j = S(x_i, t_j), \quad P_i^{(1),j} = P^{(1)}(x_i, t_j), \quad P_i^{(2),j} = P^{(2)}(x_i, t_j), \\ &i = 0, \ldots, N, \quad j = 0, \ldots, M. \end{aligned} \tag{38}$$

2.2.1 CEC Mode

Partial differential equations (16) can be approximated by the following implicit finite difference scheme:

$$\begin{aligned} \frac{S_i^{j+1} - S_i^j}{\tau} &= D_S \frac{S_{i+1}^{j+1} - 2S_i^{j+1} + S_{i-1}^{j+1}}{h^2} - \frac{V_{max}^{(1)} S_i^j}{K_M^{(1)} + S_i^j}, \\ \frac{P_i^{(1),j+1} - P_i^{(1),j}}{\tau} &= D_P^{(1)} \frac{P_{i+1}^{(1),j+1} - 2P_i^{(1),j+1} + P_{i-1}^{(1),j+1}}{h^2} \\ &\quad + \frac{V_{max}^{(1)} S_i^j}{K_M^{(1)} + S_i^j} + \frac{V_{max}^{(2)} P_i^{(2),j}}{K_M^{(2)} + P_i^{(2),j}}, \\ \frac{P_i^{(2),j+1} - P_i^{(2),j}}{\tau} &= D_P^{(2)} \frac{P_{i+1}^{(2),j+1} - 2P_i^{(2),j+1} + P_{i-1}^{(2),j+1}}{h^2} - \frac{V_{max}^{(2)} P_i^{(2),j}}{K_M^{(2)} + P_i^{(2),j}}, \\ i &= 1, \ldots, N-1, \quad j = 0, \ldots, M-1. \end{aligned} \tag{39}$$

The initial conditions (17) are approximated as follows:

$$\begin{aligned} &S_i^0 = 0, \quad i = 0, \dots, N-1, \\ &S_N^0 = S_0, \\ &P_i^{(1),0} = P_i^{(2),0} = 0, \quad i = 0, \dots, N. \end{aligned} \tag{40}$$

The boundary conditions (18)–(21) can be approximated as follows:

$$\begin{aligned} &P_0^{(1),j} = 0, \\ &S_0^j = S_1^j, \quad S_N^j = S_0, \\ &D_P^{(2)}(P_1^{(2),j} - P_0^{(2),j}) = -D_P^{(1)}(P_1^{(1),j} - P_0^{(1),j}), \\ &P_N^{(1),j} = P_N^{(2),j} = 0, \\ &j = 1, \dots, M. \end{aligned} \tag{41}$$

The density of the biosensor current can easily be calculated,

$$i_{CEC}(t_j) \approx i_{CEC,j} = n_e F D_P P_1^{(1),j}/h, \quad j = 1, \dots, M. \tag{42}$$

2.2.2 CCE Mode

In the case of CCE mode, the governing equations (28) can be approximated similarly as in the case of the CEC mode,

$$\begin{aligned} \frac{S_i^{j+1} - S_i^j}{\tau} &= D_S \frac{S_{i+1}^{j+1} - 2S_i^{j+1} + S_{i-1}^{j+1}}{h^2} - \frac{V_{max}^{(1)} S_i^j}{K_M^{(1)} + S_i^j}, \\ \frac{P_i^{(1),j+1} - P_i^{(1),j}}{\tau} &= D_P^{(1)} \frac{P_{i+1}^{(1),j+1} - 2P_i^{(1),j+1} + P_{i-1}^{(1),j+1}}{h^2} \\ &\quad + \frac{V_{max}^{(1)} S_i^j}{K_M^{(1)} + S_i^j} - \frac{V_{max}^{(2)} P_i^{(1),j}}{K_M^{(2)} + P_i^{(1),j}}, \\ \frac{P_i^{(2),j+1} - P_i^{(2),j}}{\tau} &= D_P^{(2)} \frac{P_{i+1}^{(2),j+1} - 2P_i^{(2),j+1} + P_{i-1}^{(2),j+1}}{h^2} + \frac{V_{max}^{(2)} P_i^{(1),j}}{K_M^{(2)} + P_i^{(1),j}}, \\ j = 1, \dots, N-1, &\quad j = 0, \dots, M-1. \end{aligned} \tag{43}$$

Since the initial conditions are the same as in the case of the CEC scheme, the same approximation (40) can be used.

The boundary condition (29) is approximated as follows:

$$P_0^{(2),j} = 0, \quad j = 1, \dots, M. \tag{44}$$

All other boundary conditions (19)–(21) are common for both cases of triggering: the CEC and the CCE. Thus, the same approximation can be used for the boundary conditions.

The density of the biosensor current i_{CCE} can be calculated as follows:

$$i_{CCE}(t_j) \approx i_{CCE,j} = n_e F D_P P_1^{(2),j} / h, \quad j = 1, \dots, M. \tag{45}$$

2.3 Simulated Response

Figures 3 and 4 show the concentration profiles of the substrate and products in the enzyme layer for biosensors acting in the CEC and the CCE modes calculated at the following values of the parameters:

$$\begin{aligned} &D_S = D_P = 300\,\mu\mathrm{m}^2/\mathrm{s}, \quad d = 100\,\mu\mathrm{m}, \\ &K_M^{(1)} = K_M^{(2)} = 100\,\mu\mathrm{M}, \quad V_{max}^{(1)} = V_{max}^{(2)} = 100\,\mu\mathrm{M}/s, \quad S_0 = 100\,\mu\mathrm{M}. \end{aligned} \tag{46}$$

Figures 3 and 4 show the concentration profiles at the time when the steady state as well as 50% of the steady state response has been reached. It is easy to notice in

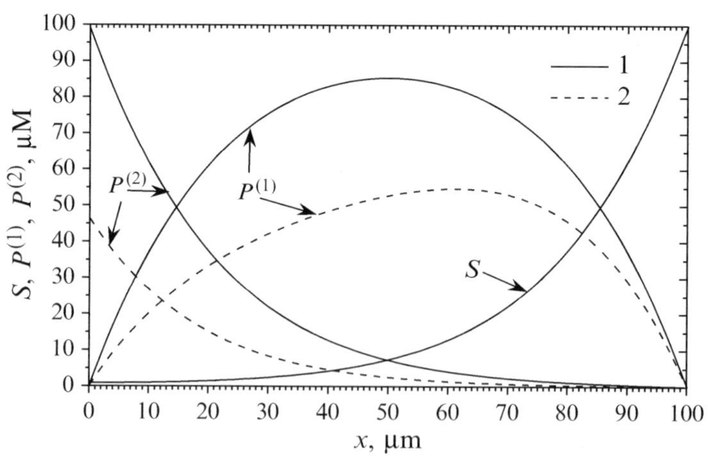

Fig. 3 Concentration profiles of the substrate (S) and products ($P^{(1)}$ and $P^{(2)}$) in the enzyme layer of the CEC biosensor at the steady state time $T = 120$ s (1) when the steady state is reached and at the half of it, $T_{0.5} = 10.5$ s (2). Model parameters are defined in (46)

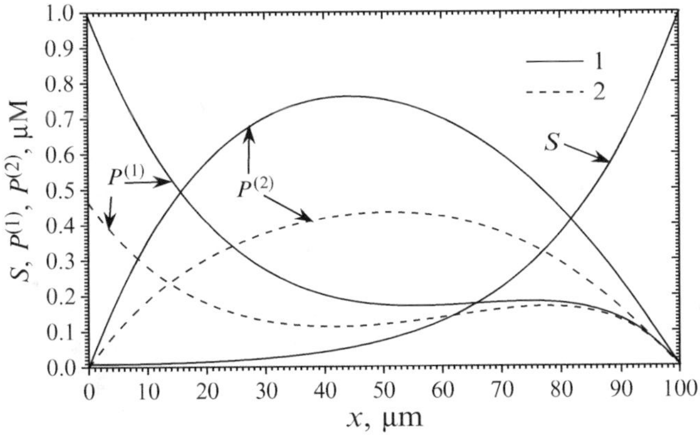

Fig. 4 Concentration profiles of the substrate and products in the enzyme layer of the CCE biosensor at $T = 120$ s (1) and $T_{0.5} = 11$ s (2). The other parameters and notation are the same as in Fig. 3

the figures that the concentrations of the substrate at steady state conditions are approximately the same for both biosensors [32]. At the time when the half of the steady state response was reached, no significant difference has also been observed in the entire enzyme layer, $x \in [0, d]$. The similarity of the substrate concentrations in the both modes (CEC and CCE) of the biosensor operation, can be explained analytically by the identity of the equations describing the dynamics of the substrate concentration.

The density of the steady state current and the time of steady state are very similar in both types of biosensors, $I_{CEC} \approx i_{CEC}(120) \approx 0.26\,\mu A/mm^2$ and $I_{CCE} \approx i_{CCE}(120) \approx 0.25\,\mu A/mm^2$. At the steady state conditions, i.e. when $\partial S/\partial t = \partial P^{(1)}/\partial t = \partial P^{(2)}/\partial t = 0$, the equality $S(x,t) + P^{(1)}(x,t) + P^{(2)}(x,t) = S_0$ is valid for all $x \in [0, d]$ when $t \to \infty$. This can be observed in both Figs. 3 and 4.

2.4 *Peculiarities of the Response*

2.4.1 Dependence of the Steady State Current on the Reactions Rates

Figures 5 and 6 show the dependence of steady state current on the activity of both enzymes (E_1 and E_2) for both action modes: the CEC and the CCE. In calculations, the maximal enzymatic rates $V_{max}^{(1)}$ and $V_{max}^{(2)}$ varied in four orders of magnitude: from 10^{-7} upto 10^{-3} M/s, the substrate concentration S_0 was equal $100\,\mu M = K_M{}^{(1)} = K_M{}^{(2)}$, the thickness of the enzyme layer was equal 100 μm. One can see in Figs. 5 and 6 that I_{CEC} as well as I_{CCE} are monotonously increasing functions of both enzymatic rates: $V_{max}^{(1)}$ and $V_{max}^{(2)}$.

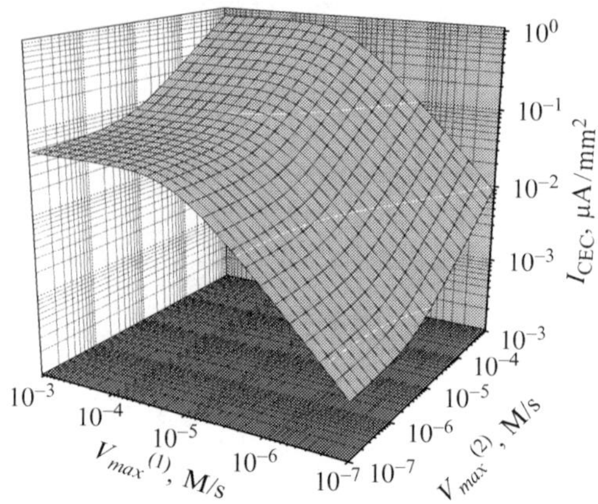

Fig. 5 The density I_{CEC} of the steady state current versus the maximal enzymatic rates $V_{max}^{(1)}$ and $V_{max}^{(2)}$ of the biosensor acting in CEC mode. The values of the other parameters are the same as in Fig. 3

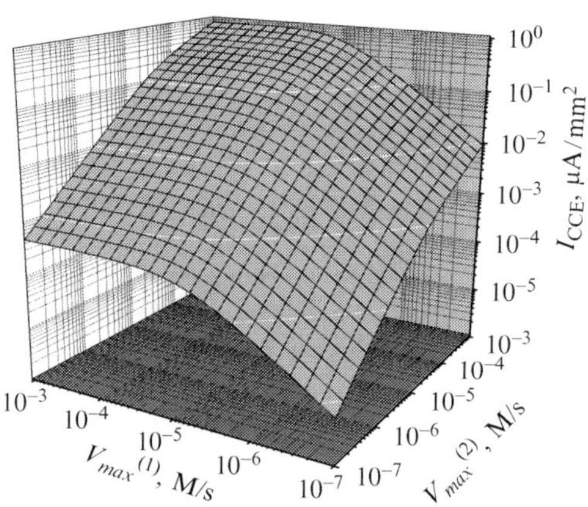

Fig. 6 The density I_{CCE} of the steady state current versus the maximal enzymatic rates $V_{max}^{(1)}$ and $V_{max}^{(2)}$ of the biosensor acting in CCE mode. The values of the model parameters are the same as in Fig. 3

In the case of the CEC mode, the activity of the enzyme E_2 ($V_{max}^{(2)} > 0$) stimulates a notable increase of the biosensor current [32]. When the enzyme membrane (layer) contains no enzyme E_2 ($V_{max}^{(2)} = 0$), the biosensor acting in CEC mode still generates the current only if $V_{max}^{(1)} > 0$. In the case of CCE mode, the appearance of an active enzyme E_2 ($V_{max}^{(2)} > 0$) is a critical factor for the biosensor current. $I_{CCE} = 0$ if $V_{max}^{(2)} = 0$ even if the activity of the enzyme E_1 is very high ($V_{max}^{(1)} \gg 0$). Because of this, at low values of $V_{max}^{(2)}$ the density I_{CCE} of the steady state current increases very sharply with the increase of $V_{max}^{(2)}$. That effect is observed in Figs. 5 and 6 and is called the surface salience. The salience of the surface I_{CCE} (Fig. 6) is more noticeable than the salience of the surface I_{CEC} (Fig. 5). Consequently, at $V_{max}^{(2)} \to 0$ and $V_{max}^{(1)} > 0$, $I_{CCE} \to 0$ in CCE mode of the biosensor operation and $I_{CEC} \to I_{CE}$ in another mode (CEC) of triggering. On the other hand, Figs. 5 and 6 show, that $I_{CCE} \approx I_{CEC}$ at very high values of the maximal enzymatic rate $V_{max}^{(2)}$.

2.4.2 Effect the Reaction Rates on the Amplification

To investigate the effect of the amplification the density I_{CE} of the stationary current has to be calculated at the same conditions as those given above. Having I_{CEC}, I_{CCE} and I_{CE} the gains G_{CEC} and G_{CCE} can be easily calculated [32]. The results of calculations are depicted in Figs. 7 and 8.

One can see in both figures that the signal gain increases with the increase of the enzymatic rate $V_{max}^{(2)}$. The increase is especially notable at high values of $V_{max}^{(2)}$. The variation of $V_{max}^{(1)}$ on the response gain is only slight. The gain varies from 15 to 19 at $V_{max}^{(1)} = 1$ mM/s in both modes of the biosensor operation: the CEC and the CCE.

When comparing the gain in the CEC mode (Fig. 7) with the gain in the CCE mode (Fig. 8), one can notice a significant difference at low values of $V_{max}^{(2)}$. The gain G_{CEC} starts to increase from about unity, while G_{CCE} at low values of $V_{max}^{(2)}$ ($V_{max}^{(2)} < 1$ μM/s)) is even less than unity. It means that in the case of low activity of the enzyme E_2, the steady state current of the biosensor acting in the CCE mode is

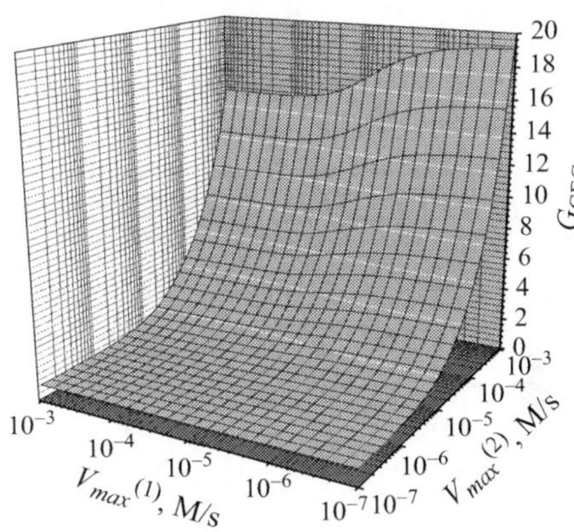

Fig. 7 The signal gain G_{CEC} versus the maximal enzymatic rates $V_{max}^{(1)}$ and $V_{max}^{(2)}$ of the biosensor acting in the CEC mode under the same conditions as in Fig. 3

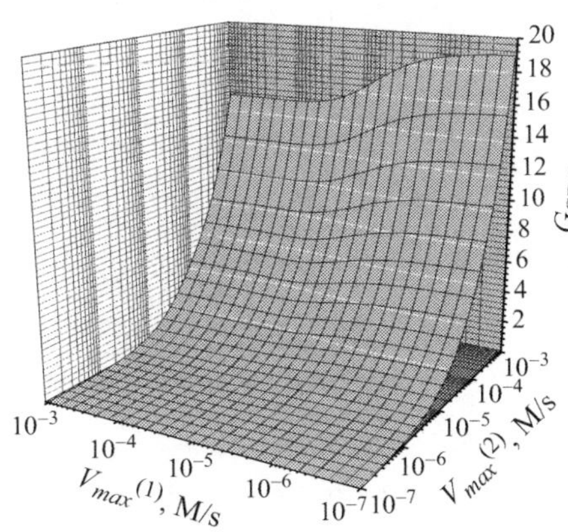

Fig. 8 The signal gain G_{CEC} versus the maximal enzymatic rates $V_{max}^{(1)}$ and $V_{max}^{(2)}$ of the biosensor acting in the CCE mode under the same conditions as in Fig. 3

even less than the steady state current of a biosensor acting in the CE mode under the same conditions. It follows from the model of the CCE biosensor that $P^{(2)}(x,t) \approx 0$ when $V_{max}^{(2)} \to 0$. Thus, when $V_{max}^{(2)} \to 0$ at all positive values of $V_{max}^{(1)}$, $G_{CCE} \to 0$ and $G_{CEC} \to 1$. On the other hand, Figs. 7 and 8 show that $G_{CEC} \approx G_{CCE}$ at high maximal enzymatic rate $V_{max}^{(2)}$, e.g. at $V_{max}^{(2)} = 1\,\text{mM/s}$.

2.4.3 The Amplification Versus the Substrate Concentration

To investigate the dependence of the signal gain on the substrate concentration S_0 the biosensor response has to be simulated changing S_0 in a wide range, e.g. from 1 μM to 100 mM [32]. Since the signal gain of trigger biosensors is only significant at relatively high maximal rate $V_{max}^{(2)}$ of the enzyme E_2 (see Figs. 7 and 8), the following two values of $V_{max}^{(2)}$ were used: 0.1 and 1 mM/s. In numerical simulation, two values of the maximal rate $V_{max}^{(1)}$ of the enzyme E_1 were the same as of $V_{max}^{(2)}$. Assuming $K_M = K_M^{(1)} = K_M^{(2)} = 100\,\mu\text{M}$, the signal gains G_{CEC} and G_{CCE} are depicted in Fig. 9 as functions of the normalized substrate concentration S_0/K_M.

One can see in Fig. 9 that the behaviour of the signal gain versus the substrate concentration is very similar in both modes of the biosensor action: the CEC and the CCE. Some noticeable difference between values of G_{CEC} and G_{CCE} is observed only at high substrate concentrations, $S_0 > K_M$. However, in the case of a higher value of $V_{max}^{(2)}$, $V_{max}^{(2)} = 1\,\text{mM/s}$, and a lower $V_{max}^{(1)}$, $V_{max}^{(1)} = 0.1\,\text{mM/s}$, no noticeable difference is observed between values of G_{CEC} (curve 5 in Fig. 9) and G_{CCE} (curve 6 in Fig. 9) in the entire domain of the substrate concentration.

Figure 9 shows the significant importance of the maximal enzymatic rate $V_{max}^{(2)}$ for both signal gains: G_{CEC} and G_{CCE}. That importance is especially perceptible at low and moderate concentrations of substrate, $S_0 < K_M$. At $S_0 < 0.1K_M$ and

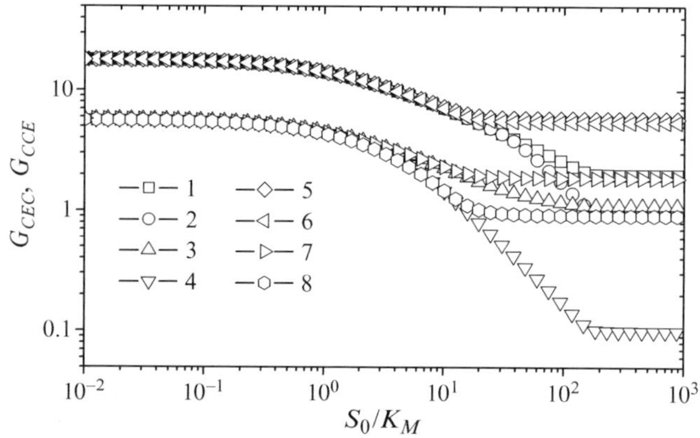

Fig. 9 The signal gains G_{CEC} (1, 3, 5, 7) and G_{CCE} (2, 4, 6, 8) versus the normalized substrate concentration S_0/K_M at the following maximal enzymatic rates $V_{max}^{(1)}$: 1 (1–4), 0.1 (5–8) and $V_{max}^{(2)}$: 1 (1, 2, 5, 6), 0.1 (3, 4, 7, 8) mM/s. The other parameters are the same as in Fig. 3

$V_{max}^{(2)} = 1\,\text{mM/s}$, due to the amplification, the steady state current increases up to about 18 times ($G_{CEC} \approx G_{CCE} \approx 18$). However, at the same S_0 and tenfold lower value of $V_{max}^{(2)}$, the gain is about three times less, $G_{CEC} \approx G_{CCE} \approx 5.7$. Consequently, at low substrate concentrations, $S_0 < 0.1K_M$, and a wide range of the maximal enzymatic rate $V_{max}^{(1)}$, the tenfold reduction of $V_{max}^{(2)}$ reduces the signal gain about three times. That property is valid for both modes of triggering: the CEC and the CCE.

An increase in the substrate concentration causes a decrease in the signal gain when S_0 becomes greater than K_M (Fig. 9), i.e. when $S_0 > K_M^{(1)}$ and $S_0 > K_M^{(2)}$. The decrease is more marked in cases of the higher values of the enzymatic rate $V_{max}^{(1)}$ (curves 1–4 in Fig. 9) rather than of the lower ones (curves 5–8). Additional calculations showed that at a very low activity of enzyme E_1 when $V_{max}^{(1)} = 0.1\,\text{mM/s}$, both signal gains practically do not vary when changing the substrate concentration in the entire domain [32].

Because of a very stable amplification of the biosensor signal at wide range of substrate concentration, the use of the biosensors acting in a trigger mode is especially reasonable at relatively low maximal enzymatic activity (rate $V_{max}^{(1)}$) of enzyme E_1 and high activity (rate $V_{max}^{(2)}$) of enzyme E_2. In the case of relatively high maximal enzymatic activity $V_{max}^{(1)}$ the signal amplification is stable only for low concentrations of the substrate.

Additional calculations showed that the signal gains vanish rapidly with the decrease of the enzymatic activity $V_{max}^{(2)}$ of enzyme E_2. For example, in the case of $V_{max}^{(2)} = 1\,\mu\text{M/s}$ the gains become less than 2 even at low substrate concentration, $G_{CEC} \approx 1.9$ and $G_{CCE} \approx 1.3$ at $S_0 = 0.01K_M$ [32]. That effect is also observed in Figs. 7 and 8. Numerical simulation confirmed the property that the tenfold reduction in $V_{max}^{(2)}$ reduces the signal gains G_{CEC} and G_{CCE} about three times at wide range also of $V_{max}^{(2)}$.

Similar dependence of the signal gain on the substrate concentration was observed in the case of an amperometric enzyme electrode acting under chemical amplification by cyclic substrate conversion discussed in Sect. 6 of first chapter of Part II. In the case of the biosensor with substrate cyclic conversion the signal gain of 26 times was observed at the maximal enzymatic rate of 1 mM/s and the thickness of 150 μm of the enzyme layer. In order to compare that gain with the gain achieved in the trigger mode, the gains G_{CEC} and G_{CCE} were calculated for the enzyme membrane of thickness 150 μm. The numerical simulation of the action of the trigger biosensors showed very similar amplification, $G_{CEC} \approx G_{CCE} \approx 22$ at $V_{max}^{(1)} = V_{max}^{(2)} = 1\,\text{mM/s}$.

2.4.4 Effect of the Enzyme Membrane Thickness on the Amplification

As has been mentioned in the previous chapters, the steady state current of membrane biosensors significantly depends on the thickness of enzyme layer. The steady state time varies even in orders of magnitude. To investigate the dependence of the signal gain of the trigger biosensors on the enzyme membrane thickness d the response of biosensors was simulated varying d from 1 upto 500 μm at different maximal enzymatic rate $V_{max}^{(1)}$ of the enzyme E_1 and rate $V_{max}^{(2)}$ of the enzyme E_2 [32].

Figure 10 shows the signal gains G_{CEC} and G_{CCE} versus the membrane thickness d at the maximal enzymatic rate $V_{max}^{(1)} = 1$ mM/s and three values of the rate $V_{max}^{(2)}$: 0.01, 0.1 and 1 mM/s. When comparing the gain G_{CEC} with the gain G_{CCE}, one can notice a valuable difference in the behaviour of the signal gains. In the case of CEC biosensor action, no notable amplification is observed for the thin enzyme membranes ($d < 10$ μm). A more distant increase in the thickness d causes an increase of the gain G_{CEC}. The thickness at which G_{CEC} starts to increase depends on the maximal enzymatic rate $V_{max}^{(2)}$.

The diffusion module is one of the principal parameters controlling the biosensor behaviour. Since the diffusion coefficients and the Michaelis–Menten constant were the same in all the numerical experiments as defined in (46), and the behaviour of the biosensors acting in the trigger modes is mainly determined by the enzymatic rate $V_{max}^{(2)}$ (Figs. 7 and 8), the thickness d_σ of the enzyme layer is introduced as a function of $V_{max}^{(2)}$ at which the diffusion module equals unity, $\sigma^2 = 1$,

$$d_\sigma\left(V_{max}^{(2)}\right) = \sqrt{\frac{D_S K_M}{V_{max}^{(2)}}}. \tag{47}$$

The following three values of $d_\sigma(V_{max}^{(2)})$ are depicted in Fig. 10: $d_\sigma(1\text{ mM/s}) \approx 5.48$ μm, $d_\sigma(0.1\text{ mM/s}) \approx 17.3$ μm and $d_\sigma(0.01\text{ mM/s}) \approx 54.8$ μm. These values were calculated at constant values of D_S and K_M defined in (46). When comparing

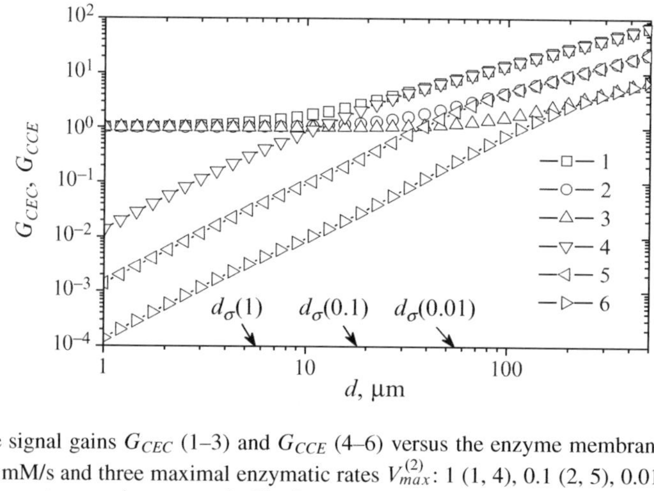

Fig. 10 The signal gains G_{CEC} (1–3) and G_{CCE} (4–6) versus the enzyme membrane thickness d at $V_{max}^{(1)} = 1$ mM/s and three maximal enzymatic rates $V_{max}^{(2)}$: 1 (1, 4), 0.1 (2, 5), 0.01 (3, 6) mM/s; The other parameters are the same as in Fig. 3

value $d_\sigma(1\,\mathrm{mM/s})$ with the membrane thickness at which the gain G_{CEC} starts to increase at corresponding $V_{max}^{(2)}$ of 1 mM/s (curve 1 in Fig. 10), one can see that the amplification becomes noticeable when the mass transport by diffusion starts to control the biosensor response. As one can see in Fig. 10, this effect is also valid for two other values of the maximal enzymatic rate $V_{max}^{(2)}$: 0.1 and 0.01 mM/s. However, this is valid only in the case of the biosensor acting in the CEC mode.

In the case of the CCE mode, the signal gain G_{CCE} permanently increases with an increase in the thickness d. G_{CCE} is approximately a linear increasing function of the enzyme layer thickness d. However, practical amplification ($G_{CCE} > 1$) takes place only in cases of relatively thick membranes ($d >\approx 2d_\sigma$). As has been noted above (see Fig. 8), the steady state current of the biosensor acting in the CCE mode may be even significantly less than the steady state current of the corresponding biosensor acting in the CE mode at the same conditions. In the case of relatively thick enzyme membrane the gain G_{CCE} is equal approximately to G_{CEC}.

Using computer simulation for different enzymatic rates $V_{max}^{(2)}$ the thickness d_G of the enzyme membrane at which $G_{CCE} = 1$ was calculated. Accepting $V_{max}^{(1)} =$ 1 mM/s the following values of were found: $d_G \approx 9\,\mu\mathrm{m}$ at $V_{max}^{(2)} = 1$ mM/s, $d_G \approx 30\,\mu\mathrm{m}$ at $V_{max}^{(2)} = 0.1$ mM/s, and $d_G \approx 90\,\mu\mathrm{m}$ at $V_{max}^{(2)} = 0.01$ mM/s. These values of the membrane thickness compare favourably with values of the thickness d_{max} (see formula (64)) at which the steady state current as a function of the membrane thickness d gains the maximum [27, 32]. Consequently, for low substrate concentrations, the thickness d_G of the enzyme membrane at which $G_{CCE} = 1$ can be rather precisely expressed as $d_G \approx 1.5d_\sigma$, where d_σ was defined in (47). Additional calculations showed that this property is valid for wide ranges of both maximal enzymatic rates, $V_{max}^{(1)}$ and $V_{max}^{(2)}$, if only the substrate concentration S_0 is less than the Michaelis–Menten constant K_M.

2.4.5 Effect of the Membrane Thickness on the Response Time

From a practical point of view it is very important to have the biosensor response time as short as possible. To compare the time of steady state amplified biosensor response with the steady state time of the response without amplification, the prolongation of the response time is introduced as the ratio of the steady state time of the trigger biosensor to the steady state time of the corresponding CE biosensor,

$$
\begin{aligned}
L_{CEC}(V_{max}^{(2)}) &= \frac{T_{CEC}(V_{max}^{(2)})}{T_{CE}} = \frac{T_{CEC}(V_{max}^{(2)})}{T_{CEC}(0)},\\
L_{CCE}(V_{max}^{(2)}) &= \frac{T_{CCE}(V_{max}^{(2)})}{T_{CE}} = \frac{T_{CCE}(V_{max}^{(2)})}{T_{CEC}(0)},
\end{aligned}
\qquad (48)
$$

where $T_{CEC}(V_{max}^{(2)})$ and $T_{CCE}(V_{max}^{(2)})$ are the steady state times of the trigger biosensors acting in the CEC and the CCE mode, respectively, calculated at maximal

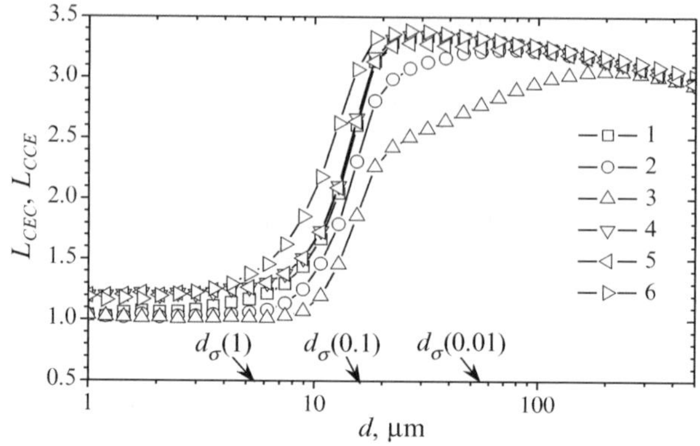

Fig. 11 The response time prolongations L_{CEC} (1–3) and L_{CCE} (4–6) versus the enzyme membrane thickness d. The parameters and the notation are the same as in Fig. 10

enzymatic rate $V_{max}^{(2)}$ of the enzyme E_2, T_{CE} is the steady state time of the corresponding CE biosensor. Since the action of the CE biosensor can be simulated as an action of the CEC biosensor accepting $V_{max}^{(2)} = 0$, $T_{CE} = T_{CEC}(0)$ is assumed.

Figure 11 shows the change of the response time versus the membrane thickness d at $V_{max}^{(1)} = 1$ mM/s and different values of $V_{max}^{(2)}$. One can see in Fig. 11 that in all of the presented cases the prolongation of the response time (L_{CEC} as well as L_{CCE}) is a non monotonous function of the thickness d of the enzyme layer. The shoulder on the curves is especially noticeable at high maximal enzymatic rates. Similar effect was noticed in the case of biosensors with substrate cyclic conversion [29] and during the oxidation of β-nicotinamide adenine dinucleotide (NADH) at poly(aniline)-coated electrodes [46].

In cases of the thin enzyme membranes ($d < 10\,\mu$m), the prolongation of the response time is insignificant. However, an increase in the membrane thickness d prolongs the response time up to 3.4 times in both modes of the triggering: the CEC and the CCE.

In the case of CEC mode, the slight influence of the maximal enzymatic rate $V_{max}^{(2)}$ on L_{CEC} can be noticed in Fig. 11, while no notable influence of $V_{max}^{(2)}$ on L_{CCE} is observed in the case of the CCE action mode. Additional calculations showed that the response time prolongation slightly depends on the substrate concentration S_0 as well as on the maximal activity $V_{max}^{(1)}$ of the enzyme E_1 [32].

2.5 Concluding Remarks

The mathematical model (16)–(21) of the biosensor operation can be used to investigate the dynamics of the response of biosensors utilizing a trigger enzymatic reaction following electrochemical and enzymatic product cyclic conversion (CEC scheme (13)–(15)), while the model (28), (17), (19)–(21), (29) can be applied as a framework to investigate the behaviour of biosensors utilizing a trigger enzymatic reaction following enzymatic and electrochemical conversion of the product (CCE scheme (25)–(27)).

The signal gains in sensitivity, G_{CEC} and G_{CCE}, of trigger biosensors are mainly determined by the maximal enzymatic activity $V_{max}^{(2)}$ of the enzyme E_2 (Figs. 7 and 7). The enzymatic activity $V_{max}^{(2)}$ is a critical factor for the biosensor current in the case of the CCE mode, $G_{CCE} \to 0$ if $V_{max}^{(2)} \to 0$. In the case of the CEC biosensor, a decrease of activity $V_{max}^{(2)}$ causes a decrease in gain G_{CEC}, however, G_{CEC} stays greater than unity, $G_{CEC} \to 1$ if $V_{max}^{(2)} \to 0$.

Both signal gains, G_{CEC} and G_{CCE}, are mostly valuable when the concentration of the substrate is less than the Michaelis–Menten constant (Fig. 9). A valuable amplification (up to dozens of times) at wide range of substrate concentration is achieved only in the case of relatively low maximal enzymatic activity of enzyme E_1 and high activity of enzyme E_2. In both biosensor action modes, the CEC and the CCE, a significant amplification of the signal is observed if the response is under the mass transport control, i.e. if $\sigma^2 \gg 1$ (Fig. 10).

In cases when the valuable amplification of the signal of a triggering biosensor is achieved, the response time is up to several times longer than the response time of the corresponding biosensor acting without triggering (Fig. 11) [32].

Multi-Layer Models of Biosensors

Contents

1 Multi-Layer Approach 140
1.1 Mathematical Model of Multi-Layer System 141
1.2 Numerical Approximation 143
1.3 Three-Layer Model 145
2 Two-Compartment Model 147
2.1 Mathematical Model 147
2.2 Transient Numerical Solution 150
2.3 Validation of Numerical Solution 152
2.4 Simulated Biosensor Responses 153
2.5 Effect of the Diffusion Layer 155
2.6 The Nernst Diffusion Layer 157
2.7 Dimensionless Model 160
2.8 Impact of the Diffusion Module 162
3 Biosensors with Outer Porous Membrane 163
3.1 Mathematical Model 164
3.2 Numerical Simulation 166
3.3 Effect of the Porous Membrane 168
3.4 Concluding Remarks 171
4 Biosensors with Selective and Outer Perforated Membranes 172
4.1 Mathematical Model 173
4.2 Numerical Simulation 176
5 Biosensors Based on Chemically Modified Electrode 178
5.1 Mathematical Model 178
5.2 Numerical Simulation 182
5.3 Dimensionless Model 184
5.4 Simulated Biosensor Action 186
5.5 Impact of the Diffusion Module 189
5.6 Impact of the Substrate Concentration 191
5.7 Concluding Remarks 192
6 Optical and Fluorescence Biosensors 193
6.1 Mathematical Model 193
6.2 Numerical Simulation 197
6.3 Simulated Biosensor Action 198
6.4 Impact of the Substrate Concentration 201

R. Baronas et al., *Mathematical Modeling of Biosensors*, Springer Series on Chemical Sensors and Biosensors 9, DOI 10.1007/978-90-481-3243-0_8,

There are various reasons for applying a multi-layer approach to the modeling of biosensors. Multi-layer models are usually used in the following cases [35, 119, 235, 236]:

- The bulk solution is assumed to be slightly-stirred or non-stirred. This assumption leads to two-compartment models [51, 60, 102, 120, 286].
- The enzyme layer is covered with an inert outer membrane [234]. The membrane stabilizes the enzyme layer and creates a diffusion limitation to the substrate, i.e. lowers the substrate concentration in the enzymatic layer and thereby prolongs the calibration curve of the biosensor [152, 166, 229, 238, 258].
- The electrode is covered with a selective membrane [236]. Selective membranes are usually impermeable to certain molecules and permeable to a desired substance. This arrangement can notably increase the biosensor selectivity. The selective layer can also protect the metal interface of the electrode [16, 47, 91, 157].
- In multienzyme systems, enzymes are often immobilized separately in different active layers packed in a sandwich-like multi-layer arrangement [13, 14, 15, 113, 184, 242]. This approach seems to be a rather fast and cheap method to design biosensors for different purposes.

1 Multi-Layer Approach

According to a multi-layer approach, a biosensor is considered as an electrode, having several layers each with a single enzyme sandwich-likely applied onto the electrode surface [234, 235].

In each enzyme layer the substrate (S) combines reversibly with an enzyme (E_k) to form a complex (ES_k). The complex then dissociates into a product (P) and the enzyme is released,

$$S + E_k \rightleftarrows ES_k \rightarrow E_k + P, \quad k = 1, \ldots, K, \tag{1}$$

where K is the number of the layers.

Assuming the quasi steady state approximation, the concentration of the intermediate complex (ES_k) does not change and may be neglected when simulating the biochemical behaviour of biosensors [229, 234, 235, 258]. Consequently, consider in each enzyme layer a scheme where the substrate (S) binds to the enzyme (E_k) and is converted to the product (P),

$$S \xrightarrow{E_k} P, \quad k = 1, \ldots, K. \tag{2}$$

1.1 Mathematical Model of Multi-Layer System

Coupling the enzyme-catalyzed reaction in each enzyme layer with the one-dimensional-in-space diffusion leads to the following equations ($t > 0$):

$$\frac{\partial S^{(k)}}{\partial t} = D_S^{(k)} \frac{\partial^2 S^{(k)}}{\partial x^2} - \frac{V_{max}^{(k)} S^{(k)}}{K_M^{(k)} + S^{(k)}},$$
$$\frac{\partial P^{(k)}}{\partial t} = D_P^{(k)} \frac{\partial^2 P^{(k)}}{\partial x^2} + \frac{V_{max}^{(k)} S^{(k)}}{K_M^{(k)} + S^{(k)}}, \tag{3}$$
$$x \in (a_{k-1}, a_{k-1} + d_k), \quad a_k = a_{k-1} + d_k, \quad k = 1, \dots, K,$$

where x and t stand for space and time, $S^{(k)}(x,t)$ and $P^{(k)}(x,t)$ are the concentrations of the substrate and of the product in k-th layer, d_k is the thickness of the k-th enzyme layer, $D_S^{(k)}$, $D_P^{(k)}$ are the diffusion coefficients, $V_{max}^{(k)}$ and $K_M^{(k)}$ denote the maximal enzymatic rate and the Michaelis constant for the enzyme E_k, a_k stands for the interface separating two adjacent layers, $a_0 = 0$. The closed interval $[a_0, a_K]$ (as well as $[0, a_K]$) covers all the enzyme layers.

Let $x = a_0 = 0$ represent the electrode surface, while $x = a_K$ represents the boundary between the sandwich-like enzyme membrane and the bulk solution. According to the following initial conditions, the biosensor operation starts when the substrate appears over the surface of the upper layer ($t = 0$):

$$S^{(k)}(x,0) = 0, \quad x \in [a_{k-1}, a_k], \quad k = 1, \dots, K-1,$$
$$S^{(K)}(x,0) = 0, \quad x \in [a_{K-1}, a_K), \tag{4}$$
$$S^{(K)}(a_K,0) = S_0,$$

where S_0 is the concentration of the substrate in the bulk solution.

Initially, no product appears in the entire domain,

$$P^{(k)}(x,0) = 0, \quad x \in [a_{k-1}, a_k], \quad k = 1, \dots, K. \tag{5}$$

At the electrode surface ($x = a_0 = 0$), the boundary conditions depend on the electric activity of the substances. Accepting amperometry leads to the following conditions at the electrode surface ($t > 0$):

$$D_S^{(1)} \left. \frac{\partial S^{(1)}}{\partial x} \right|_{x=0} = 0,$$
$$P^{(1)}(0,t) = 0. \tag{6}$$

The problem solutions at two adjacent layers have to be continuous ($t > 0$)

$$\begin{aligned} S^{(k)}(a_k, t) &= S^{(k+1)}(a_k, t), \\ P^{(k)}(a_k, t) &= P^{(k+1)}(a_k, t), \quad k = 1, \dots, K-1. \end{aligned} \tag{7}$$

On the boundary between two adjacent layers the mass conservation relations are additionally defined

$$\begin{aligned} D_S^{(k)} \frac{\partial S^{(k)}}{\partial x}\bigg|_{x=a_k} &= D_S^{(k+1)} \frac{\partial S^{(k+1)}}{\partial x}\bigg|_{x=a_k}, \\ D_P^{(k)} \frac{\partial P^{(k)}}{\partial x}\bigg|_{x=a_k} &= D_P^{(k+1)} \frac{\partial P^{(k+1)}}{\partial x}\bigg|_{x=a_k}, \quad k = 1, \dots, K-1. \end{aligned} \tag{8}$$

The matching conditions (7) and (8) mean that the fluxes of the substrate and product through the $(k+1)$-th layer are equal to the corresponding fluxes entering the k-th layer. The partitions of the substrate and the product in the k-th layer versus the $(k+1)$-th layer are assumed to be equal [201, 235].

When the bulk solution is well-stirred outside and in the powerful motion, the concentration of the substrate as well as the product over the top layer enzyme surface remains constant during the biosensor operation ($t > 0$),

$$\begin{aligned} S^{(K)}(a_K, t) &= S_0, \\ P^{(K)}(a_K, t) &= 0. \end{aligned} \tag{9}$$

Due to conditions (7), the concentrations of the substrate (S) and of the reaction product (P), can be defined for the entire interval $x \in [a_0, a_K]$ as follows ($t \geq 0$):

$$S(x,t) = \begin{cases} S^{(1)}(x,t), & x \in [a_0, a_1], \\ S^{(2)}(x,t), & x \in (a_1, a_2], \\ \dots \\ S^{(K)}(x,t), & x \in (a_{K-1}, a_K], \end{cases} \tag{10}$$

$$P(x,t) = \begin{cases} P^{(1)}(x,t), & x \in [a_0, a_1], \\ P^{(2)}(x,t), & x \in (a_1, a_2], \\ \dots \\ P^{(K)}(x,t), & x \in (a_{K-1}, a_K]. \end{cases} \tag{11}$$

Both concentration functions (S and P) are now continuous in the entire domain $x \in [a_0, a_K]$.

The biosensor current depends upon the flux of the product at the electrode surface,

$$i(t) = n_e F D_P^{(1)} \frac{\partial P^{(1)}}{\partial x}\bigg|_{x=0}, \tag{12}$$

where $i(t)$ is the density of the anodic current at time t, n_e is the number of electrons involved in a charge transfer at the electrode surface, and F is the Faraday constant.

The system (3)–(9) approaches a steady state as $t \to \infty$,

$$I = \lim_{t\to\infty} i(t). \tag{13}$$

where I is the density of the steady state biosensor current.

1.2 Numerical Approximation

Equations (3)–(9) form K initial boundary value problems formulated on each layer. Due to the matching conditions (8) these problems can not be solved independently of each other [234, 235].

In the case of the first-order rate of the enzymatic reactions ($S_0 \ll K_M^{(k)}$ for $\forall k : k = 1, \ldots, K$), the problem (3)–(9) can be solved analytically. Schulmeister solved this problem analytically by using polygon approximations of the boundary conditions (8) [234].

To find a numerical solution of the problem (3)–(9) in the domain $[a_0, a_K] \times [0, T]$, a discrete grid has to be introduced. Since the thicknesses of the enzyme layers may differ even in orders of magnitude [280], it is reasonable to use an individual step size for each layer. Assuming the step size h_k constant through k-th layer, each layer can be uniformly discretized, $k = 1, \ldots, K$,

$$\begin{aligned} \omega_h^{(k)} &= \{x_i^{(k)} : x_i^{(k)} = a_{k-1} + ih_k,\ i = 0, 1, \ldots, N_k,\ h_k N_k = d_k\}, \\ \omega_\tau &= \{t_j : t_j = j\tau,\ j = 0, 1, \ldots, M;\ \tau M = T\}, \end{aligned} \tag{14}$$

where T is the time of the biosensor operation.

Let us assume the following notation:

$$\begin{aligned} S_i^{(k),j} &= S^{(k)}(x_i^{(k)}, \tau j), \\ P_i^{(k),j} &= P^{(k)}(x_i^{(k)}, \tau j), \\ i &= 0, \ldots, N_k, \quad j = 0, \ldots, M, \quad k = 1, \ldots, K. \end{aligned} \tag{15}$$

The governing equations (3), the initial conditions (4), (5) and the boundary conditions (6)–(9) can be approximated using finite differences as discussed in Sect. 3 of first chapter of Part II. Both types of the finite difference schemes, explicit and implicit, are applicable. The matching conditions (8) are approximated as follows [222, 235]:

$$
\begin{aligned}
&D_S^{(k)} \frac{S_{N_k}^{(k),j} - S_{N_k-1}^{(k),j}}{h_k} = D_S^{(k+1)} \frac{S_1^{(k+1),j} - S_0^{(k+1),j}}{h_{k+1}}, \\
&D_P^{(k)} \frac{P_{N_k}^{(k),j} - P_{N_k-1}^{(k),j}}{h_k} = D_P^{(k+1)} \frac{P_1^{(k+1),j} - P_0^{(k+1),j}}{h_{k+1}}, \\
&k = 1, \ldots, K-1.
\end{aligned}
\tag{16}
$$

A union of $\omega_h^{(1)}, \ldots, \omega_h^{(K)}$ forms a discretization ω_h for the entire interval $[a_0, a_K]$,

$$
\omega_h = \bigcup_{k=1}^{K} \omega_h^{(k)}. \tag{17}
$$

The ordered set ω_h contains $N = (N_1 \times N_2 \times \ldots \times N_K) + 1$ points. Let $x_0, x_1, \ldots, x_N$ be an ordered sequence of all points consisting ω_h, $x_0 = a_0 = 0$. In the discrete presentation ω_h of the interval $[a_0, a_K]$, $K - 1$ points correspond to the interfaces separating adjacent media. Let $bound(k)$ be an index of a point in ω_h corresponding to a boundary between k-th and $(k+1)$-th enzyme layers, $k = 1, \ldots, K-1$,

$$
bound(k) = \sum_{i=1}^{k} N_k. \tag{18}
$$

The following relation maps the concentrations of the substrate and the product for each layer on the entire domain ($j = 1, \ldots, M$):

$$
S_i^j = \begin{cases}
S^{(1)}(x_i^{(1)}, \tau j), & 0 \le i \le bound(1), \\
S^{(2)}(x_{i-bound(1)}^{(2)}, \tau j), & bound(1) < i \le bound(2), \\
\ldots \\
S^{(K)}(x_{i-bound(K-1)}^{(K)}, \tau j), & bound(K-1) < i \le bound(K),
\end{cases} \tag{19}
$$

$$
P_i^j = \begin{cases}
P^{(1)}(x_i^{(1)}, \tau j), & 0 \le i \le bound(1), \\
P^{(2)}(x_{i-bound(1)}^{(2)}, \tau j), & bound(1) < i \le bound(2), \\
\ldots \\
P^{(K)}(x_{i-bound(K-1)}^{(K)}, \tau j), & bound(K-1) < i \le bound(K).
\end{cases} \tag{20}
$$

Using (19) and (20) the difference equations can be solved to find approximate values of the concentrations $S(x,t)$ and $P(x,t)$ (see descriptions (10) and (11)) in the domain $[a_0, a_K] \times [0, T]$.

Having the numerical solution of the problem, the density of the biosensor current at time $t = t_j$ is calculated by

$$i(t_j) \approx i_j = n_e F D_P^{(1)} P_1^j / h_1, \quad j = 1, \ldots, M. \tag{21}$$

1.3 Three-Layer Model

Figures 1 and 2 show the steady state concentration profiles of the substrate S and the product P calculated applying the multi-layer approach to a three-layer model. To see some interaction between the layers, the biosensor response has to be simulated at different values of the maximal enzymatic rate keeping the bulk concentration constant. All three enzyme layers were of the same thickness, and the

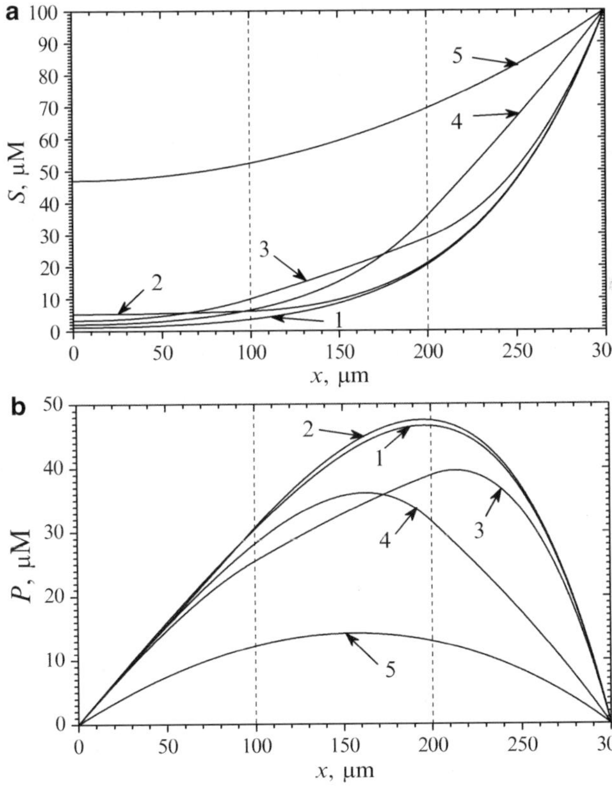

Fig. 1 Steady state concentration profiles of the substrate (**a**) and the product (**b**) in three enzyme layers at the following maximal enzymatic rates $V_{max}^{(1)}$: 10 (1, 3, 4), 1 (2, 5), $V_{max}^{(2)}$: 10 (1, 2, 4), 1 (3, 5), $V_{max}^{(3)}$: 10 (1–3), 1 (4, 5) µM/s. *Dashed lines* show boundaries between adjacent layers

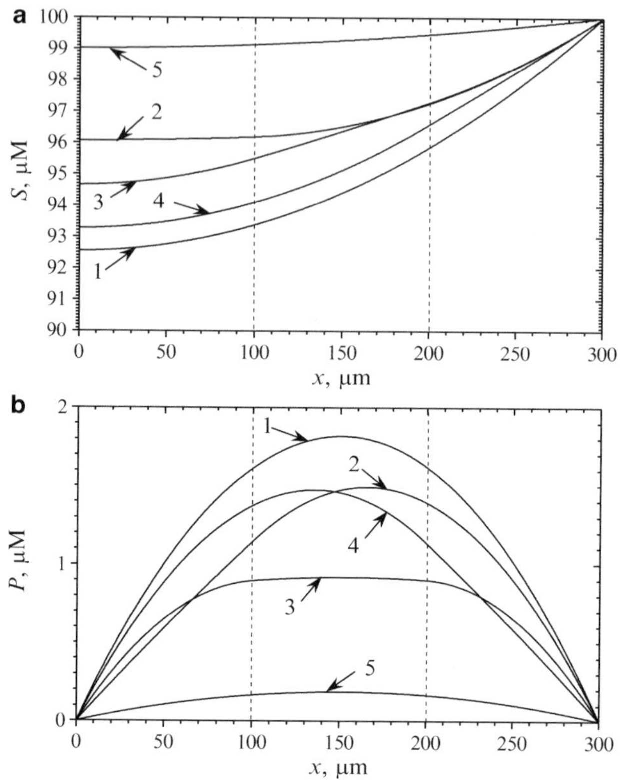

Fig. 2 Steady state concentration profiles of the substrate (**a**) and the product (**b**) in three enzyme layers at the following maximal enzymatic rates $V_{max}^{(1)}$: 0.1 (1, 3, 4), 0.01 (2, 5), $V_{max}^{(2)}$: 0.1 (1, 2, 4), 0.01 (3, 5), $V_{max}^{(3)}$: 0.1 (1–3), 0.01 (4, 5) μM/s. *Dashed lines* show boundaries between adjacent layers

diffusion coefficients as well as the Michaelis constants were also identical for all three layers,

$$\begin{aligned} &D_S^{(k)} = D_P^{(k)} = 300\,\mu\text{m}^2/\text{s}, \quad K_M^{(k)} = 100\mu\text{M}, \quad d_k = 100\,\mu\text{m}, \\ &k = 1, 2, 3, \quad S_0 = 100\,\mu\text{M}. \end{aligned} \tag{22}$$

Figure 1 shows the steady state concentration profiles in the case when a three-layer biosensor acts under limitation of the diffusion (diffusion module $\sigma^2 > 1$), while Fig. 2 corresponds to the limitation of the enzyme kinetics ($\sigma^2 < 1$). When comparing the shape of corresponding curves in the figures, one can see different positions of peaks.

The shape of curves 1–4 in Fig. 1b differs notably, however, the gradients of the product concentrations at the electrode surface ($x = 0$) are comparable. The density of the biosensor current in these four cases differs less than 5%. One can observe a very similar situation in Fig. 2b when comparing curves 1 and 4. In both Figs. 1b and 2b, curves 1 and 4 correspond to different enzymatic activity in the distal (from the electrode) layer and the same enzymatic activities in other two layers. Consequently, the decrease in enzyme activity in a farther layer more considerably effects the biosensor response than in the nearest layer.

In Figs. 1 and 2, curves 1 and 5 show the concentration profiles for biosensors with three practically identical layers, i.e. in all three layers even the maximal enzymatic rates are the same. In such cases of identical layers, a multi-layer biosensor becomes practically identical to a mono-layer biosensor with the thickness equal to the sum of the thicknesses of all the layers of the corresponding multi-layer biosensor. This feature of models can be applied to testing the numerical solution of multi-layer model by comparing the solution with the solution of the corresponding mono-layer model (see Sect. 3.3 of first chapter of Part II).

2 Two-Compartment Model

Starting from 70ths, two-compartment models have been extensively investigated in the modeling of biosensors [47, 51, 102, 120]. In a two-compartment model, the bulk solution region is usually considered as a second layer. That model can be considered as a particular case of the multi-layer model [235].

Some authors have considered models with an infinite bulk solution [51, 102]. Below a two-compartment model with a finite compartment corresponding to the bulk solution is considered [235].

If the bulk solution is well-stirred and in powerful motion, then the diffusion layer remains at a constant thickness of the enzyme layer. In such a case, a one-layer model is usually applied as discussed in first chapter of Part II. The concentration of the substrate as well as of the product over the enzyme surface (bulk solution/membrane interface) remains constant while the biosensor keeps in touch with the substrate. However, if the bulk solution is slightly stirred or not stirred at all, then the concentrations of the substrate and the product over the enzyme surface depend on the diffusion of the species in the bulk solution. Because of this the model to be considered consists of three regions: the enzyme layer where the enzyme reaction as well as the mass transport by diffusion takes place, the diffusion limiting region where only the mass transport by diffusion takes place, and the convective region, where the analyte concentration is maintained constant.

2.1 Mathematical Model

A scheme where the substrate (S) binds into the enzyme (E) and is converted to the product (P) has been already discussed in first chapter of Part II,

$$S \xrightarrow{E} P \tag{23}$$

An amperometric biosensor is considered as an electrode, having a layer of mono-enzyme (enzyme membrane) applied onto the electrode surface. Assuming the symmetrical geometry of the electrode and the homogeneous distribution of the immobilized enzyme in the enzyme membrane, the dynamics of the concentrations of the substrate as well as the product in the enzyme layer can be described by the reaction-diffusion system [26, 235],

$$\begin{aligned} \frac{\partial S_e}{\partial t} &= D_{S_e} \frac{\partial^2 S_e}{\partial x^2} - \frac{V_{max} S_e}{K_M + S_e}, \\ \frac{\partial P_e}{\partial t} &= D_{P_e} \frac{\partial^2 P_e}{\partial x^2} + \frac{V_{max} S_e}{K_M + S_e}, \quad x \in (0, d), \quad t > 0, \end{aligned} \tag{24}$$

where x and t stand for space and time, respectively, $S_e(x,t)$ and $P_e(x,t)$ are the concentrations of the substrate and the reaction product in the enzyme layer, respectively, d is the thickness of the enzyme membrane, D_{S_e} and D_{P_e} are the diffusion coefficients, V_{max} is the maximal enzymatic rate and K_M is the Michaelis constant (see Sect. 1 of first chapter of Part II).

In the second layer only the mass transport by diffusion takes place,

$$\begin{aligned} \frac{\partial S_b}{\partial t} &= D_{S_b} \frac{\partial^2 S_b}{\partial x^2}, \\ \frac{\partial P_b}{\partial t} &= D_{P_b} \frac{\partial^2 P_b}{\partial x^2}, \quad x \in (d, d+\delta), \quad t > 0, \end{aligned} \tag{25}$$

where δ is the thickness of the external diffusion layer, $S_b(x,t)$ and $P_b(x,t)$ are the concentrations of the substrate and the reaction product in the diffusion layer, respectively, D_{S_b}, D_{P_b} are the diffusion coefficients of the species in the bulk solution.

Let $x = 0$ represent the electrode surface, while $x = d$ represents the boundary layer between the enzyme membrane and the bulk solution. The biosensor operation starts when some substrate appears in the bulk solution ($t = 0$),

$$\begin{aligned} &S_e(x,0) = 0, \quad P_e(x,0) = 0, \quad x \in [0, d], \\ &S_b(x,0) = 0, \quad P_b(x,0) = 0, \quad x \in [d, d+\delta), \\ &S_b(d+\delta, 0) = S_0, \quad P_b(d+\delta, 0) = 0, \end{aligned} \tag{26}$$

where S_0 is the concentration of the substrate in the bulk solution.

On the boundary between two subregions having different diffusion coefficients the matching conditions are defined ($t > 0$),

$$\begin{aligned} D_{S_e} \left.\frac{\partial S_e}{\partial x}\right|_{x=d} &= D_{S_b} \left.\frac{\partial S_b}{\partial x}\right|_{x=d}, \quad S_e(d,t) = S_b(d,t), \\ D_{P_e} \left.\frac{\partial P_e}{\partial x}\right|_{x=d} &= D_{P_b} \left.\frac{\partial P_b}{\partial x}\right|_{x=d}, \quad P_e(d,t) = P_b(d,t). \end{aligned} \tag{27}$$

The external diffusion layer of thickness δ remains unchanged with time. Away from it the concentration of the substrate as well as of the product remains constant ($t > 0$),

$$S_b(d+\delta,t) = S_0, \quad P_b(d+\delta,\, t) = 0. \tag{28}$$

At the electrode surface ($x = 0$), due to the amperometry, the potential is chosen to keep zero concentration of the reaction product,

$$P_e(0,t) = 0, \quad D_{S_e}\frac{\partial S_e}{\partial x}\bigg|_{x=0} = 0, \quad t > 0. \tag{29}$$

The system (24)–(29) approaches a steady state as $t \to \infty$. The density $i(t)$ of the biosensor current at time t can be obtained explicitly from the Faraday and the Fick laws [235],

$$i(t) = n_e F D_{P_e}\frac{\partial P_e}{\partial x}\bigg|_{x=0}, \qquad I = \lim_{t\to\infty} i(t), \tag{30}$$

where n_e is the number of electrons involved in a charge transfer at the electrode surface, F is the Faraday constant and I is the density of the steady state biosensor current.

For $\delta \to 0$, the solution of the two-compartment model approaches the solution of the corresponding one-layer model discussed in first chapter of Part II.

The two-compartment mathematical model (24)–(29) of an amperometric biosensor would become the two-compartment model of the corresponding potentiometric biosensor if the boundary condition $P_e(0,t) = 0$ was replaced with the following boundary condition:

$$D_{P_e}\frac{\partial P_e}{\partial x}\bigg|_{x=0} = 0, \quad t > 0. \tag{31}$$

The initial condition (26) indicates that biosensor operation starts when the substrate of concentration S_0 touches the distal boundary of the diffusion layer, i.e. the substrate enters to diffusion layer. These conditions model the infusion of the substrate into the bulk solution. Sometimes a biosensor is instantly immersed into the bulk solution containing the substrate of concentration S_0 [26]. In this case initial conditions (26) have to be replaced with

$$\begin{aligned}
&S_e(x,0) = 0, \quad x \in [0,\, d), \quad S_e(d,0) = S_0,\\
&P_e(x,0) = 0, \quad x \in [0,\, d],\\
&S_b(x,0) = S_0, \quad P_b(x,0) = 0, \quad x \in [d,\, d+\delta].
\end{aligned} \tag{32}$$

Let us notice that both initial conditions, (26) and (32), lead to the same steady state solution [10,56]. Only a transient solution is effected by the initial conditions.

The thickness δ of the external diffusion layer depends upon the nature and intensity of the stirring of the buffer solution. The less intense stirring corresponds to the thicker diffusion layer. A non-perfect stirring is usually modeled by a relatively thick external diffusion layer. To model a biosensor action in a non-stirred buffer solution some authors use a non-leakage (zero flux) boundary condition on the boundary $d + \delta$,

$$D_{S_b} \frac{\partial S_b}{\partial x}\bigg|_{x=d+\delta} = 0, \quad D_{P_b} \frac{\partial P_b}{\partial x}\bigg|_{x=d+\delta} = 0, \quad t > 0, \tag{33}$$

instead of conditions (28), and the initial condition (32) instead of (26) [22,120,235].

2.2 Transient Numerical Solution

At the transient conditions the problem (24)–(29) can be solved numerically using the finite difference technique. To simulate the biosensor action a discrete grid has to be introduced. In a common case, different step sizes can be used for different compartments. For simplicity a constant step size in the entire domain can be applied. To find a finite difference solution of the problem in the domain $[0, d + \delta] \times [0, T]$ a uniform discrete grid $\omega_h \times \omega_\tau$ is introduced,

$$\begin{aligned} \omega_h &= \{x_i : x_i = ih, i = 0, \dots, N_d, \dots, N;\ hN_d = d,\ hN = d + \delta\}, \\ \omega_\tau &= \{t_j : t_j = j\tau, j = 0, \dots, M;\ \tau M = T\}. \end{aligned} \tag{34}$$

As in a general multi-layer model introduced in (10) and (11), the concentrations of both species, S and P, can be defined in the entire domain $x \in [0, d + \delta]$ as follows ($t \geq 0$):

$$\begin{aligned} S(x,t) &= \begin{cases} S_e(x,t), & x \in [0, d], \\ S_b(x,t), & x \in (d, d + \delta], \end{cases} \\ P(x,t) &= \begin{cases} P_e(x,t), & x \in [0, d], \\ P_b(x,t), & x \in (d, d + \delta]. \end{cases} \end{aligned} \tag{35}$$

Both concentration functions, S and P, are continuous in the entire domain $x \in [0, d + \delta]$. The concentrations of both species are defined on the discrete grid (34),

$$\begin{aligned} &S_i^j = S(x_i, t_j), \quad P_i^j = P(x_i, t_j), \quad i_j = i(t_j), \\ &i = 0, \dots, N, \quad j = 0, \dots, M. \end{aligned} \tag{36}$$

To find approximate values of S and P on $\omega_h \times \omega_\tau$, an implicit finite difference scheme can be used replacing the differential equations (24) and (25) with the following difference equations:

$$\frac{S_i^{j+1}-S_i^j}{\tau}=D_{S_e}\frac{S_{i+1}^{j+1}-2S_i^{j+1}+S_{i-1}^{j+1}}{h^2}-\frac{V_{max}S_i^j}{K_M+S_i^j},$$
$$\frac{P_i^{j+1}-P_i^j}{\tau}=D_{P_e}\frac{P_{i+1}^{j+1}-2P_i^{j+1}+P_{i-1}^{j+1}}{h^2}+\frac{V_{max}S_i^j}{K_M+S_i^j},\tag{37}$$
$$i=1,\ldots,N_d-1,\quad j=0,\ldots,M-1,$$

$$\frac{S_i^{j+1}-S_i^j}{\tau}=D_{S_b}\frac{S_{i+1}^{j+1}-2S_i^{j+1}+S_{i-1}^{j+1}}{h^2},$$
$$\frac{P_i^{j+1}-P_i^j}{\tau}=D_{P_b}\frac{P_{i+1}^{j+1}-2P_i^{j+1}+P_{i-1}^{j+1}}{h^2},\tag{38}$$
$$i=N_d+1,\ldots,N-1,\quad j=0,\ldots,M-1.$$

The initial conditions (26) are approximated by

$$S_i^0=0,\quad i=0,\ldots,N-1,$$
$$S_N^0=S_0,\tag{39}$$
$$P_i^0=0,\quad i=0,\ldots,N.$$

The matching (27) and the boundary conditions (28) and (29) are approximated as follows:

$$D_{S_e}\frac{S_{N_d}^j-S_{N_d-1}^j}{h}=D_{S_b}\frac{S_{N_d+1}^j-S_{N_d}^j}{h},$$
$$D_{P_e}\frac{P_{N_d}^j-P_{N_d-1}^j}{h}=D_{P_b}\frac{P_{N_d+1}^j-P_{N_d}^j}{h},\tag{40}$$
$$S_N^j=S_0,\quad P_N^j=0,$$
$$D_{S_e}\frac{S_1^j-S_0^j}{h}=0,\quad P_0^j=0,\quad j=1,\ldots,M.$$

The resulting systems of linear algebraic equations are solved efficiently because of the tridiagonality of their matrices.

Having a numerical solution of the problem, the density of the biosensor current at time $t=t_j$ can be easily calculated by

$$i_j\approx i(t_j)=n_eFD_{P_e}(P_1^j-P_0^j)/h,\quad j=0,\ldots,M.\tag{41}$$

In the case of momentary immersion of a biosensor into a bulk solution, the initial conditions (26) are approximated by

$$S_i^0=0,\quad i=0,\ldots,N_d-1,$$
$$S_i^0=S_0,\quad i=N_d,\ldots,N,\tag{42}$$
$$P_i^0=0,\quad i=0,\ldots,N.$$

In a numerical simulation, the biosensor response time T and the half time $T_{0.5}$ of the steady state can be calculated as in the case of single-layer model, i.e. by (36) and (25), respectively.

2.3 Validation of Numerical Solution

The existing exact analytical solution of the reaction–diffusion problem (24)–(29) can be applied to validate the numerical solution.

At so low concentrations of the substrate as $S_0 \ll K_M$, the non-linear reaction rate in (24) reduces to the first order reaction rate $V_{max} S_e / K_M$. At these conditions the two-compartment model can be solved analytically. The stationary concentrations of the substrate and the product in both layers have been presented by Schulmeister [235] ($t \to \infty$),

$$
\begin{aligned}
S_e(x,t) &= S_0 \frac{D_{S_b} \cosh(\sigma x/d)}{D_{S_b} \cosh(\sigma) + D_{S_e} \sigma(\delta/d) \sinh(\sigma)}, \\
S_b(x,t) &= S_0 \frac{D_{S_b} \cosh(\sigma) + D_{S_e} \sigma((x-d)/d) \sinh(\sigma)}{D_{S_b} \cosh(\sigma) + D_{S_e} \sigma(\delta/d) \sinh(\sigma)},
\end{aligned} \tag{43}
$$

$$
\begin{aligned}
P_e(x,t) = {}& \frac{S_0}{d+\delta} \left(d + \delta \frac{D_{S_b} - D_{S_e} \sigma \sinh(\sigma)/\cosh(\sigma)}{D_{S_b} + D_{S_e} \sigma(\delta/d) \sinh(\sigma)/\cosh(\sigma)} \right) \\
& \times \left[\left(\frac{\sigma D_{S_e} \delta \sinh(\sigma)}{d \cosh(\sigma)} + \frac{D_{S_e} D_{P_b}}{D_{P_e}} \left(1 - \frac{1}{\cosh(\sigma)} \right) \right) \frac{x}{D_{P_b} d + D_{P_e} \delta} \right. \\
& \left. + \frac{D_{S_e}}{D_{P_e}} \frac{1 - \cosh(\sigma x/d))}{\cosh(\sigma)} \right], \\
P_b(x,t) = {}& \frac{S_0}{d+\delta} \left(d + \delta \frac{D_{S_b} - D_{S_e} \sigma \sinh(\sigma)/\cosh(\sigma)}{D_{S_b} + D_{S_e} \sigma(\delta/d) \sinh(\sigma)/\cosh(\sigma)} \right) \\
& \times \left[\left(\frac{\sigma D_{S_e} \sinh(\sigma)}{\cosh(\sigma)} + D_{S_e} \left(1 - \frac{1}{\cosh(\sigma)} \right) \right) / \left(D_{P_b} d + D_{P_e} \delta \right) \right] \\
& \times (d + \delta - x).
\end{aligned} \tag{44}
$$

where σ^2 is the dimensionless diffusion module or the Damköhler number which has been defined for the single-layer single-enzyme biosensors in Sect. 1.3 of first chapter of Part II,

$$
\sigma^2 = \frac{V_{max} d^2}{K_M D_{S_e}}. \tag{45}
$$

The density of the steady state current reads [235]

$$I = n_e F D_{P_e} \frac{S_0}{d+\delta}\left(d + \delta \frac{D_{S_b} - \sigma D_{S_e}\sinh(\sigma)/\cosh(\sigma)}{D_{S_b} + \sigma(\delta/d) D_{S_e}\sinh(\sigma)/\cosh(\sigma)}\right) \times \left(\frac{\sigma D_{S_e}\delta \sinh(\sigma)}{d\cosh(\sigma)} + \frac{D_{S_e} D_{P_b}}{D_{P_e}}\left(1 - \frac{1}{\cosh(\sigma)}\right)\right) / \left(D_{P_b} d + D_{P_e}\delta\right). \quad (46)$$

2.4 Simulated Biosensor Responses

Figure 3 shows the concentration profiles of the substrate S and the product P calculated from the two-compartment model (24)–(29) at the following values of the model parameters:

$$D_{S_e} = D_{P_e} = 300\,\mu\text{m}^2/\text{s}, \quad D_{S_b} = D_{P_b} = 600\,\mu\text{m}^2/\text{s},$$

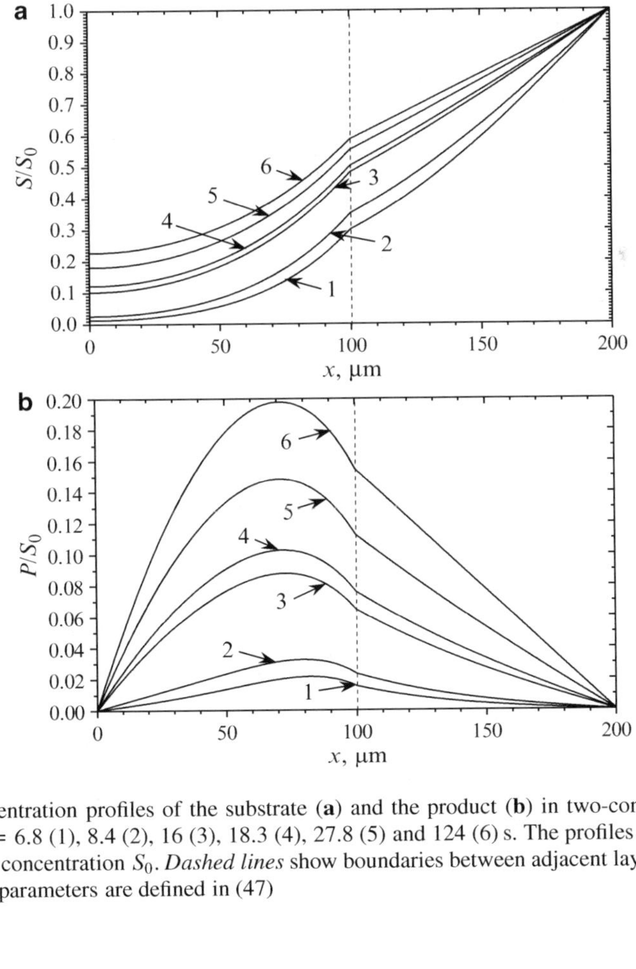

Fig. 3 Concentration profiles of the substrate (**a**) and the product (**b**) in two-compartments obtained for t = 6.8 (1), 8.4 (2), 16 (3), 18.3 (4), 27.8 (5) and 124 (6) s. The profiles are normalized with the bulk concentration S_0. *Dashed lines* show boundaries between adjacent layers. The values of the model parameters are defined in (47)

$$K_M = S_0 = 100\mu\text{M}, \quad V_{max} = 10\,\mu\text{M/s}. \quad d = \delta = 100\,\mu\text{m}. \quad (47)$$

In Fig. 3, curves 6 show the concentration profiles at the steady state conditions. Curves 4 correspond to the concentrations at time $T_{0.5}$ of a half of the steady state. The other curves show the concentrations at the intermediate values of time: $T_{0.05} = 6.8$ (curve 1), $T_{0.1} = 8.4$ (2), $T_{0.25} = 16$ (3), $T_{0.75} = 27.8$ (5) s. It should be borne in mind that T_α is the time needed to reach $\alpha\%$ of the stationary current (see Sect. 2.4 of first chapter of Part II for details). One can see in Fig. 3 linear curves 6 at $x \in [d, d + \delta]$. At the steady state conditions the concentrations approach straight line because of linearity of governing equations (25).

The evolution of the density $i(t)$ of the biosensor current accepting different initial conditions is presented in Fig. 4. Figure 4a shows the dynamics of the biosensor current in the case of infusion modeled by initial conditions (26), while 4b corresponds to the immersion modeled by initial conditions (32). For both modes of the

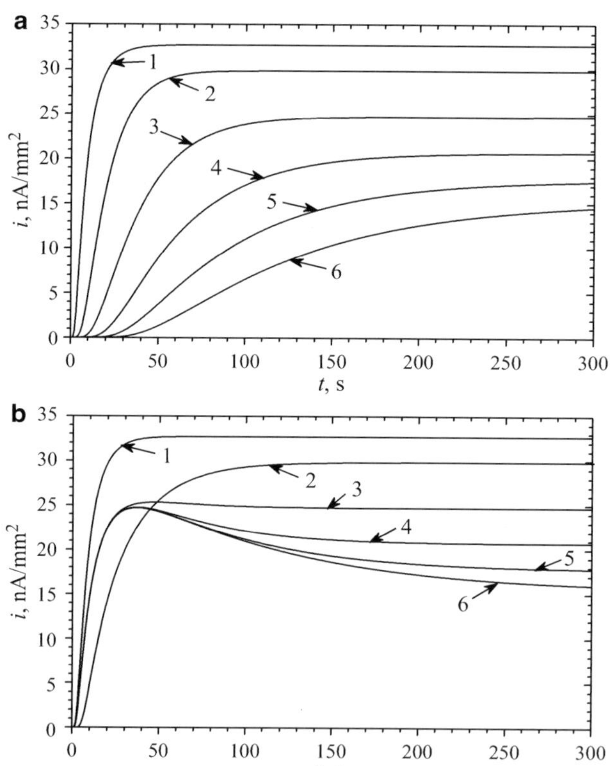

Fig. 4 The dynamics of the biosensor current at six values of the thickness δ: 0 (1), 100 (2), 200 (3), 300 (4), 400 (5) and 500 (6) μm in the cases of infusion (**a**, initial conditions (26)) and immersion (**b**, (32)). The other parameters are the same as in Fig. 3

biosensor operation, the response was modeled for biosensors at different thickness δ of the external diffusion layer keeping the enzyme membrane thickness constant.

Although the shape of curves in Fig. 4a notably differs from these in Fig. 4b, the corresponding steady state currents are identical. In the case of the infusion (Fig. 4a), $i(t)$ is a monotonous increasing function for all six thicknesses of the external diffusion layer. In the case of the immersion (Fig. 4b), $i(t)$ is a non-monotonous function for relatively thick external diffusion layers (curves 3–6). As has been mentioned above, the distinction in the initial conditions (26) and (32) makes no influence on the steady state response. Both curves marked by 1 are identical because both of them correspond to zero thickness δ of the diffusion layer.

In cases of the non-monotonous density $i(t)$ of the biosensor current (curves 3–6 in Fig. 4b), the maximal current is notably less (about 23%) than the steady state (maximal) current when the diffusion layer is neglected, $\delta = 0$ (curve 1). No notable difference is observed between the maximal currents in all cases of non-monotonous current density $i(t)$ (curves 3–6). Similar effect was also observed in the case of amperometric biosensors acting with a substrate cyclic conversion [30].

As one can see in Fig. 4 an increase in the thickness δ of the external diffusion layer notably decreases the steady state current and prolongs the response time. At the thickness $\delta = 5d = 500\,\mu\text{m}$ (curve 6) the density of the steady state current ($I = 15.4\,\text{nA/mm}^2$) is more than two times less than the density of the steady state current ($I = 32.7\,\text{nA/mm}^2$) at zero thickness of the diffusion layer. The half ($T_{0.5}$) of these two steady state responses were achieved at 113 and 7.7 s, respectively. However, additional numerical experiments showed that in some cases an increase in the thickness δ can even increase the steady state current. The effect of the external diffusion layer on the biosensor response is investigated in detail in the following sections.

2.5 *Effect of the Diffusion Layer*

To investigate the dependence of the steady state biosensor current on the relative thickness of the diffusion layer, a dimensionless ratio γ is introduced. The relative thickness γ of the diffusion layer is the ratio of the thickness δ of the external diffusion layer to the thickness d of the enzyme layer, $\gamma = \delta/d$, $\gamma \geq 0$.

Since the steady state current is very sensitive to the thickness of the enzyme layer, the steady state biosensor current has to be normalized to evaluate the effect of the ratio γ on the biosensor response. The normalized steady state biosensor current I_N is expressed by the steady state current at the thickness δ of the diffusion layer divided by the steady state current assuming the zero thickness d of the diffusion layer

$$I_N(d,\delta) = \frac{I(d,\delta)}{I(d,0)}, \quad d > 0,\ \delta \geq 0, \tag{48}$$

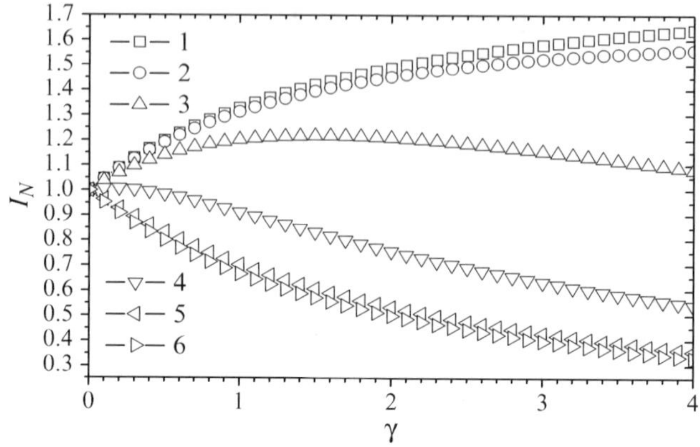

Fig. 5 The normalized maximal biosensor current I_N versus ratio $\gamma = \delta/d$ at six values of the thickness d of the enzyme layer: 10 (1), 20 (2), 50 (3), 100 (4), 200 (5) and 500 (6) µm. The other parameters are the same as in Fig. 3

where $I(d, \delta)$ is the density of the steady state current as defined by (30) and calculated at given thickness d of the enzyme membrane and the thickness δ of the diffusion layer.

The biosensor response versus the dimensionless ratio $\gamma = \delta/d$ was calculated at very different values of the membrane thickness d. The results obtained at values of the model parameters (47) are depicted in Fig. 5.

One can see in Fig. 5 that the steady state biosensor current notably decreases with an increase in the ratio γ in cases when the enzyme membrane thickness d is equal or greater than 100 µm (curves 4–6). That decrease is non-linear. In cases of relatively thin enzyme membranes ($d \leq 20$ µm, curves 1 and 2), the biosensor current increases with the increase in the ratio γ. In the case of $d = 50$ µm, the density I of the steady state current is a non-monotonous function of the ratio γ.

When analyzing the dependence of the enzyme membrane thickness d on the behaviour of the stationary current as a function of the ratio γ, the diffusion module σ^2 is calculated. Accepting the values of the model parameters defined in (47), σ^2 equals unity at $d \approx 55$ µm. This thickness correlates well with the thickness at which I_N is a non-monotonous function of γ (curve 3 in Fig. 5). Additional calculations of the response changing the maximal enzymatic rate V_{max} confirm a statement that the density I of the stationary current is a monotonous decreasing function of the dimensionless ratio γ if the biosensor response is distinct under the diffusion control ($\sigma^2 \gg 1$). The stationary current is a monotonous increasing function of γ when the enzyme kinetics controls the biosensor response ($\sigma^2 \ll 1$).

The Biot number Bi is another widely used dimensionless ratio of the internal mass transfer resistance to the external one [10, 104, 108, 165, 284],

$$Bi = \frac{d/D_{S_e}}{\delta/D_{S_b}} = \frac{D_{S_b} d}{D_{S_e} \delta}. \tag{49}$$

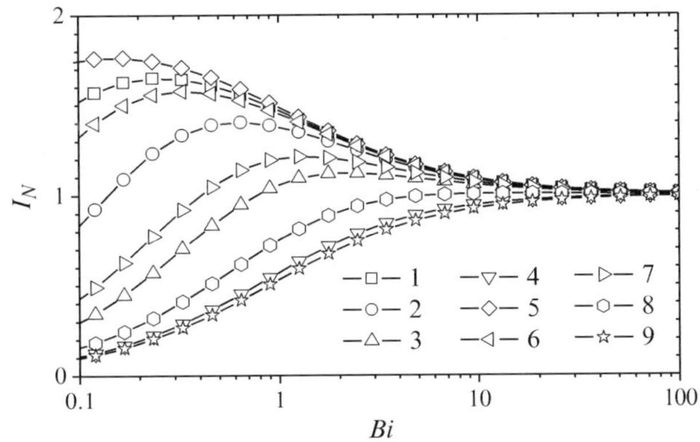

Fig. 6 The normalized steady state current I_N versus the Biot number Bi at $V_{max} = 100$ (1–4), $V_{max} = 10\,\mu\text{M/s}$ (5–9), and nine diffusion module σ: 0.18 (5), 0.29 (1), 0.37 (6), 0.58 (2), 0.91 (7), 1.15 (3), 1.83 (8), 2.89 (4), 3.65 (9). The other parameters are the same as in Fig. 3

The dependence of the normalized stationary current on the Bio number Bi has been investigated [116],

$$I_N(Bi) = \frac{I(Bi)}{I(\infty)}, \tag{50}$$

where $I(Bi)$ is the density of the steady state current calculated at the given Biot number Bi. $I(\infty)$ corresponds to the biosensor response for the zero thickness of the external diffusion layer, $\delta = 0$.

The biosensor response versus the Biot number Bi was investigated at different values of the maximal enzymatic rate V_{max} and the membrane thickness d. The results of the calculations obtained at two values of V_{max}: 10 and 100 μM/s, and various values of the thickness d are depicted in Fig. 6.

As one can see in Fig. 6, the steady state biosensor current is a monotonous increasing function of the Biot number Bi when the response is under the diffusion control ($\sigma >\approx 1.5$) [27]. I_N is a non-monotonous function of Bi when the enzyme kinetics controls the biosensor response ($\sigma <\approx 1.5$). In cases when $\sigma \approx 1$ the steady state biosensor current varies slightly. At all values of the diffusion module σ, the steady state current varies slightly when $Bi > 10$.

2.6 The Nernst Diffusion Layer

The thickness δ of the external diffusion layer depends upon the nature and the intensity of stirring of the buffer solution. Usually, the more intense stirring corresponds to the thinner diffusion layer. That diffusion layer is also known as the Nernst diffusion layer [73, 158, 199, 267]. According to the Nernst approach, a layer of thickness δ remains stagnant. Away from it the solution is in motion and uniform in concentration. The thickness of the diffusion layer remains unchanged with time [2, 164, 165]. The thickness of the Nernst diffusion layer practically does not depend upon the enzyme membrane thickness.

In the case when the bulk solution is stirred by rotation of the enzyme electrode, the thickness δ of the external diffusion layer can be calculated as follows [73, 158]:

$$\delta = 1.61 D^{1/3} \omega^{-1/2} \nu^{1/6} \tag{51}$$

where D is the diffusion coefficient of the species, ω is the rotation rate in Hz, ν is the viscosity in m^2/s. According to (51), the thickness δ is inversely dependent on the square root of the rotation rate ω. Theoretically, δ can be as minimal as desirable. Practically, the thickness of the Nernst diffusion layer can be minimized up to $\delta = 2\,\mu m$ by increasing the rotation speed [134, 158]. In another frequently used case when the solution is stirred in a magnetic stirrer, it is difficult to achieve the thickness δ less than about $20\,\mu m$.

In the case when an analyte is well-stirred and in powerful motion, the mass transport by diffusion outside the enzyme membrane quite often is neglected [56, 234, 235]. However, in practice, the zero thickness of the Nernst layer can not be achieved. The biosensor model taking into consideration the Nernst diffusion layer describes the biosensor action more precisely than another model where the Nernst diffusion layer is neglected. In addition, the Nernst diffusion layer of thickness δ may be neglected for a biosensor having membrane thickness d only if the steady state current calculated, considering the Nernst layer, is approximately the same as in the case when the Nernst diffusion layer is neglected. Consequently, the Nernst diffusion layer may be neglected if $I(d,\delta) \approx I(d,0)$, i.e. $I_N(d,\delta) \approx 1$, where $I(d,\delta)$ is the density of the stationary current calculated at the given values of d and δ, I_N is the normalized stationary current as defined by (48).

The relative error of the biosensor response as a function of the Biot number is described as follows,

$$R(Bi) = \frac{|I(Bi) - I(\infty)|}{I(Bi)}. \tag{52}$$

$R(Bi)$ can be called the relative error of the use of the mathematical model where the diffusion layer of thickness δ is neglected at the Biot number Bi. This function may also be regarded as a level of reliability of the mathematical model where the Nernst diffusion layer is not taken into account.

Let us investigate the conditions when the Nernst diffusion layer may be neglected to simulate the response of amperometric biosensors accurately. To investigate the effect of the Nernst diffusion layer on the biosensor response when the analyte is well stirred the relative error R has to be calculated at practically minimal thickness of the diffusion layer. Since the effect of the diffusion layer on the biosensor response significantly depends upon the diffusion module, the normalized response was calculated changing in a wide range both: the maximal enzymatic rate V_{max} and the membrane thickness d. Figure 7 shows the results of the calculation for two values of the minimal thickness δ of the Nernst diffusion layer, corresponding to two types of stirring: by rotating electrode (a, $\delta = 2\,\mu m$) and by magnetic stirrer (b, $\delta = 20\,\mu m$).

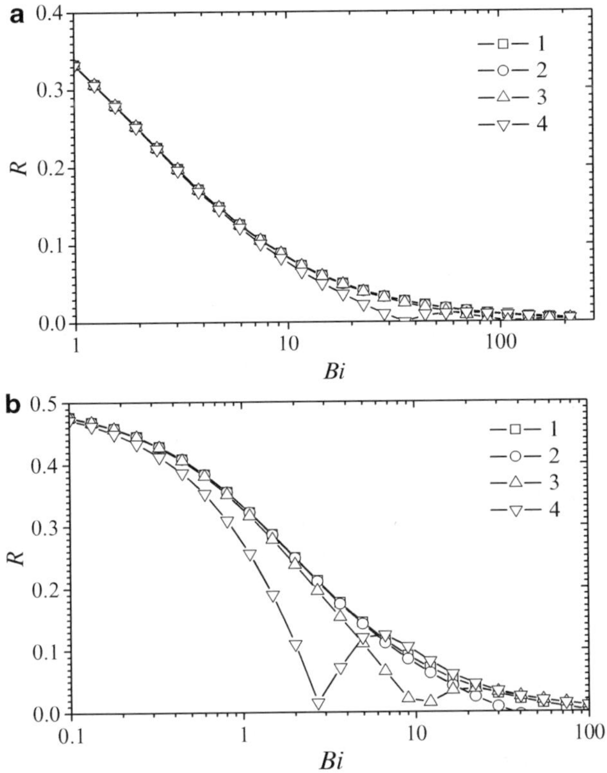

Fig. 7 The relative error R versus the Biot number Bi at two thicknesses δ of the Nernst layer: 2 (**a**) and 20 (**b**) μm. $V_{max} = 0.1$ (1), 1 (2), 10 (3) and 100 (4) μM/s, the other parameters are the same as in Fig. 3

As one can see in Fig. 7, the effect of the Nernst layer mainly decreases with the increase in the Biot number Bi as well as in the membrane thickness d.

Figure 7 shows that the Nernst diffusion layer should be taken into consideration in all the cases when the enzyme membrane is thinner than about 25δ ($Bi = 50$). If the Nernst diffusion layer is neglected in modeling of the biosensor response when $Bi = 1$ then the simulated density I of the steady state current differs even more than 30% ($R > 0.3$) from the true stationary current. In all cases when the Biot number is greater than about 50 ($d > 25\delta$), the relative error is less than 2%. The effect of the Nernst diffusion layer becomes slight only in cases when the Biot number is greater than about 50. At higher values of maximal enzymatic activity V_{max} (curves 3 and 4), the relative errors are slightly lower than at lower activity (curves 1 and 2). The Nernst diffusion layer should be taken into consideration when an analytical system based on an amperometric biosensor acts under the conditions when the Biot number is less than about 50 (Fig. 7). On the other hand, the Nernst diffusion layer may be neglected when Biot number is grater than about 50. When the diffusion layer is neglected, a two compartment model reduces to a notably simpler one layer model.

2.7 Dimensionless Model

Each additional layer to be considered in the modeling of the biosensor action includes some additional parameters into a mathematical model. The two compartment model (24)–(29) in comparison with the corresponding one layer model contains three additional parameters: the thickness δ of the diffusion layer as well as the diffusion coefficients D_{S_b} and D_{P_b}. Increasing the number of the model parameters complicates the investigation of the behaviour of the biosensor response. Because of this it is very important to find the main governing parameters of the mathematical model. Introducing dimensionless parameters is a widely used technique to reduce the number of model parameters [2, 67, 73, 113, 164, 219].

In order to define the main governing parameters of the two compartment model (24)–(29) the dimensional variables (x and t) and unknown concentrations (S_e, P_e, S_b and P_b) are replaced with the following dimensionless parameters:

$$\begin{aligned}
&\hat{x} = \frac{x}{d}, \quad \hat{t} = \frac{D_{S_e} t}{d^2}, \\
&\hat{S}_e = \frac{S_e}{K_M}, \quad \hat{P}_e = \frac{P_e}{K_M}, \quad \hat{S}_b = \frac{S_b}{K_M}, \quad \hat{P}_b = \frac{P_b}{K_M},
\end{aligned} \tag{53}$$

where $\hat{x}$ is the dimensionless distance from the electrode surface, $\hat{t}$ stands for the dimensionless time, and $\hat{S}_e$, $\hat{P}_e$, $\hat{S}_b$, $\hat{P}_b$ are the dimensionless concentrations. Having defined dimensionless variables and unknowns, the following dimensionless parameters characterizes the domain geometry and the substrate concentration in the bulk:

$$\hat{\delta} = \frac{\delta}{d}, \quad \hat{S}_0 = \frac{S_0}{K_M}, \quad \hat{T} = \frac{T D_{S_e}}{d^2}, \quad \hat{T}_{0.5} = \frac{T_{0.5} D_{S_e}}{d^2}, \tag{54}$$

where $\hat{\delta}$ is the dimensionless thickness of the external diffusion layer, $\hat{S}_0$ is the dimensionless substrate concentration in the bulk solution, $\hat{T}$ is the dimensionless response time, $\hat{T}_{0.5}$ is the half time of the steady state. The dimensionless thickness of enzyme membrane equals one.

The governing equations (24) in the dimensionless coordinates are expressed as follows:

$$\begin{aligned}
&\frac{\partial \hat{S}_e}{\partial \hat{t}} = \frac{\partial^2 \hat{S}_e}{\partial \hat{x}^2} - \sigma^2 \frac{\hat{S}_e}{1 + \hat{S}_e}, \\
&\frac{\partial \hat{P}_e}{\partial \hat{t}} = \frac{D_{P_e}}{D_{S_e}} \frac{\partial^2 \hat{P}_e}{\partial \hat{x}^2} + \sigma^2 \frac{\hat{S}_e}{1 + \hat{S}_e}, \quad \hat{x} \in (0, 1), \quad \hat{t} > 0,
\end{aligned} \tag{55}$$

where σ^2 is the dimensionless diffusion module as defined in (45).

The governing equations (25) take the following form:

$$\begin{aligned}\frac{\partial \hat{S}_b}{\partial \hat{t}} &= \frac{D_{S_b}}{D_{S_e}} \frac{\partial^2 \hat{S}_b}{\partial \hat{x}^2},\\ \frac{\partial \hat{P}_b}{\partial \hat{t}} &= \frac{D_{P_b}}{D_{S_e}} \frac{\partial^2 \hat{P}_b}{\partial \hat{x}^2}, \quad \hat{x} \in (1,\, 1+\hat{\delta}), \quad \hat{t} > 0.\end{aligned} \tag{56}$$

The initial conditions (26) transform to the following conditions:

$$\begin{aligned}&\hat{S}_e(\hat{x},0) = 0, \quad \hat{P}_e(\hat{x},0) = 0, \quad \hat{x} \in [0,\, 1],\\ &\hat{S}_b(\hat{x},0) = 0, \quad \hat{P}_b(\hat{x},0) = 0, \quad \hat{x} \in [1,\, 1+\hat{\delta}),\\ &\hat{S}_b(1+\hat{\delta},\, 0) = \hat{S}_0, \quad \hat{P}_b(1+\hat{\delta},0) = 0,\end{aligned} \tag{57}$$

The matching (27) and the boundary (28), (29) conditions are rewritten as follows ($\hat{t} > 0$):

$$\begin{aligned}&\left.\frac{\partial \hat{S}_e}{\partial \hat{x}}\right|_{\hat{x}=1} = \left.\frac{D_{S_b}}{D_{S_e}} \frac{\partial \hat{S}_b}{\partial \hat{x}}\right|_{\hat{x}=1}, \quad \hat{S}_e(1,\hat{t}) = \hat{S}_b(1,\hat{t}),\\ &\left.\frac{\partial \hat{P}_e}{\partial \hat{x}}\right|_{\hat{x}=1} = \left.\frac{D_{P_b}}{D_{P_e}} \frac{\partial \hat{P}_b}{\partial \hat{x}}\right|_{\hat{x}=1}, \quad \hat{P}_e(1,\hat{t}) = \hat{P}_b(1,\hat{t}),\end{aligned} \tag{58}$$

$$\begin{aligned}&\hat{S}_b(1+\hat{\delta},\hat{t}) = \hat{S}_0, \quad \hat{P}_b(1+\hat{\delta},\, \hat{t}) = 0,\\ &\hat{P}_e(0,\hat{t}) = 0, \quad \left.\frac{\partial \hat{S}_e}{\partial \hat{x}}\right|_{\hat{x}=0} = 0.\end{aligned} \tag{59}$$

The dimensionless current (flux) $\hat{i}$ and the corresponding dimensionless stationary current $\hat{I}$ are defined as follows:

$$\begin{aligned}\hat{i}(\hat{t}) &= \left.\frac{\partial \hat{P}_e}{\partial \hat{x}}\right|_{\hat{x}=0} = \frac{i(t)\, d}{n_e F D_{P_e} K_M},\\ \hat{I} &= \lim_{\hat{t}\to\infty} \hat{i}(\hat{t}).\end{aligned} \tag{60}$$

Assuming the same diffusion coefficients for both species (the substrate and the product), only the following dimensionless parameters remain in the dimensionless mathematical model (55)–(60): $\hat{\delta}$ – the thickness of the diffusion layer, $\hat{S}_0$ – the substrate concentration in the bulk solution, σ^2 – the diffusion module, and D_{rel} – the ratio of the external diffusivity to the internal diffusivity, $D_{rel} = D_{S_b}/D_{S_e} = D_{P_b}/D_{P_e}$. In all the calculations, D_{rel} was equal to 2 as defined in (47). The advent of the diffusion module σ^2 is one of the most important outcomes of defining a dimensionless model of the biosensor action. The diffusion module is the main parameter expressing all internal characteristics of the biosensor.

2.8 Impact of the Diffusion Module

To investigate the effect of two main parameters of the two compartment model, the diffusion module σ^2 and the Biot number Bi, on the biosensor response, the biosensor action was simulated at different values of Bi, changing the module σ^2 in a few orders of magnitude. Figure 8a shows the dependence of the steady state dimensionless current $\hat{I}$ on σ^2, while Fig. 8b shows the corresponding dependence of the sensitivity B_S. The diffusion module σ^2 varied by changing exponentially the maximal enzymatic rate V_{max} from 0.1 up to 1,000 μM/s. Values of all other parameters were kept constant as defined in (47).

As one can see in Fig. 8a that at low values the diffusion module ($\sigma^2 < 1$) when the enzyme kinetics controls the biosensor response, the stationary dimensionless current $\hat{I}$ is approximately a linear function of σ^2. At these conditions the relative thickness Bi of the external diffusion layer effects the dimensionless current $\hat{I}$ only slightly. When the biosensor response is considerably controlled by the diffusion

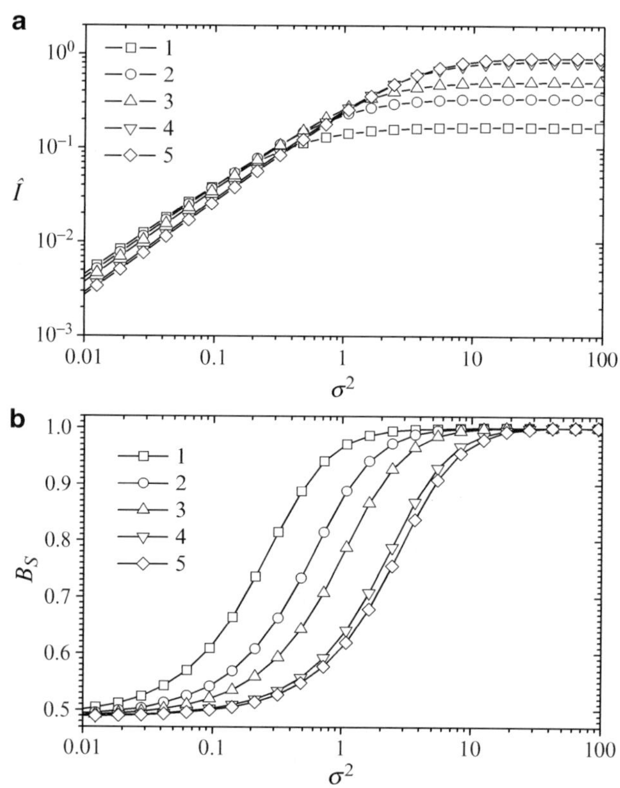

Fig. 8 The steady state dimensionless current $\hat{I}$ (**a**) and the sensitivity B_S (**b**) versus the diffusion module σ^2 at five values of the Biot number Bi: 0.2 (1), 0.5 (2), 1 (3), 5 (4) and 10 (5). The other parameters are the same as in Fig. 3

($\sigma^2 > 10$) the stationary current practically does not depend on σ^2. However, the effect of the Biot number on the current $\hat{I}$ is much more significant namely at $\sigma^2 > 10$. One can also see in Fig. 8 a longer linear range of curves at greater values of the Biot number rather than at lover ones.

Figure 8b shows the dependence of the biosensor sensitivity on the diffusion module σ^2. The sensitivity B_S monotonously increases with an increase in module σ^2. The shape of all the curves looks very similar. However, the smaller Biot number corresponds to a greater sensitivity B_S. This is especially noticeable at moderate values of σ^2, $0.1 < \sigma^2 < 10$. Let us recall that a smaller Bi corresponds to a thicker external diffusion layer. The thicker external diffusion layer creates a greater diffusion limitation to the substrate. Figure 8b well illustrates a widely known and a very useful feature of biosensors that an additional diffusion limitation on the substrate increases their sensitivity and prolongs their calibration curve [152, 166, 194, 255].

3 Biosensors with Outer Porous Membrane

Some drawbacks, like a relatively short linear range of the calibration curve, an instability, a low specificity due to the interfering compounds of the biosensors can be completely or partially solved by an application of an additional the inert outer membrane on the surface of the biosensors [90, 184]. Cellulose acetate, polyurethane, latex and a number of other membranes have been used to protect the surface of the electrodes from electrochemically active compounds, like uric acid, ascorbic acid, free amino acids, paracetamol and a number of other electrochemically active compounds of natural and artificial nature [8, 230, 256]. The outer porous membrane can also create a diffusion limitation to the substrate, i.e. to lower the substrate concentration in the enzymatic layer and thereby prolong the calibration curve of the biosensor [152, 166, 254].

Here an amperometric biosensor is considered as a flat electrode deposited with a mono-layer of enzyme and covered with an inert porous membrane [37, 117, 229]. Figure 9 shows the principal structure of the biosensor, where d_e is the thickness of

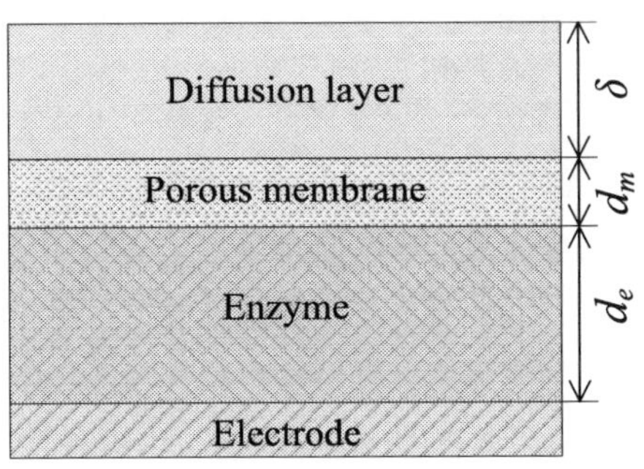

Fig. 9 A principal structure of a biosensor with an outer porous membrane. The figure is not to scale

the enzyme layer immobilized onto the surface of the electrode, d_m is the thickness of the porous membrane, and δ is the thickness of the external diffusion layer.

3.1 Mathematical Model

The model to be considered consists of two main regions: the enzyme layer where the enzyme reaction as well as the mass transport by diffusion takes place and an outer porous membrane where only the mass transport by diffusion takes place. Assuming relatively thin layers of the enzyme and the porous membrane, an additional diffusion limiting region where only the mass transport by diffusion takes place, has to be considered even in modeling of a well-stirred buffer solution. If the bulk solution is slightly stirred or not stirred at all, then the layer of external diffusion especially has to be taken into account. The analyte concentration is maintained constant farther from the diffusion layer.

3.1.1 Governing Equations

Assuming the homogeneous distribution of the immobilized enzyme in the enzyme membrane of the uniform thickness, the dynamics of the concentrations of the substrate as well as product in the enzyme layer can be described by a system of the reaction–diffusion equations [26, 37, 235],

$$\begin{aligned} \frac{\partial S_e}{\partial t} &= D_{S_e}\frac{\partial^2 S_e}{\partial x^2} - \frac{V_{max} S_e}{K_M + S_e}, \\ \frac{\partial P_e}{\partial t} &= D_{P_e}\frac{\partial^2 P_e}{\partial x^2} + \frac{V_{max} S_e}{K_M + S_e}, \quad x \in (0,\, d_e), \quad t > 0, \end{aligned} \tag{61}$$

where x and t stand for space and time, respectively, $S_e(x,t)$ and $P_e(x,t)$ are the concentrations of the substrate and reaction product, respectively, d_e is the thickness of the enzyme membrane, D_{S_e} and D_{P_e} are the diffusion coefficients, V_{max} is the maximal enzymatic rate and K_M is the Michaelis constant (see Sect. 2.1).

Assuming the porous membrane as a periodic media, the homogenization process can been applied to the membrane domain [18, 22, 82, 271]. After this, the porous membrane is modeled as a diffusion layer with an effective (averaging) diffusion coefficient,

$$\begin{aligned} \frac{\partial S_m}{\partial t} &= D_{S_m}\frac{\partial^2 S_m}{\partial x^2}, \\ \frac{\partial P_m}{\partial t} &= D_{P_m}\frac{\partial^2 P_m}{\partial x^2}, \quad x \in (d_e,\, d_e + d_m), \quad t > 0, \end{aligned} \tag{62}$$

where d_m is the thickness of the porous membrane, $S_m(x,t)$ and $P_m(x,t)$ are the concentrations of the substrate and the reaction product, respectively, D_{S_m}, D_{P_m} are the effective diffusion coefficients of the species in the membrane.

In the homogeneous external region, only the mass transport by diffusion of the substrate as well as of the product takes place,

$$\begin{aligned} \frac{\partial S_b}{\partial t} &= D_{S_b}\frac{\partial^2 S_b}{\partial x^2}, \\ \frac{\partial P_b}{\partial t} &= D_{P_b}\frac{\partial^2 P_b}{\partial x^2}, \quad x \in (d_e + d_m,\, d_e + d_m + \delta), \quad t > 0, \end{aligned} \tag{63}$$

where δ is the thickness of the external diffusion layer, $S_b(x,t)$ and $P_b(x,t)$ are the concentrations of the substrate and reaction product, respectively, D_{S_b}, D_{P_b} are the diffusion coefficients of the species in the bulk solution.

3.1.2 Initial and Boundary Conditions

Let $x = 0$ represent the electrode surface. The simulation of the biosensor operation starts when some substrate appears in the bulk solution ($t = 0$),

$$\begin{aligned} &S_e(x,0) = 0, \quad P_e(x,0) = 0, \quad x \in [0,\, d_e], \\ &S_m(x,0) = 0, \quad P_m(x,0) = 0, \quad x \in [d_e,\, d_e + d_m], \\ &S_b(x,0) = 0, \quad P_b(x,0) = 0, \quad x \in [d_e + d_m,\, d_e + d_m + \delta), \\ &S_b(d_e + d_m + \delta,\, 0) = S_0, \quad P_b(d_e + d_m + \delta, 0) = 0, \end{aligned} \tag{64}$$

where S_0 is the concentration of the substrate in the bulk solution.

Due to the amperometry, the substrate is assumed as an electro-inactive substance, while the product is an electro-active substance. The electrode potential is chosen to keep zero concentration of the product at the electrode surface ($t > 0$),

$$P_e(0,t) = 0, \quad D_{S_e}\left.\frac{\partial S_e}{\partial x}\right|_{x=0} = 0. \tag{65}$$

On the boundary between the enzyme layer and the porous membrane as well as between the porous membrane and the external diffusion layer the matching conditions are defined ($t > 0$),

$$\begin{aligned} D_{S_e}\left.\frac{\partial S_e}{\partial x}\right|_{x=d_e} &= D_{S_m}\left.\frac{\partial S_m}{\partial x}\right|_{x=d_e}, \quad S_e(d_e,t) = S_m(d_e,t), \\ D_{P_e}\left.\frac{\partial P_e}{\partial x}\right|_{x=d_e} &= D_{P_m}\left.\frac{\partial P_m}{\partial x}\right|_{x=d_e}, \quad P_e(d_e,t) = P_m(d_e,t), \end{aligned} \tag{66}$$

$$
\begin{aligned}
&D_{S_m} \frac{\partial S_m}{\partial x}\bigg|_{x=d_e+d_m} = D_{S_b} \frac{\partial S_b}{\partial x}\bigg|_{x=d_e+d_m}, \\
&S_m(d_e + d_m, t) = S_b(d_e + d_m, t), \\
&D_{P_m} \frac{\partial P_m}{\partial x}\bigg|_{x=d_e+d_m} = D_{P_b} \frac{\partial P_b}{\partial x}\bigg|_{x=d_e+d_m}, \\
&P_m(d_e + d_m, t) = P_b(d_e + d_m, t).
\end{aligned}
\tag{67}
$$

The external diffusion layer of thickness δ remains unchanged with time. Away from it the concentration of the substrate and the product remain constant ($t > 0$),

$$
S_b(d_e + d_m + \delta, t) = S_0, \quad P_b(d_e + d_m + \delta, t) = 0. \tag{68}
$$

3.1.3 Biosensor Response

The system (61)–(68) approaches a steady state as $t \to \infty$. The density $i(t)$ of the biosensor current at time t as well as the corresponding steady state current I is defined as in the two compartment model,

$$
i(t) = n_e F D_{P_e} \frac{\partial P_e}{\partial x}\bigg|_{x=0}, \qquad I = \lim_{t\to\infty} i(t), \tag{69}
$$

where n_e is the number of electrons involved in the charge transfer at the electrode surface, F is the Faraday constant.

3.2 *Numerical Simulation*

At the transient conditions the initial boundary value problem (61)–(68) can be solved numerically using the finite difference technique. Since the governing equations, initial as well as boundary conditions are of the same type as in a two-compartment model discussed in Sect. 2.2, the model equations can be approximated with the difference equations by using the same technique. For building the discrete grid we also refer to the above Sects. 1.2 and 2.2.

For $\delta \to 0$, the solution of the model approaches the solution of the corresponding two-compartment discussed in Sect. 2. Similarly, vanishing the thickness of the porous membrane, $d_m \to 0$, the solution of three-compartment model (61)–(68) also approaches the solution of two-compartment model with the governing equations (61) and (63). These features can be applied to validation of the numerical solution of the three-compartment model by using analytical solutions presented in the previous section.

As in a general multi-layer model as well as in a two-compartment model, the concentrations of both species, S and P, can be defined continuously in the entire domain $x \in [0, d_e + d_m + \delta]$ as follows ($t \geq 0$):

$$S(x,t) = \begin{cases} S_e(x,t), & x \in [0, d_e], \\ S_m(x,t), & x \in (d_e, d_e + d_m], \\ S_b(x,t), & x \in (d_e + d_m, d_e + d_m + \delta], \end{cases}$$
$$P(x,t) = \begin{cases} P_e(x,t), & x \in [0, d], \\ P_m(x,t), & x \in (d_e, d_e + d_m], \\ P_b(x,t), & x \in (d_e + d_m, d_e + d_m + \delta]. \end{cases} \tag{70}$$

Figure 10 shows the simulated concentration profiles of the substrate S and the product P calculated from the three-compartment model (61)–(68) at the following values of the model parameters:

$$\begin{aligned} &D_{S_e} = D_{P_e} = D_e = 300\,\mu\text{m}^2/\text{s}, \quad D_{S_m} = D_{P_m} = D_m = 30\,\mu\text{m}^2/\text{s}, \\ &D_{S_b} = D_{P_b} = D_b = 600\,\mu\text{m}^2/\text{s}, \quad d_e = 4\,\mu\text{m}, \quad d_m = \delta = 2\,\mu\text{m}, \\ &K_M = 100\mu\text{M}, \quad S_0 = 100\mu\text{M}, \quad V_{max} = 10\,\mu\text{M/s}. \end{aligned} \tag{71}$$

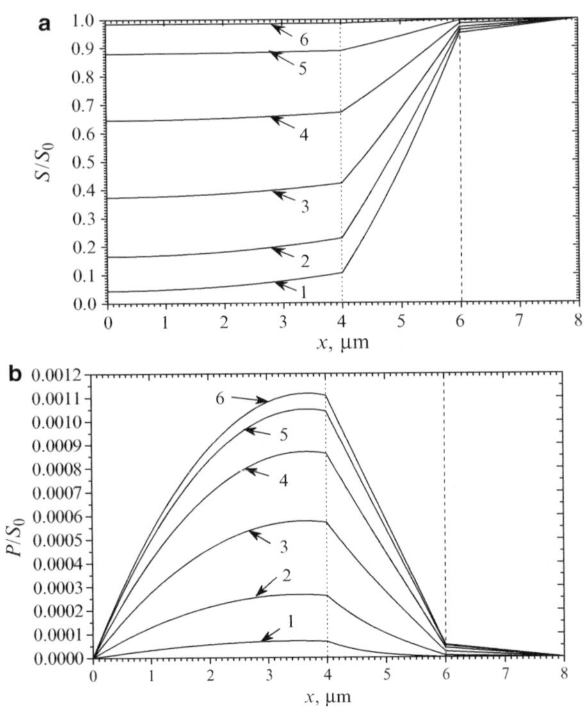

Fig. 10 Concentration profiles of the substrate (**a**) and the product (**b**) obtained at $t = 0.05$ (1), 0.1 (2), 0.2 (3), 0.4 (4), 0.8 (5) and 4.14 (6, steady state time) s. The profiles are normalized with the bulk concentration S_0. *Dot lines* show the boundaries between the enzyme layer and the porous membrane, *dashed lines* show the boundaries between the porous membrane and the external diffusion layer. Values of the model parameters are defined in (71)

In Fig. 10, the concentration profiles are depicted at the steady state (curves 6) and the intermediate (curves 1–5) conditions. The steady state condition was achieved at time $T = 4.14\,\text{s}$. The half of the steady state was achieved at 0.19 s ($T_{0.5} = 0.19\,\text{s}$). Curves 3 correspond approximately to the half time $T_{0.5}$ of the steady state. One can see in Fig. 10 linear curves 6 at $x \in (d_e, d_e + d_m + \delta)$. At the steady state conditions, the concentrations approach straight lines because of the linearity of the governing equations (62) and (63).

3.3 *Effect of the Porous Membrane*

Using numerical simulation, the influence of the permeability and of the thickness of the outer porous membrane on the biosensor response has been investigated [37]. In terms of the mathematical model (61)–(68), the permeability as well as the porosity is expressed by the diffusion coefficients D_{S_m} and D_{P_m}.

3.3.1 Effect of the Membrane Permeability

When calculating the concentrations shown in Fig. 10, the diffusion coefficients D_{S_m} and D_{P_m} for the porous membrane were tenfold less than the corresponding coefficients for the enzyme, $D_m = D_{S_m} = D_{P_m} = D_{S_e}/10 = D_{P_e}/10 = D_e/10$.

To investigate the dependence of the biosensor response on the diffusivity (permeability) of the outer porous membrane, the biosensor response was simulated at constant thickness $d_m = 2\,\mu\text{m}$ of the outer membrane changing the diffusion coefficient D_m ($D_m = D_{S_m} = D_{P_m}$) from 600 to $6\,\mu\text{m}^2/\text{s}$, i.e. from D_b to $D_b/100$, where $D_b = D_{S_b} = D_{P_b}$.

The simulated biosensor current was normalized with respect to the stationary current calculated at theoretically maximal value D_m of the diffusivity of the outer membrane to be analyzed,

$$I_{D_m} = \frac{I(D_m)}{I(D_b)}, \tag{72}$$

where $I(D_m)$ is the density of the steady state current calculated at the diffusivity D_m of the outer porous membrane. The maximal limiting value $D_m = D_b$ of diffusivity corresponds to a limiting case when the diffusivity in the membrane matches the diffusivity in the bulk. Practically, the diffusivity $D_m = D_b$ of the outer membrane corresponds to the extreme case when the porous membrane becomes just a diffusion layer. Calculations were done at three maximal enzymatic rates V_{max}: 10, 100, 1,000 $\mu\text{M/s}$, combined with the three values of the substrate concentration S_0 in bulk: 10, 100, 1,000 $\mu\text{M/s}$. When normalizing these values of S_0 with respect to the Michalis constant $K_M = 100\,\mu\text{M/s}$ the following dimensionless concentrations of the substrate are obtained: 0.1, 1 and 10. Figure 11a represents the dependence of the normalized steady state current I_{D_m} on the diffusivity D_m, while Fig. 11b represents the corresponding dependence of the biosensor sensitivity B_S.

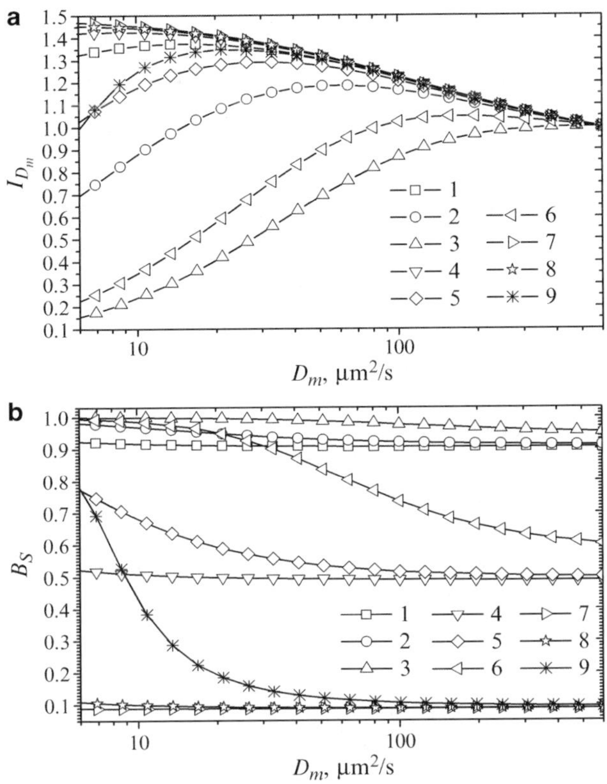

Fig. 11 The normalized steady state current I_{D_m} (**a**) and the sensitivity B_S (**b**) versus the diffusivity D_m of the outer membrane at three maximal enzymatic rates V_{max}: 10 (1, 4, 7), 100 (2, 5, 8), 1,000 (3, 6, 9) μM/s, and three substrate concentrations S_0: 10 (1–3), 100 (4–6), 1,000 (7–9) μM. Values of the other parameters are defined in (71)

As one can see in Fig. 11a the biosensor current notably depends on the diffusivity of the porous membrane. A decrease in the diffusivity corresponds to a decrease in the permeability and, usually, to a decrease in the porosity. The effect of the diffusivity D_m entirely depends on the maximal enzymatic rate V_{max} and the substrate concentration S_0. At moderate values of the enzyme activity V_{max} (10 and 100 μM/s) and relatively high values of the concentration S_0 (100 and 1,000 μM, curves 4, 7 and 8), the stationary current monotonously increases when D_m changes hundredfold from 600 down to 6 μm^2/s. Due to the increase in the diffusivity D_m, the stationary current can increase half as much again. However, at a high enzyme activity V_{max} (1 mM/s) and low concentration S_0 (1 mM, curve 3), the stationary current monotonously decreases when D_m changes from 600 to 6 μm^2/s. In all other cases of V_{max} and S_0, I_{D_m} is a non-monotonous function at $D_m \in [6, 600]$ (μm^2/s). Since at the theoretically minimal limiting value of $D_m = 0$ no current upstarts, the normalized current I_{D_m} is a non-monotonous function of D_m changing D_m from the maximal limiting value (D_b) to the minimal one (0).

Figure 11b shows a well known feature of the biosensors that the biosensor sensitivity is higher at lower substrate concentrations rather than at higher ones. However, in cases of a high enough enzymatic activity V_{max}, the sensitivity B_S can be notably increased by decreasing the diffusivity D_m of the outer membrane even at a relatively high substrate concentration S_0 (curves 5, 6, 9 in Fig. 11b). This is a very important advantage of the outer membranes in extending the region of the biosensor application [152, 166, 230].

3.3.2 Effect of the Membrane Thickness

To investigate the effect of the thickness d_m of the outer porous membrane on the biosensor response the steady state current changing the thickness d_m has to be calculated at different values of the maximal enzymatic rate V_{max} and the substrate concentration S_0. To evaluate the effect of the membrane thickness on the biosensor response the biosensor stationary current is normalized with respect to the steady state current of the corresponding biosensor having no outer membrane,

$$I_{d_m} = \frac{I(d_m)}{I(0)}, \tag{73}$$

where $I(d_m)$ is the density of the steady state current calculated at the thickness d_m of the outer porous membrane.

Figure 12a shows the dependence of the steady state biosensor response on the thickness d_m of the outer membrane at different values of the maximal enzymatic rate V_{max}, the substrate concentration S_0 keeping the membrane diffusivity $D_m = 0.1\,D_e = 30\,\mu\text{m}^2/\text{s}$ and the other parameters constant as defined in (71). Figure 12b shows the effect of the membrane thickness on the biosensor sensitivity at the same values of the parameters.

One can see in Fig. 12a that the shape of the normalized steady state current I_{d_m} (as well as of the non-normalized one I) is very sensitive to changes of the maximal enzymatic rate V_{max} and the substrate concentration S_0.

I_{d_m} is a monotonous increasing function of the outer membrane thickness d_m at moderate values of the enzyme activity V_{max} (10 and 100 μM/s) and relatively high values of the concentration S_0 (100 and 1,000 μM, curves 4, 7 and 8) when d_m changes from 0 to 8 μm. I_{d_m} is a monotonous decreasing function of the thickness d_m at a high value of V_{max} (1 mM/s) and a low value of S_0 (10 μM) (curve 3). Similar behaviour of the biosensor response was observed when analyzing the effect of the thickness of diffusion layer for the amperometric biosensors when applying a two-compartment model [26] (see also Sect. 2.5).

Although the shapes of curves in Fig. 12 notably differ from those in Fig. 11, the effect of the diffusivity D_m of the porous membrane is very similar to that of the membrane thickness d_m. A decrease in diffusivity D_m influences the steady

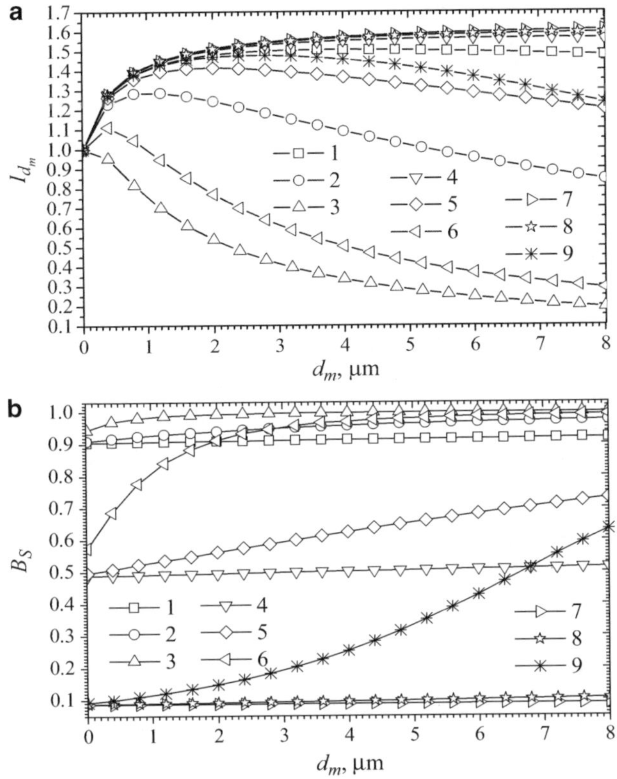

Fig. 12 The normalized steady state current I_{d_m} (**a**) and the sensitivity B_S (**b**) versus the thickness d_m of the outer membrane. The parameters and notations are the same as in Fig. 11

state current as well as the sensitivity similarly to an increase in thickness D_m of the membrane. To make the similarities more apparent, the normalized steady state current I_{d_m} and the sensitivity B_S (Fig. 11) were replotted in Fig. 13 as functions of the inverse thickness $1/d_m$ of the outer membrane.

One can see that the shapes of curves in Fig. 13 are approximately the same as in Fig. 11. When preparing highly sensitive biosensors, the increase of the outer membrane can be effectively combined with the decrease of the membrane diffusivity. The membrane diffusivity usually decreases with the decrease in the porosity.

3.4 Concluding Remarks

Assuming the porous membrane of a uniform thickness as a periodic media, the membrane can be modeled as a diffusion layer with an effective diffusion coefficient depending on the porosity of the membrane [18, 271].

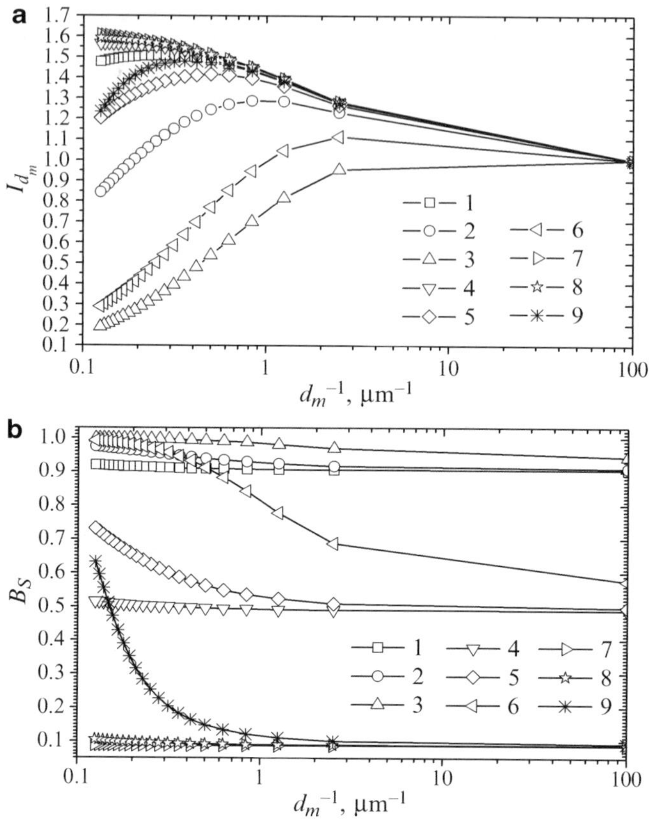

Fig. 13 The normalized steady state current I_{d_m} (**a**) and the sensitivity B_S (**b**) versus the inverse thickness d_m^{-1} of the outer membrane. The parameters and notations are the same as in Fig. 11

The outer porous membrane creates a diffusion limitation to the substrate, i.e. lowers the substrate concentration in the enzymatic layer and thereby prolongs the calibration curve of the biosensor (see Fig. 11, 12 and 13) [152, 166, 230, 254].

The diffusivity of the porous membrane influences the steady state biosensor current as well as the sensitivity very similarly to the increase in the membrane thickness (see Fig. 11 and 13).

4 Biosensors with Selective and Outer Perforated Membranes

A practical biosensor contains a multi-layer enzyme membrane [16, 47, 91]. The electrode acting as a transducer of the biosensor is rather often covered by a selective membrane, following a layer of immobilized enzyme and an outer membrane. The outer membrane can be prepared from a dialysis membrane or, in modern biosensors, from a perforated nucleopore type membrane [8]. The selective membrane is used to increase the biosensors selectivity [50, 205, 226]. The selective membrane is permeable for the selected species only, usually, for the product of the biochemical reaction.

The modeling of biosensors with perforated membranes has been performed by Schulmeister and Pfeiffer [236]. The model did not take into account the geometry of the holes in the membranes. It included the diffusion coefficients having a limited physical sense, and as the authors acknowledged "its quantitative value is limited" [236]. However, this assumption allowed formulating the mathematical model of the biosensor operation in a one dimensional space. The perforation of the outer membrane was modeled as the permeability (porosity or diffusivity) of the porous membrane discussed in Sect. 3. This section concentrates on the modeling of the selective membrane. By changing the input parameters the output results are numerically analyzed with a special emphasis on the influence of the selective membrane to the response of the biosensors.

Figure 14 shows the principal structure of a biosensor to be considered, where d_s is thickness of the selective membrane, d_e is thickness of the enzyme layer, d_p is thickness of the perforated membrane, and δ is the thickness of the external diffusion layer.

4.1 Mathematical Model

The thickness of the selective membrane as well as of the perforated membrane of the biosensor is assumed to be much less than its length and width. In the biosensor, the selective and the perforated membranes are of the uniform thickness. Figure 14 shows the profile of a one dimensional unit of the biosensor, where $d_s = b_1 - b_0$ stands for the thickness of the selective membrane, $d_e = b_2 - b_1$ is the thickness of the enzyme layer, $d_p = b_3 - b_2$ is the thickness of the perforated membrane, and $\delta = b_4 - b_3$ is the thickness of the external diffusion (Nernst) layer.

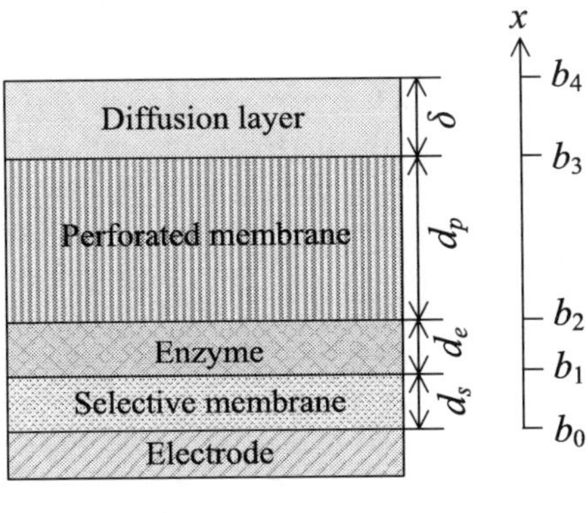

Fig. 14 The principal structure of the biosensor with selective and outer perforated membranes. The figure is not to scale

In the enzyme region the mass transport by diffusion and the enzyme-catalyzed reaction are considered,

$$\mathrm{S} + \mathrm{E} \rightleftarrows \mathrm{ES} \rightarrow \mathrm{E} + \mathrm{P} \tag{74}$$

In this scheme the substrate (S) combines reversibly with an enzyme (E) to form a complex (ES), which then dissociates into a product (P) and the enzyme is regenerated. Assuming the quasi steady state approximation, the concentration of the intermediate complex (ES) does not change and may be neglected when simulating the biochemical behaviour of analytical systems [229, 258]. In the selective membrane only the mass transport by diffusion of the reaction product takes place.

4.1.1 Governing Equations

The dynamics of the concentrations of the substrate as well as the product in the enzyme layer can be described by a one-dimensional-in-space system of the reaction–diffusion equations [35, 236]. In the selective membrane only the mass transport by diffusion of the reaction product takes place,

$$\frac{\partial P_1}{\partial t} = D_{P_1} \frac{\partial^2 P_1}{\partial x^2}, \quad x \in (b_0, b_1), \quad t > 0, \tag{75}$$

where x and t stand for space and time, respectively, $P_1(x,t)$ is the concentration of the reaction product in the selective membrane, $d_s = b_1 - b_0$ is the thickness of the enzyme membrane, D_{P_1} is the diffusion coefficient.

In the enzyme layer the mass transport by diffusion and the enzyme-catalyzed reaction are considered,

$$\begin{aligned} \frac{\partial S_2}{\partial t} &= D_{S_2} \frac{\partial^2 S_2}{\partial x^2} - \frac{V_{max} S_2}{K_M + S_2}, \\ \frac{\partial P_2}{\partial t} &= D_{P_2} \frac{\partial^2 P_2}{\partial x^2} + \frac{V_{max} S_2}{K_M + S_2}, \quad x \in (b_1, b_2), \quad t > 0, \end{aligned} \tag{76}$$

where $S_2(x,t)$ and $P_2(x,t)$ are the concentrations of the substrate and reaction product, respectively, $d_e = b_2 - b_1$ is the thickness of the enzyme membrane, V_{max} is the maximal enzymatic rate and K_M is the Michaelis constant.

Assuming the perforated membrane as a periodic media, a homogenization process can been applied to the membrane domain [18,22,82]. After this, the perforated membrane is modeled as a diffusion layer with effective diffusion coefficients. So, the dynamics of concentrations in the perforated and the external diffusion layer are described by the systems of the diffusion equations,

$$\begin{aligned} \frac{\partial S_j}{\partial t} &= D_{S_j} \frac{\partial^2 S_j}{\partial x^2}, \\ \frac{\partial P_j}{\partial t} &= D_{P_j} \frac{\partial^2 P_j}{\partial x^2}, \quad x \in (b_{j-1}, b_j), \quad t > 0, \quad j = 3, 4, \end{aligned} \tag{77}$$

where $S_j(x,t)$ and $P_j(x,t)$ are the concentrations of the substrate and the reaction product, respectively, $d_p = b_3 - b_2$ is the thickness of the perforated membrane, $\delta = b_4 - b_3$ is the thickness of the external diffusion layer, $j = 3, 4$.

4.1.2 Initial and Boundary Conditions

Let $x = 0$ represent the electrode surface. The simulation of the biosensor operation starts when some substrate appears in the bulk solution ($t = 0$),

$$\begin{aligned} &P_i(x,0) = 0, \quad x \in [b_{i-1}, b_i], \quad i = 1,2,3,4, \\ &S_j(x,0) = 0, \quad x \in [b_{j-1}, b_j], \quad j = 2,3, \\ &S_4(x,0) = 0, \quad x \in [b_3, b_4), \quad S_4(b_4,0) = S_0, \end{aligned} \tag{78}$$

where S_0 is the concentration of the substrate in the bulk solution.

Due to the amperometry, the electrode potential is chosen to keep zero concentration of the product at the electrode surface ($t > 0$),

$$P_1(b_0,t) = 0. \tag{79}$$

On the boundary between the selective membrane and the enzyme layer the matching conditions for the product and the non-leakage condition for the substrate are defined ($t > 0$),

$$\begin{aligned} &D_{P_1} \frac{\partial P_1}{\partial x}\bigg|_{x=b_1} = D_{P_2} \frac{\partial P_2}{\partial x}\bigg|_{x=b_1}, \quad P_1(b_1,t) = P_2(b_1,t), \\ &D_{S_2} \frac{\partial S_2}{\partial x}\bigg|_{x=b_1} = 0. \end{aligned} \tag{80}$$

On the other boundaries between the adjacent layers the matching conditions are applied ($t > 0$),

$$\begin{aligned} &D_{S_j} \frac{\partial S_j}{\partial x}\bigg|_{x=b_j} = D_{S_{j+1}} \frac{\partial S_{j+1}}{\partial x}\bigg|_{x=b_j}, \quad S_j(b_j,t) = S_{j+1}(b_j,t), \\ &D_{P_j} \frac{\partial P_j}{\partial x}\bigg|_{x=b_j} = D_{P_{j+1}} \frac{\partial P_{j+1}}{\partial x}\bigg|_{x=b_j}, \quad P_j(b_j,t) = P_{j+1}(b_j,t), \\ &j = 2,3. \end{aligned} \tag{81}$$

The external diffusion layer of thickness δ remains unchanged with time. Away from it the concentrations of the substrate and the product remain constant ($t > 0$),

$$S_4(b_4,t) = S_0, \quad P_4(b_4, t) = 0. \tag{82}$$

4.1.3 Biosensor Response

The system (75)–(82) approaches a steady state as $t \to \infty$. The biosensor current is proportional to the gradient of the concentration P_1 of the reaction product at the electrode surface. The density $i(t)$ of the current at time t as well as the steady state current I is defined as usual,

$$i(t) = n_e F D_{P_1} \left.\frac{\partial P_1}{\partial x}\right|_{x=b_0}, \qquad I = \lim_{t\to\infty} i(t), \tag{83}$$

where n_e is the number of the electrons involved in a charge transfer at the electrode surface, F is the Faraday constant.

4.2 Numerical Simulation

The initial boundary value problem (75)–(82) can be solved numerically using the finite difference technique [56, 222]. Since the governing equations, the initial and the boundary conditions are of the same type as in the two-compartment model as well as in the modeling of a biosensor with the porous membrane discussed in the previous Sects. 2.2 and 3, the model equations can be approximated with the difference equations by using the same technique.

For $d_s \to 0$, the solution of the model approaches the solution of the corresponding three-layer model of the biosensor with the outer porous membrane discussed in Sect. 3. This can be applied to validate the numerical solution of the four-compartment model (75)–(82).

As for the model of the biosensor with an outer porous membrane, the concentrations of both species, S and P, can be defined continuously in the entire domain $x \in [b_0, b_4]$ as follows ($t \geq 0$):

$$S(x,t) = \begin{cases} S_2(x,t), & x \in [b_1, b_2], \\ S_3(x,t), & x \in (b_2, b_3], \\ S_4(x,t), & x \in (b_3, b_4], \end{cases} \tag{84}$$

$$P(x,t) = \begin{cases} P_1(x,t), & x \in [b_0, b_1], \\ P_2(x,t), & x \in (b_1, b_2], \\ P_3(x,t), & x \in (b_2, b_3], \\ P_4(x,t), & x \in (b_3, b_4]. \end{cases} \tag{85}$$

Both concentration functions (S and P) are now continuous in the entire domain $x \in [b_0, b_4]$.

In the simulation of the biosensor action, the diffusion coefficient for the selective membrane was accepted hundredfold less and the diffusion coefficient for the perforated membrane – tenfold less than the diffusion coefficient for the enzyme layer. Figure 15 shows the simulated concentration profiles of the substrate S and the product P calculated from the four-compartment model (75)–(82) at the following values of the model parameters:

$$\begin{aligned}
&D_{P_1} = 3\,\mu\text{m}^2/\text{s}, \quad D_{S_2} = D_{P_2} = 300\,\mu\text{m}^2/\text{s},\\
&D_{S_3} = D_{P_3} = 30\,\mu\text{m}^2/\text{s}, \quad D_{S_4} = D_{P_4} = 600\,\mu\text{m}^2/\text{s},\\
&d_s = b_1 - b_0 = b_1 = 2\,\mu\text{m}, \quad d_e = b_2 - b_1 = 4\,\mu\text{m},\\
&d_p = b_3 - b_2 = 8\,\mu\text{m}, \quad \delta = b_4 - b_3 = 2\,\mu\text{m},\\
&K_M = 100\,\mu\text{M}, \quad S_0 = 100\,\mu\text{M}, \quad V_{max} = 1\,\mu\text{M/s}.
\end{aligned} \tag{86}$$

Figure 15 shows the concentration profiles at the steady state (curves 5) and the intermediate (curves 1–4) conditions. The steady state condition was achieved at

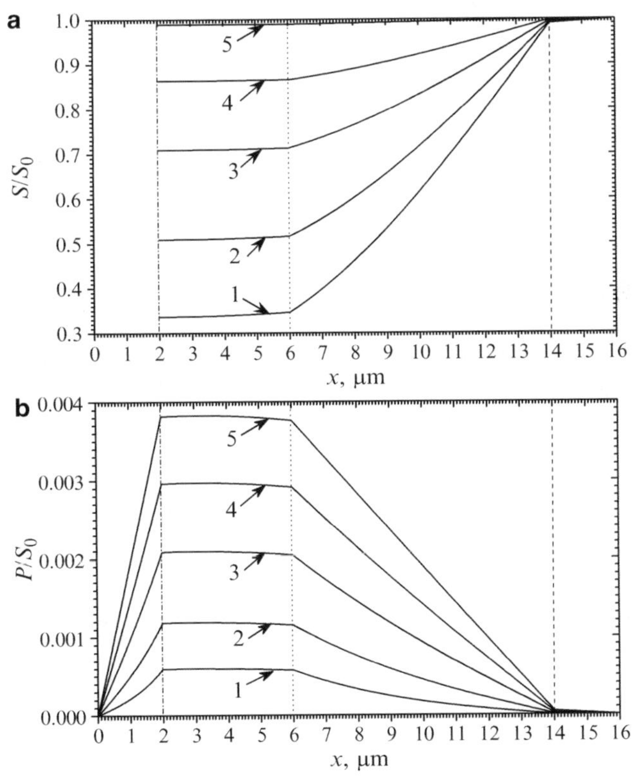

Fig. 15 Concentration profiles of the substrate (**a**) and the product (**b**) obtained at $t = 1.1$ (1), 1.67 (2), 2.66 (3), 4.12 (4) and 10.04 (5) s. The profiles are normalized with the bulk concentration S_0. *Dash–dot* lines show the boundary between the selective membrane and the enzyme layer ($x = 2\,\mu$m), *dot lines* – the boundary between the enzyme layer and the perforated membrane ($x = 6\,\mu$m), and *dashed lines* – the boundary between the perforated membrane and the external diffusion layer ($x = 14\,\mu$m). The values of the model parameters are defined in (86)

time $T = 10.04$ s. The half of the steady state was achieved at 2.66 s ($T_{0.5} = 2.66$ s). Curves 3 correspond to the half time $T_{0.5}$ of the steady state. The other curves show the concentrations at the intermediate values of time: $T_{0.1} = 1.1(1)$, $T_{0.25} = 1.67$ (2), $T_{0.75} = 4.12$ (4) s.

The response peculiarities of the biosensor with the selective and the perforated membranes are also discussed in farther Sect. 4 of fourth chapter of Part II.

5 Biosensors Based on Chemically Modified Electrode

A large part of commercially available and disposable biosensors is prepared by screen-printing technology [64, 107, 224, 258]. They usually contain chemically modified (CM) graphite together with an enzyme [89, 213, 257]. At a CM electrode (CME), electrocatalysis is accomplished by an immobilized redox substance acting as an electron transfer mediator between the graphite electrode and the reaction substrate [182, 190, 195, 253].

To improve the efficiency of the biosensors design and to optimize their configuration the mathematical modeling has rather been widely used [3,9,71,87,247,276]. Mathematical modeling has also been successfully applied to specific sensors based on the CME [45, 64, 67, 146, 167, 198, 287].

This section presents a model allowing an effective computer simulation of the amperometric biosensors based on the CME [42]. Using this model the influence of the physical and the kinetic parameters on the biosensor response is investigated. An ordered ping-pong scheme of the enzyme catalyzed substrate conversion in the presence of the mediator is considered. The CME is considered as an electrode containing a relatively thin layer of the low soluble mediator and covered with an enzyme membrane. The developed model approaches the two-compartment model. In order to define the main governing parameters of the mathematical model the corresponding dimensionless model is derived. By changing the input parameters the output results are numerically analyzed at transition and steady state conditions [42].

5.1 Mathematical Model

An ordered ping-pong scheme of the enzyme (E) catalyzed substrate (S) conversion in the presence of the mediator (M) is considered,

$$\mathrm{E_{ox}} + \mathrm{S} \underset{k_{-1}}{\overset{k_1}{\rightleftharpoons}} \mathrm{ES} \overset{k_2}{\rightarrow} \mathrm{E_{red}} + \mathrm{P_1} \tag{87a}$$

$$\mathrm{E_{red}} + \mathrm{M} \overset{k_3}{\rightarrow} \mathrm{E_{ox}} + \mathrm{P} \tag{87b}$$

where E_{ox}, E_{red} and ES are oxidized enzyme, reduced enzyme and the enzyme substrate, respectively, P and P_1 are the reaction products.

The reaction takes place on a chemically modified electrode (CME). The CME is considered as an electrode containing a relatively thin layer of the low soluble mediator and covered with an enzyme membrane. The developed model approaches the two-compartment model and involves three regions: the enzyme layer where the enzymatic reaction as well as the mass transport by diffusion take place, the diffusion limiting region where only the mass transport by diffusion takes place and the convective region where the analyte concentration is maintained constant.

5.1.1 Governing Equations

Assuming the quasi steady state approximation, the concentration of the intermediate complex (ES) does not change and is usually neglected when simulating the biochemical behaviour of the biosensors [229, 258]. Additionally assuming the symmetrical geometry of the electrode and the homogeneous distribution of the immobilized enzyme in the enzyme layer of a uniform thickness, the mass transport and the kinetics in the enzyme layer can be expressed by the system of reaction–diffusion equations ($t > 0$),

$$\frac{\partial S_e}{\partial t} = D_{S_e} \frac{\partial^2 S_e}{\partial x^2} - V(M_e, S_e), \tag{88a}$$

$$\frac{\partial M_e}{\partial t} = D_{M_e} \frac{\partial^2 M_e}{\partial x^2} - V(M_e, S_e), \tag{88b}$$

$$\frac{\partial P_e}{\partial t} = D_{P_e} \frac{\partial^2 P_e}{\partial x^2} + V(M_e, S_e), \quad 0 < x < d, \tag{88c}$$

where x stands for space, t stands for time, $S_e(x,t)$, $M_e(x,t)$, $P_e(x,t)$ are the concentrations of the substrate, the mediator and the reaction product, respectively, d is thickness of the enzyme layer, D_{S_e}, D_{M_e} and D_{P_e} are the diffusion coefficients for the substrate, mediator and reaction product, respectively, and $V(M_e, S_e)$ is the quasi steady state enzyme reaction rate of the ordered ping-pong scheme (87). According to the scheme

$$\frac{E_t}{V} = \frac{1}{k_{cat}} + \frac{1}{k_{red} S_e} + \frac{1}{k_{ox} M_e}, \tag{89}$$

where E_t is the total concentration of the enzyme, k_{cat} is the catalytic constant of the ES conversion, $k_{cat} = k_2$, k_{ox} is a constant of the enzyme interaction with the mediator, $k_{ox} = k_3$, k_{red} is an apparent bimolecular constant of the enzyme and the substrate interaction, $k_{red} = k_1 k_2/(k_{-1} + k_2)$. The total sum E_t of the concentrations of all the enzyme forms is assumed to be constant in the entire enzyme layer, $E_t = E_{ox} + E_{red} + E_s$, where E_{ox}, E_{red}, E_s are the concentrations of E_{ox}, E_{red}, ES, respectively. From (89) the following non-linear expression of the reaction rate is obtained:

$$V(M_e, S_e) = \frac{E_t k_{cat} k_{red} k_{ox} M_e S_e}{k_{red} k_{ox} M_e S_e + k_{cat} k_{ox} M_e + k_{cat} k_{red} S_e}. \tag{90}$$

Outside the enzyme layer only the mass transport by diffusion takes place. The outer mass transport obeys a finite diffusion regime,

$$\frac{\partial S_b}{\partial t} = D_{S_b}\frac{\partial^2 S_b}{\partial x^2}, \tag{91a}$$

$$\frac{\partial M_b}{\partial t} = D_{M_b}\frac{\partial^2 M_b}{\partial x^2}, \tag{91b}$$

$$\frac{\partial P_b}{\partial t} = D_{P_b}\frac{\partial^2 P_b}{\partial x^2}, \quad d < x < d+\delta, \tag{91c}$$

where $S_b(x,t)$, $M_b(x,t)$ and $P_b(x,t)$ are the concentrations of the substrate, the mediator and the product, respectively, in the diffusion layer, δ is the thickness of the diffusion layer, D_{S_b}, D_{M_b}, D_{P_b} are the diffusion coefficients.

According to the Nernst approach the layer ($d < x < d+\delta$) remains unchanged with time. Away from it the solution is in motion and uniform in concentration [2, 73, 158, 164, 199, 267].

5.1.2 Initial Conditions

Let $x = 0$ represent the surface of the CME, while $x = d$ – the boundary between the enzyme membrane and the buffer solution. The biosensor operation starts when some substrate appears in the bulk solution. This occurs in the initial conditions ($t = 0$),

$$S_e(x,0) = 0, \quad 0 \le x \le d, \tag{92a}$$

$$M_e(x,0) = \begin{cases} M_0, & x = 0, \\ 0, & 0 < x \le d, \end{cases} \tag{92b}$$

$$P_e(x,0) = 0, \quad 0 \le x \le d, \tag{92c}$$

$$S_b(x,0) = \begin{cases} 0, & d \le x < d+\delta, \\ S_0, & x = d+\delta, \end{cases} \tag{92d}$$

$$M_b(x,0) = 0, \quad d \le x \le d+\delta, \tag{92e}$$

$$P_b(x,0) = 0, \quad d \le x \le d+\delta, \tag{92f}$$

where M_0 is the concentration of the mediator at the boundary between the electrode and the enzyme layer, S_0 is the concentration of the substrate in the bulk solution.

5.1.3 Boundary and Matching Conditions

On the boundary between two regions having different diffusivities, the matching conditions are defined ($t > 0$),

$$D_{S_e} \frac{\partial S_e}{\partial x}\bigg|_{x=d} = D_{S_b} \frac{\partial S_b}{\partial x}\bigg|_{x=d}, \quad S_e(d,t) = S_b(d,t), \tag{93a}$$

$$D_{M_e} \frac{\partial M_e}{\partial x}\bigg|_{x=d} = D_{M_b} \frac{\partial M_b}{\partial x}\bigg|_{x=d}, \quad M_e(d,t) = M_b(d,t), \tag{93b}$$

$$D_{P_e} \frac{\partial P_e}{\partial x}\bigg|_{x=d} = D_{P_b} \frac{\partial P_b}{\partial x}\bigg|_{x=d}, \quad P_e(d,t) = P_b(d,t). \tag{93c}$$

These conditions mean that the fluxes of the substrate, the mediator and the product through the stagnant external layer are equal to the corresponding fluxes entering the surface of the enzyme membrane. The partition of the substrate, mediator and product in the membrane versus the bulk is assumed to be equal.

In the bulk solution the concentrations of the substrate, the mediator and the product remain constant ($t > 0$),

$$S_b(d+\delta,t) = S_0, \tag{94a}$$

$$M_b(d+\delta,t) = 0, \tag{94b}$$

$$P_b(d+\delta,t) = 0. \tag{94c}$$

The concentration P_e of the reaction product at the electrode surface ($x = 0$) is being permanently reduced to zero due to the electrode polarization. Following the scheme (87), the substrate is an electro-inactive substance. The concentration of the mediator covering the electrode surface is kept constant. This is described by the following boundary conditions ($t > 0$):

$$D_{S_e} \frac{\partial S_e}{\partial x}\bigg|_{x=0} = 0, \tag{95a}$$

$$M_e(0,t) = M_0, \tag{95b}$$

$$P_e(0,t) = 0. \tag{95c}$$

The constant concentration M_0 of the mediator on the electrode can be achieved by permanent dissolution of adsorbed mediator. The direct measurements show that M_0 can be as low as 10^{-6} M [126].

5.1.4 Biosensor Response

The measured current is accepted as a response of an amperometric biosensor in physical experiments. The anodic current is directly proportional to the flux of the reaction product at the electrode surface [229, 258], i.e. on the border $x = 0$. Since the total current is also directly proportional to the area of the electrode surface, the total current is normalized with the area of that surface. The density $i(t)$ of the biosensor current at time t can be obtained explicitly from the Faraday and the Fick laws,

$$i(t) = n_e F D_{P_e} \frac{\partial P_e}{\partial x}\bigg|_{x=0}, \tag{96}$$

where n_e is the number of electrons involved in a charge transfer at the electrode surface, and F is the Faraday constant.

The system (88), (91)–(95) approaches a steady state as $t \to \infty$,

$$I = \lim_{t\to\infty} i(t), \tag{97}$$

where I is assumed as the density of the steady state biosensor current.

5.2 *Numerical Simulation*

The concentrations S, M and P of the substrate, the mediator and the reaction product, respectively, can be defined for the entire interval $x \in [0, d + \delta]$ as follows $(t \geq 0)$:

$$U(x,t) = \begin{cases} U_e(x,t), & x \in [0,d], \\ U_b(x,t), & x \in (d, d+\delta], \end{cases} \qquad U = S, M, P. \tag{98}$$

The concentration functions (S, M and P) are continuous in the entire interval $x \in [0, d + \delta]$.

The initial boundary value problem (88), (91)–(95) can be solved numerically by applying the finite difference technique [56,222]. Since the governing equations, the initial and the boundary conditions are of the same type as in a two-compartment model and as in modeling a biosensor with porous membrane discussed in previous sections, the model equations can be approximated to the difference equations by using the same technique.

The adequacy of the mathematical model of the biosensor can be evaluated using a known analytical solution of the two-compartment model of amperometric biosensors (see Sect. 2). As one can see from the reaction rate V introduced by (90) the kinetics of the biochemical reaction significantly depends on the ratio of the substrate and the mediator concentrations. Let us introduce the dimensionless ratio Σ of the substrate (S_0) to the mediator (M_0) concentrations combining them with the rates of the corresponding reactions (87),

$$\Sigma = \frac{S_0 k_{red}}{M_0 k_{ox}}. \tag{99}$$

At relatively low concentrations of the substrate when $\Sigma \ll 1$ ($S_0 k_{red} \ll M_0 k_{ox}$), the reaction rate $V(M, S)$ introduced by (90) reduces as follows:

$$V(M,S) \approx \frac{E_t k_{cat} k_{red} S}{k_{red} S + k_{cat}}. \tag{100}$$

Consequently, in this case the mediator concentration does not affect the product concentration, and the governing equation (88b) can be neglected when simulating the biosensor response.

Assuming (100), the governing equations (88a), (88c), (91a), (91c), together with the initial conditions (92a), (92c), (92d), (92f) and the boundary conditions (93a), (93c), (94a), (94c), (95a), (95c) form an initial boundary value problem which can be solved analytically in cases when the reaction function (100) approaches a linear function [235]. At so low concentrations of the substrate as $S_0 \ll k_{cat}/k_{red}$, the reaction rate $V(M, S)$ reduces further to $E_t k_{red} S$. Assuming $V(M, S) \approx E_t k_{red} S$, the density I of the steady state current can be calculated as follows [235] (also see Sect. 2.3):

$$\begin{aligned} I &= n_e F D_{P_e} S_0 \\ &\times \frac{1}{d+\delta}\left(d + \delta \frac{D_{S_b} - \sigma_{red} D_{S_e} \sinh(\sigma_{red})/\cosh(\sigma_{red})}{D_{S_b} + \sigma_{red} D_{S_e} (\delta/d) \sinh(\sigma_{red})/\cosh(\sigma_{red})}\right) \\ &\times \left(\frac{\sigma_{red} D_{S_e} \delta}{d} \times \frac{\sinh(\sigma_{red})}{\cosh(\sigma_{red})} + \frac{D_{S_e} D_{pd}}{D_{P_e}}\left(1 - \frac{1}{\cosh(\sigma_{red}}\right)\right) / \\ &\qquad \left(D_{P_b} d + D_{P_e} \delta\right), \end{aligned} \tag{101}$$

$$\sigma_{red}^2 = k_{red} q, \quad q = \frac{E_t d^2}{D_{S_e}}. \tag{102}$$

The dimensionless factor σ_{red}^2 is known as the diffusion module or the Damköhler number [10]. The diffusion module compares the rate $(E_t k_{red})$ of the enzyme reaction with the diffusion rate (D_{S_e}/d^2).

At relatively low concentrations of the mediator when $\Sigma \gg 1$ $(M_0 k_{ox} \ll S_0 k_{red})$, the reaction rate $V(M, S)$ reduces to

$$V(M, S) \approx \frac{E_t k_{cat} k_{ox} S}{k_{ox} M + k_{cat}}. \tag{103}$$

In this case the substrate concentration may be neglected when simulating the biosensor response.

Assuming (103), the governing equations (88b), (88c), (91b), (91c), together with the initial conditions (92b), (92c), (92e), (92f) and the boundary conditions (93b), (93c), (94b), (94c), (95b), (95c) form an initial boundary value problem which can be solved analytically in cases when the reaction function (103) approaches a linear function [64]. At concentrations of the mediator as low as $M_0 \ll k_{cat}/k_{ox}$ and the zero thickness of the external diffusion layer, $\delta = 0$, the density I of the steady state current can be calculated as follows [64]:

$$I = n_e F D_{P_e} M_0 \frac{1}{d} \left(\sigma_{ox} \coth(\sigma_{ox}) - 1\right), \tag{104}$$

$$\sigma_{ox}^2 = k_{ox} q. \tag{105}$$

The number q introduced by (102) incorporates the diffusion rate (D_{S_e}/d^2) and the total concentration E_t of the enzyme. q includes all the parts of the diffusion module except the constant k_{red} of the enzyme–substrate interaction and the constant k_{ox} of the enzyme–mediator interaction. Assuming constant values of k_{red} as well as of k_{ox}, the number q can be used as a reduced diffusion module instead of two module σ_{red}^2 and σ_{ox}^2.

It is rather well known that an ordinary enzyme electrode acts under diffusion limitation when the diffusion module is much greater than unity [26, 235]. If the diffusion module is significantly less than unity then the enzyme kinetics predominates in the biosensor response.

In the case of CM electrode, the kinetics of the enzymatic reaction was expressed by two rates: k_{red} and k_{ox}. These two rates of the reactions (87) lead to two diffusion module: σ_{red}^2 and σ_{ox}^2. Assuming $k_{red} < k_{ox}$ and taking into consideration definitions (102) and (105), one can state that the biosensor acts under limitation of the enzyme–mediator interaction when $q \ll 1/k_{ox}$ ($\sigma_{ox}^2 \ll 1$). If $q \gg 1/k_{red}$ ($\sigma_{red}^2 \gg 1$) then the response is under control of the mass transport by diffusion. At intermediate values of q ($1/k_{ox} < q < 1/k_{red}$) the biosensor acts under mixed limitation of the diffusion and the enzyme–substrate interaction.

5.3 *Dimensionless Model*

In order to define the main governing parameters of the mathematical model the following dimensionless parameters are introduced:

$$
\begin{gathered}
\hat{x} = \frac{x}{d}, \quad \hat{t} = \frac{tD_{S_e}}{d^2}, \quad \hat{\delta} = \frac{\delta}{d}, \quad \hat{T} = \frac{TD_{S_e}}{d^2}, \quad \hat{T}_{0.5} = \frac{T_{0.5}D_{S_e}}{d^2}, \\
\hat{S} = \frac{k_{red}S}{k_{cat}}, \quad \hat{M} = \frac{k_{ox}M}{k_{cat}}, \quad \hat{P} = \frac{k_{ox}P}{k_{cat}}, \\
\hat{S}_0 = \frac{k_{red}S_0}{k_{cat}}, \quad \hat{M}_0 = \frac{k_{ox}M_0}{k_{cat}},
\end{gathered} \tag{106}
$$

where S, P and M are the concentrations introduced by (98), $\hat{x}$ is the dimensionless distance from the electrode surface, $\hat{t}$ stands for the dimensionless time, $\hat{\delta}$ is the dimensionless thickness of the diffusion layer, and $\hat{S}$, $\hat{M}$, $\hat{P}$, $\hat{S}_0$, $\hat{M}_0$ are the dimensionless concentrations. The dimensionless thickness of enzyme membrane equals one.

The governing equations (88) in dimensionless coordinates are expressed as follows:

$$
\frac{\partial \hat{S}}{\partial \hat{t}} = \frac{\partial^2 \hat{S}}{\partial \hat{x}^2} - \sigma_{red}^2 \frac{\hat{M}\hat{S}}{\hat{M}\hat{S} + \hat{M} + \hat{S}}, \tag{107a}
$$

$$
\frac{\partial \hat{M}}{\partial \hat{t}} = \frac{D_{M_e}}{D_{S_e}} \frac{\partial^2 \hat{M}}{\partial \hat{x}^2} - \sigma_{ox}^2 \frac{\hat{M}\hat{S}}{\hat{M}\hat{S} + \hat{M} + \hat{S}}, \tag{107b}
$$

$$\frac{\partial \hat{P}}{\partial \hat{t}} = \frac{D_{P_e}}{D_{S_e}} \frac{\partial^2 \hat{P}}{\partial \hat{x}^2} + \sigma_{ox}^2 \frac{\hat{M}\hat{S}}{\hat{M}\hat{S} + \hat{M} + \hat{S}}, \quad 0 < \hat{x} < 1, \quad \hat{t} > 0. \tag{107c}$$

The governing equations (91) yield to the following equations:

$$\frac{\partial \hat{S}}{\partial \hat{t}} = \frac{D_{S_b}}{D_{S_e}} \frac{\partial^2 \hat{S}}{\partial \hat{x}^2}, \tag{108a}$$

$$\frac{\partial \hat{M}}{\partial \hat{t}} = \frac{D_{M_b}}{D_{S_e}} \frac{\partial^2 \hat{M}}{\partial \hat{x}^2}, \tag{108b}$$

$$\frac{\partial \hat{P}}{\partial \hat{t}} = \frac{D_{P_b}}{D_{S_e}} \frac{\partial^2 \hat{P}}{\partial \hat{x}^2}, \quad 1 < \hat{x} < 1 + \hat{\delta}, \quad \hat{t} > 0. \tag{108c}$$

The initial conditions (92) transform to the following conditions:

$$\hat{S}(\hat{x}, 0) = \begin{cases} 0, & 0 \le \hat{x} < 1 + \hat{\delta}, \\ \hat{S}_0, & \hat{x} = 1 + \hat{\delta}, \end{cases} \tag{109a}$$

$$\hat{M}(\hat{x}, 0) = \begin{cases} \hat{M}_0, & \hat{x} = 0, \\ 0, & 0 < \hat{x} \le 1 + \hat{\delta}, \end{cases} \tag{109b}$$

$$\hat{P}(\hat{x}, 0) = 0, \quad 1 \le \hat{x} \le 1 + \hat{\delta}. \tag{109c}$$

The matching (93) and the boundary (94)–(95) conditions are rewritten as follows ($\hat{t} > 0$):

$$\left.\frac{\partial \hat{S}}{\partial \hat{x}}\right|_{\hat{x}=1} = \frac{D_{S_b}}{D_{S_e}} \left.\frac{\partial \hat{S}}{\partial \hat{x}}\right|_{\hat{x}=1}, \tag{110a}$$

$$\left.\frac{\partial \hat{M}}{\partial \hat{x}}\right|_{\hat{x}=1} = \frac{D_{M_b}}{D_{M_e}} \left.\frac{\partial \hat{M}}{\partial \hat{x}}\right|_{\hat{x}=1}, \tag{110b}$$

$$\left.\frac{\partial \hat{P}}{\partial \hat{x}}\right|_{\hat{x}=1} = \frac{D_{P_b}}{D_{P_e}} \left.\frac{\partial \hat{P}}{\partial \hat{x}}\right|_{\hat{x}=1}, \tag{110c}$$

$$\hat{S}(1 + \hat{\delta}, \hat{t}) = \hat{S}_0, \tag{111a}$$

$$\hat{M}(1 + \hat{\delta}, \hat{t}) = 0, \tag{111b}$$

$$\hat{P}(1 + \hat{\delta}, \hat{t}) = 0. \tag{111c}$$

$$\left.\frac{\partial \hat{S}}{\partial \hat{x}}\right|_{\hat{x}=0} = 0, \tag{112a}$$

$$\hat{M}(0, \hat{t}) = \hat{M}_0, \tag{112b}$$

$$\hat{P}(0, \hat{t}) = 0. \tag{112c}$$

The dimensionless current (flux) $\hat{i}$ and the corresponding dimensionless stationary current $\hat{I}$ are defined as follows:

$$\hat{i}(\hat{t}) = \left.\frac{\partial \hat{P}}{\partial \hat{x}}\right|_{\hat{x}=0} = \frac{i(t)k_{ox}d}{n_e F D_{P_e} k_{cat}}, \quad \hat{I} = \lim_{\hat{t}\to\infty} \hat{i}(\hat{t}), \tag{113}$$

Assuming the same diffusion coefficients for all three species, only the following dimensionless parameters remain in the dimensionless mathematical model (107)–(112): $\hat{\delta}$ – the thickness of the diffusion layer, $\hat{S}_0$ – the substrate concentration in the bulk solution, $\hat{M}_0$ – the mediator concentration at the electrode surface, σ_{ox}^2 and σ_{red}^2 – the diffusion modules, and D_{rel} – the ratio of the external diffusivity to the internal diffusivity, $D_{rel} = D_{S_b}/D_{S_e} = D_{M_b}/D_{M_e} = D_{P_b}/D_{P_e}$. In all the calculations D_{rel} was equal to 2. As has been mentioned above, it is reasonable to use the reduced diffusion module q instead of two the modules: σ_{ox} and σ_{red}.

5.4 Simulated Biosensor Action

Figures 16 and 17 show the profiles of concentrations of the substrate, the mediator and the product in the enzyme membrane ($x \in (0, d)$, $\hat{x} \in (0, 1)$) as well as in the external diffusion layer ($x \in (d, d+\delta)$, $\hat{x} \in (1, 1+\hat{\delta})$) at $d = 100\,\mu$m, $\delta = 300\,\mu$m. The dynamics of the biosensor current is presented in Fig. 18. The biosensor action was simulated for two concentrations (0.01 and 1 M) of the substrate (S_0) as well as for two concentrations (10^{-5} and 10^{-3} M) of the mediator (M_0). The corresponding dimensionless concentrations of the substrate ($\hat{S}_0$) as well

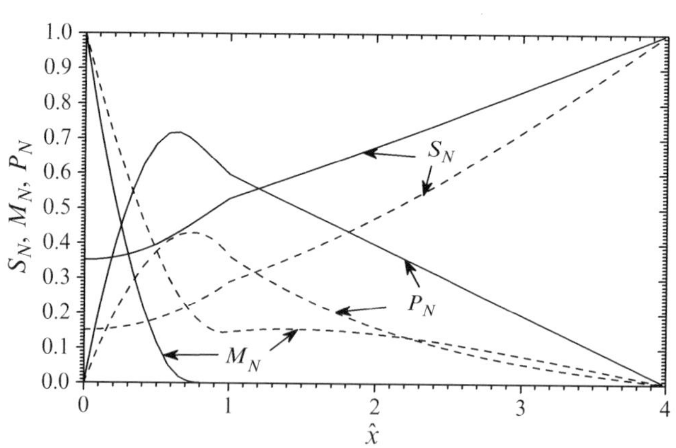

Fig. 16 The profiles of the normalized concentrations of the substrate (S_N), the mediator (M_N) and the product (P_N) in the enzyme layer $\hat{x} \in (0, 1)$ and in the diffusion layer $\hat{x} \in (1, 4)$ at an approximate steady state dimensionless time $\hat{T} = 5.73$ (*solid lines*) and at the dimensionless half-time $\hat{T}_{0.5} = 1.86$ (*dashed lines*), $\hat{S}_0 = 0.1$, $\hat{M}_0 = 10$.s The values of the other parameters are defined in (114)

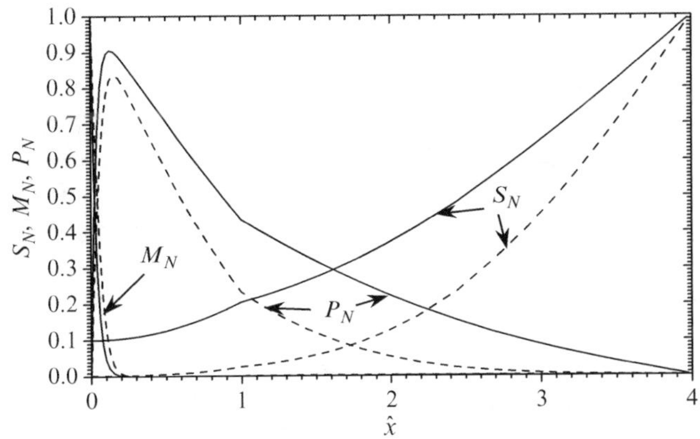

Fig. 17 The profiles of the normalized concentrations of the substrate (S_N), the mediator (M_N) and the product (P_N) in the enzyme and the diffusion layers at an approximate steady state dimensionless time $\hat{T} = 1.2$ (*solid lines*) and the dimensionless half-time $\hat{T}_{0.5} = 0.435$ (*dashed lines*), $\hat{S}_0 = 10$, $\hat{M}_0 = 0.1$. The other parameters are the same as in Fig. 16

as of the mediator ($\hat{M}_0$) are 0.1 and 10. The values of all the other parameters were constant in the numerical simulation,

$$\begin{aligned} &D_{S_e} = D_{M_e} = D_{P_e} = 300\,\mu\text{m}^2/\text{s}, \\ &D_{S_b} = 2D_{S_e}, \quad D_{M_b} = 2D_{M_e}, \quad D_{P_b} = 2D_{P_e}, \\ &k_{cat} = 10^3\,\text{s}^{-1}, \quad k_{red} = 10^4\,\text{M}^{-1}/\text{s}, \quad k_{ox} = 10^7\,\text{M}^{-1}/\text{s}, \\ &E_t = 3\,\mu\text{M}, \quad n_e = 1. \end{aligned} \tag{114}$$

In Figs. 16 and 17, the concentration profiles were normalized as follows:

$$\begin{aligned} S_N &= \hat{S}/\hat{S}_0 = S/S_0, \\ M_N &= \hat{M}/\hat{M}_0 = M/M_0, \\ P_N &= \hat{P}/\hat{M}_0 = P/M_0. \end{aligned} \tag{115}$$

In Figs. 16 and 17, the concentration profiles were plotted at the time $\hat{T}$ when the process reaches steady state and the time $\hat{T}_{0.5}$ when 50% of the steady state current has been reached. At values (114) of the parameters, the time t in seconds is converted to the dimensionless time $\hat{t}$ by $\hat{t} = 0.03t$.

As one can see in Fig. 16, there is rather a long shoulder in the profile of the mediator concentration at $1.3 < \hat{x} < 2.3$. The shoulder appears in the case of a relatively high concentration M_0 of the mediator and a low concentration S_0 of the substrate. At these conditions ($M_0 \gg S_0$, $\Sigma \ll 1$), the rate of the enzymatic reaction depends practically only on the substrate concentration as defined in (100). In the beginning of the biosensor action, there is no substrate in the enzyme membrane, and the mediator diffuses fast from the electrode surface along the enzyme membrane and even to the bulk solution. The enzymatic reaction starts only when some substrate touches the enzyme. Due to a relatively high concentration of the mediator, the reaction progresses rapidly and the concentration of the mediator inside the enzyme near the border also reduces rapidly. Consequently, for a short time the mediator concentration inside the enzyme becomes slightly lower than outside the membrane. No similar effect can be noticed in Fig. 17 which shows the concentration profiles of the opposite case of ($S_0 \gg M_0$, $\Sigma \gg 1$). Additional numerical experiments approved that a shoulder in the profile of the mediator concentration appears only in cases when $\Sigma \ll 1$.

Figure 18 shows the dynamics of the current calculated at different concentrations of the substrate and mediator. Particularly, the dynamics of the response at $\hat{S}_0 = 0.1$ and $\hat{M}_0 = 10$, at which Fig. 16 shows the profiles of the concentrations of the species. The profiles of the concentrations depicted in Fig. 17 correspond to curve 3 in Fig. 18.

One can see in Fig. 18 that the biosensor current is affected by both concentrations: S_0 and M_0. The current grows notably faster at higher concentration S_0 (curves 3 and 4) of the substrate rather than at lower one (curves 1 and 2). The effect of concentration M_0 of the mediator on the biosensor response becomes notable with some delay. The mediator diffuses very quickly from the CME into the enzyme layer in a sufficient for the reaction amount, while the substrate has to diffuse across the Nernst diffusion and enzyme layers. Therefore, at the very beginning of the biosensor operation, the biosensor acts under the limitation of the substrate diffusion.

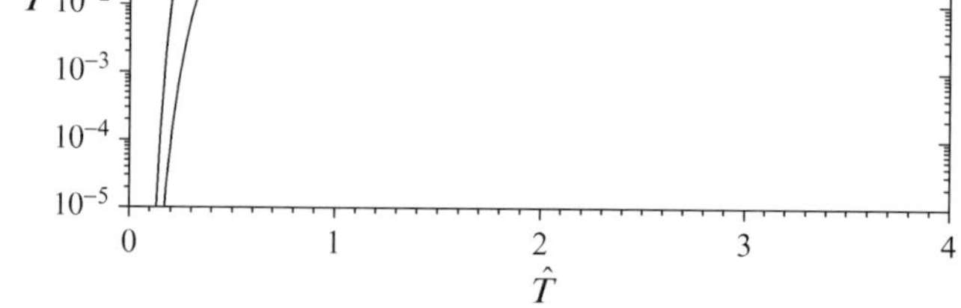

Fig. 18 The dynamics of the dimensionless biosensor current $\hat{i}(\hat{t})$ at two concentrations of the substrate $\hat{S}_0$: 0.1 (1, 2), 10 (3, 4) and two concentrations of the mediator $\hat{M}_0$: 0.1 (1, 3), 10 (2, 4). The other parameters are the same as in Fig. 16

5.5 Impact of the Diffusion Module

The dimensionless model (107)–(112) contains two diffusion modules: σ_{red}^2 and σ_{ox}^2. The reduced diffusion module q is a common part of σ_{red}^2 and σ_{ox}^2 (see (101) and (105)). At constant rates k_{red} and k_{ox} of the reactions (87), it is reasonable to use the reduced diffusion module q instead of the two modules: σ_{red}^2 and σ_{ox}^2. To investigate the effect of the diffusion module q on the biosensor response, the biosensor action was simulated at different concentrations of the substrate and the mediator changing the enzyme layer thickness.

Figure 19 shows the dependence of the steady state dimensionless current I_S on the module q, while Fig. 20 shows the corresponding dependence of the sensitivity B_S. The calculations were made at three concentrations of the substrate (S_0) and three concentrations of the mediator (M_0) changing exponentially the thickness d_e of the enzyme layer from 0.3 μm up to 1.5 mm. Values of all other parameters were assumed constant as defined in (114). Let us notice that accepting these values of the parameters, σ_{ox}^2 becomes equal to unity when $q = 10^{-7}$ Ms, and $\sigma_{red}^2 = 1$ at $q = 10^{-4}$ Ms.

As one can see in Fig. 19 at small values of the diffusion module, $\sigma_{ox}^2 < 1$, the dimensionless current $\hat{I}$ is approximately a linearly increasing function of q as well as of d^2. At large values of q, $\sigma_{red}^2 \gg 1$, $\hat{I}$ becomes a non-monotonous function of q (curves 1 and 5). To see the behaviour of the biosensor response versus the diffusion module the results of calculations were re-plotted in Fig. 21 in terms of the dimensional density I of the steady state current.

Figure 21 shows clearly the non-monotony of the density I of the steady state current versus the module q. As one can see in Fig. 21, increasing q from $1/k_{ox}$ ($\sigma_{ox}^2 = 1$) up to $1/k_{red}$ ($\sigma_{red}^2 = 1$), the density I of the steady state current changes only slightly. At greater values of q, $\sigma_{red}^2 > 1$, the density I of the steady state current monotonously decreases.

The complex effect of the diffusion module on the biosensor response can also be seen in Fig. 20. In cases when CME acts under the limitation of the enzyme-mediator interaction ($\sigma_{ox}^2 < 1$, $q < 10^{-7}$ Ms), the biosensor sensitivity B_S

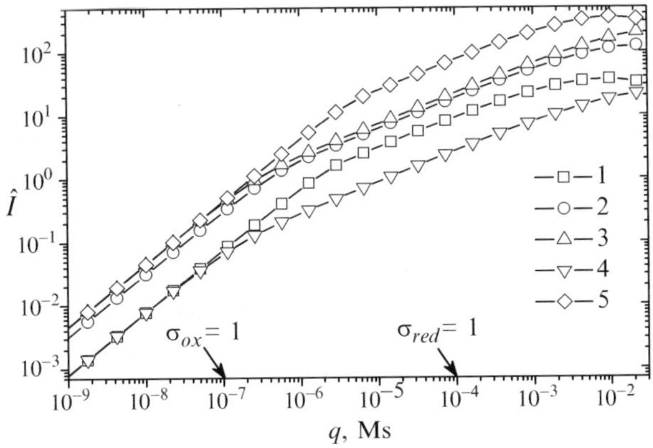

Fig. 19 The steady state dimensionless current $\hat{I}$ versus the reduced diffusion module q at different concentrations of the substrate and the mediator, $\hat{S}_0$: 0.1 (1), 1 (2, 4, 5), 10 (3), $\hat{M}_0$: 0.1 (4), 1 (1–3), 10 (5). The other parameters are the same as in Fig. 16

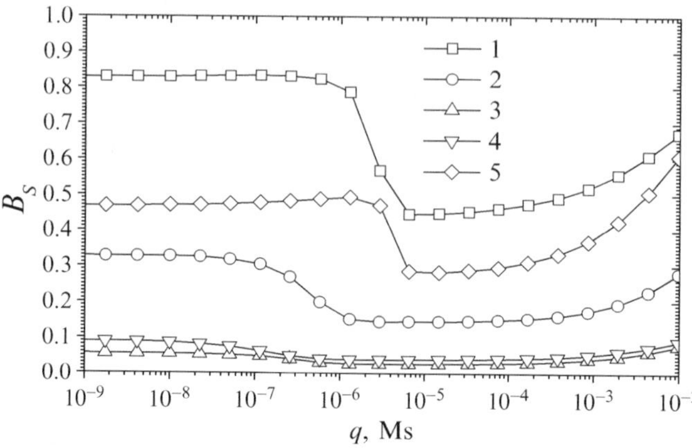

Fig. 20 The biosensor sensitivity B_S versus the reduced diffusion module q. The parameters and notation are the same as in Fig. 19

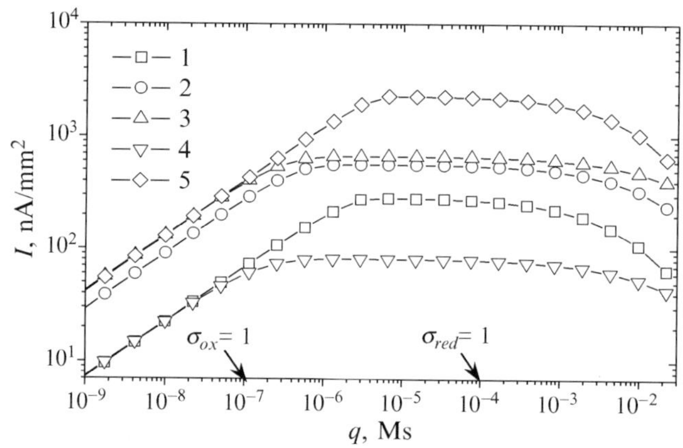

Fig. 21 The density I of the steady state biosensor current versus the reduced diffusion module q. All the parameters and the notation are the same as in Fig. 19

practically does not depend on the diffusion module. It means that at these conditions the biosensor sensitivity is very resistant to changes in the thickness d of the enzyme layer as well as in the total concentration E_t of the enzyme. This resistance notably decreases at higher values of the diffusion module. The sensitivity B_S changes even non-monotonously when q increases from 10^{-7} to 10^{-4} Ms, i.e. when $\sigma_{ox}^2 > 1$ and $\sigma_{red}^2 < 1$. In cases when the CME acts under the control of the mass transport ($\sigma_{red}^2 > 1, q > 10^{-4}$ Ms), the biosensor sensitivity slightly increases with the increase in q. The diffusion module especially affects the sensitivity in cases of low substrate concentration (curve 1) and of high concentration of the mediator (curve 5). This can also be observed in Fig. 21.

Figures 19–21 show the linearity of the biosensor response in cases when the CME acts under the limitation of the enzyme–mediator interaction ($\sigma_{ox} < 1$, $q < 10^{-7}$ Ms).

5.6 *Impact of the Substrate Concentration*

The dependence of the biosensor response on the dimensionless ratio Σ of the substrate and mediator concentrations is depicted in Figs. 22 and 23. The biosensor responses were simulated at different values of the diffusion module by changing the substrate concentration S_0 in the bulk solution and keeping the mediator concentration M_0 constant.

One can see in Fig. 22 a linear range of the calibration curve up to $\Sigma \approx 0.1$ ($\hat{S}_0 \approx 0.1$, $S_0 \approx 10$ mM). The dependence of the steady state current on the ratio Σ is noticeably affected by the diffusion module. The current is directly proportional to σ^2_{red} as well as to σ^2_{ox}. At low values of the diffusion module, the tenfold increase in σ^2_{red} also increases the steady state dimensionless current $\hat{I}$ approximately tenfold (curves 1–3). However, at $\sigma^2_{red} \geq 1$ (curves 4–6), the effect of the diffusion modules on the density I of the steady state current notably decreases. When σ^2_{red} increases from 10 (curve 5) to 100 (curve 6), the current $\hat{I}$ increases only about 2–3 times.

Figure 23 shows that the biosensor sensitivity notably decreases with an increase in the ratio Σ of the substrate and mediator concentrations at all values of the diffusion module. In general, the fact that the effect of the sensitivity decreases increasing the substrate concentration is rather well known [229,258]. As usual, the biosensors

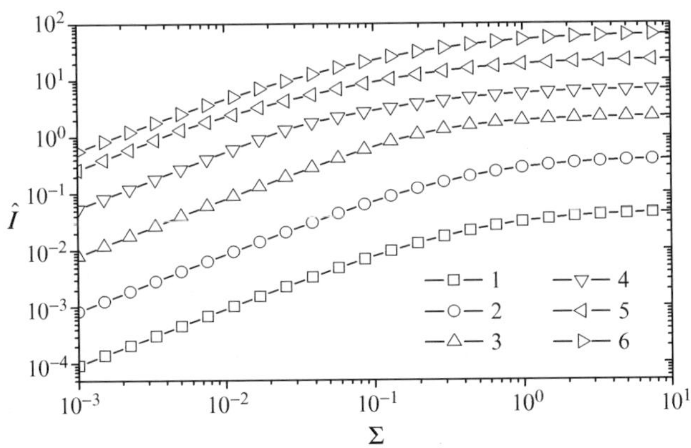

Fig. 22 The steady state dimensionless current $\hat{I}$ versus the ratio Σ of the substrate and the mediator concentrations at different values of the diffusion module σ^2_{red}: 10^{-4} (1), 10^{-3} (2), 10^{-2} (3), 0.1 (4), 1 (5), 10 (6), keeping constant concentration $\hat{M}_0 = 1$ of the mediator. $\sigma^2_{ox} = 10^3\,\sigma^2_{red}$, the other parameters are the same as in Fig. 16

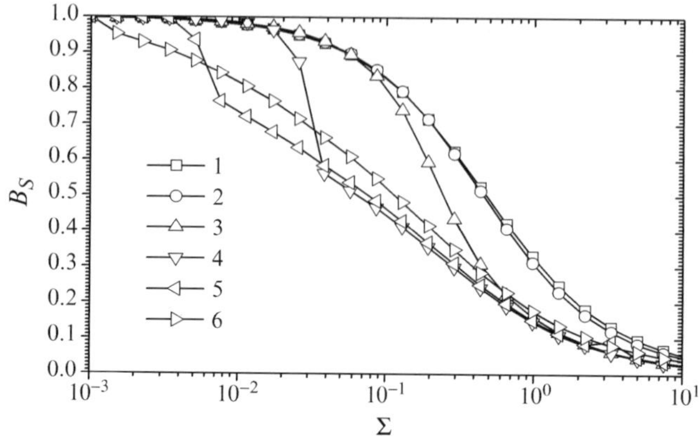

Fig. 23 The biosensor sensitivity B_S versus the ratio Σ of the substrate and the mediator concentrations at different values of the diffusion module. The parameters and notation are the same as in Fig. 22

are highly sensitive at very low concentrations of the substrate ($\Sigma \leq 10^{-2}$) and they are of very low sensitivity at high concentrations of the substrate ($\Sigma > 1$). This effect can also be noticed in Fig. 22.

One can see no notable difference between the shapes of curves 1 and 2 in Figs. 22 and 23. So, in case when $\sigma_{ox}^2 \leq 1$ ($\sigma_{red}^2 \leq 10^{-3}$, curves 1 and 2) the diffusion module practically has no influence on the biosensor sensitivity. When $\sigma_{ox}^2 > 1$ and $\sigma_{red}^2 \leq 1$ (curves 3–5) the sensitivity B_S decreases with increase in the diffusion module. The diffusion module especially affects the biosensor sensitivity at moderate concentrations of the substrate ($0.01 \leq \Sigma \leq 1$). When the response is under the diffusion control ($\sigma_{red}^2 > 1$, curve 6) the sensitivity slightly increases. This was clearly shown in Fig. 20.

5.7 *Concluding Remarks*

The mathematical model (88), (91)–(95) of an amperometric biosensor based on a chemically modified electrode can be successfully used to investigate the kinetic peculiarities of the biosensor response. The corresponding dimensionless mathematical model (107)–(112) can be used as a framework for numerical investigation of the impact of model parameters on the biosensor action and to optimize the biosensor configuration.

The biosensor current grows notably faster at higher substrate concentrations in the bulk solution than at lower ones (Fig. 18). At the very beginning of the operation, the biosensor acts under the limitation of the substrate diffusion from the bulk solution to the electrode.

The value of the diffusion module substantially determines the behaviour of the response and sensitivity of the biosensor. The steady state biosensor current is a non-monotonous function of the diffusion module (Fig. 21). In all cases when $\sigma_{ox}^2 \leq 1$, the diffusion module practically has no influence on the biosensor sensitivity. When $\sigma_{ox}^2 > 1$ and $\sigma_{red}^2 \leq 1$, the sensitivity changes non-monotonously with the diffusion module. When the response is fully under the diffusion control ($\sigma_{red}^2 > 1$), the sensitivity slightly increases with increase in the diffusion module (Figs. 20 and 23).

6 Optical and Fluorescence Biosensors

Optical biosensors are based on the measurement of absorbed or emitted light resulting from the biochemical reaction [53, 68, 161]. Optical biosensors are known to be suitable for environmental, clinical and industrial purposes [275]. Those devices allow real-time analysis of molecular interactions without labelling requirements [156]. Optical biosensors have been used to study interactions involving a wide range of interacting partners, from drugs and viruses to peptides, proteins, oligonucleotides, carbohydrates, and lipids [202, 204, 223, 249, 278].

The understanding of the kinetic peculiarities of biosensors is of crucial importance for their design. To improve the productivity as well as the efficiency of the biosensors design and to optimize the biosensors configuration a model of real biosensors should be build [71, 87]. Mathematical modeling in the design of optical biosensors has been applied only in individual cases [183, 215].

This section presents a model of peroxidase-based optical biosensors [40, 93]. The developed model is based on non-stationary reaction–diffusion equations [10, 62, 122]. By changing the input parameters the output results were numerically analyzed at transition and steady state conditions.

6.1 Mathematical Model

The reaction scheme of the optical biosensor involves hydrogen peroxide (H_2O_2) reaction with peroxidase (E) to form compound I (cmpI) and water (H_2O) with the constant reaction rate k_1. The compound I interacts with the substrate (S) to form a product (P) and a free enzyme (E) assuming the constant reaction rate k_2,

$$\mathrm{E} + \mathrm{H_2O_2} \xrightarrow{k_1} \mathrm{cmpI} + \mathrm{H_2O} \tag{116a}$$

$$\mathrm{cmpI} + \mathrm{S} \xrightarrow{k_2} \mathrm{E} + \mathrm{P} \tag{116b}$$

The product (P) absorbs light and therefore the response of the biosensor increases during the reaction as the product forms. The concentration of the analyte (S) can be directly determined by the absorption of the product (P) [263].

6.1.1 Governing Equations

Assuming the symmetrical geometry of the biosensor and the homogeneous distribution of an immobilized enzyme, the mass transport and the reaction kinetics in the enzyme layer can be described by the following system of the reaction–diffusion equations ($0 < x < d, t > 0$),

$$\frac{\partial S_e}{\partial t} = D_{S_e}\frac{\partial^2 S_e}{\partial x^2} - k_2 C S_e, \tag{117a}$$

$$\frac{\partial P_e}{\partial t} = D_{P_e}\frac{\partial^2 P_e}{\partial x^2} + k_2 C S_e, \tag{117b}$$

$$\frac{\partial H_e}{\partial t} = D_{H_e}\frac{\partial^2 H_e}{\partial x^2} - k_1 E H_e, \tag{117c}$$

$$\frac{\partial E}{\partial t} = -k_1 E H_e + k_2 C S_e, \tag{117d}$$

$$\frac{\partial C}{\partial t} = k_1 E H_e - k_2 C S_e, \tag{117e}$$

where x and t stand for space and time, $S_e(x,t)$, $P_e(x,t)$, $H_e(x,t)$, $E(x,t)$, $C(x,t)$ are the substrate, the product, hydrogen peroxide, peroxidase and compound I concentrations in the enzyme layer, d is the thickness of the enzyme layer, and D_{S_e}, D_{P_e}, D_{H_e} are the diffusion coefficients. The enzyme and the formed compound I are immobilized and therefore there are no diffusion terms in the enzyme and compound I equations.

Outside the enzyme layer only the mass transport by diffusion of the substrate, the product and the hydrogen peroxide takes place. The external mass transport obeys a finite diffusion regime ($d < x < d + \delta, t > 0$),

$$\frac{\partial S_b}{\partial t} = D_{S_b}\frac{\partial^2 S_b}{\partial x^2}, \tag{118a}$$

$$\frac{\partial P_b}{\partial t} = D_{P_b}\frac{\partial^2 P_b}{\partial x^2}, \tag{118b}$$

$$\frac{\partial H_b}{\partial t} = D_{H_b}\frac{\partial^2 H_b}{\partial x^2}, \tag{118c}$$

where δ is the thickness of the diffusion layer, $S_b(x,t)$, $P_b(x,t)$, $H_b(x,t)$ are the substrate, the product and hydrogen peroxide concentrations in the diffusion layer, and D_{S_b}, D_{P_b}, D_{H_b} are the diffusion coefficients.

6.1.2 Initial and Boundary Conditions

Let $x = 0$ represent the plate surface, while $x = d$ is the boundary between the enzyme layer and the buffer solution.

The biosensor operation starts when some substrate appears in the bulk solution. This is used in the initial conditions ($t = 0$)

$$
\begin{aligned}
&S_e(x,0) = P_e(x,0) = C(x,0) = 0, \quad 0 \le x \le d,\\
&H_e(x,0) = H_0, \quad E(x,0) = E_0, \quad 0 \le x \le d,\\
&P_b(x,0) = 0, \quad H_b(x,0) = H_0, \quad d \le x \le d + \delta,\\
&S_b(x,0) = 0, \quad d \le x < d + \delta,\\
&S_b(d + \delta, 0) = S_0,
\end{aligned}
\tag{119}
$$

where E_0 stands for the initial concentration of the enzyme in the enzyme layer, H_0 is the hydrogen peroxide concentration in the bulk solution as well as in the enzyme layer, and S_0 is the substrate concentration in the bulk solution.

The diffusion layer ($d < x < d + \delta$) may be treated as the Nernst diffusion layer [267]. According to the Nernst approach a layer of thickness δ remains unchanged with time. It was assumed that away from it the solution is uniform in concentration.

In the bulk solution the concentrations of the substrate, the product and hydrogen peroxide remain constant ($t > 0$),

$$
S_b(d + \delta, t) = S_0, \quad P_b(d + \delta, t) = 0, \quad H_b(d + \delta, t) = H_0. \tag{120}
$$

Assuming the impenetrable and unreactive plate surface, the mass flux of the species must vanish at this boundary,

$$
\left.\frac{\partial S_e}{\partial x}\right|_{x=0} = \left.\frac{\partial P_e}{\partial x}\right|_{x=0} = \left.\frac{\partial H_e}{\partial x}\right|_{x=0} = 0. \tag{121}
$$

On the boundary between two regions having different diffusivities, the matching conditions are defined ($t > 0$)

$$
D_{S_e} \left.\frac{\partial S_e}{\partial x}\right|_{x=d} = D_{S_b} \left.\frac{\partial S_b}{\partial x}\right|_{x=d}, \quad S_e(d,t) = S_b(d,t), \tag{122a}
$$

$$
D_{P_e} \left.\frac{\partial P_e}{\partial x}\right|_{x=d} = D_{P_b} \left.\frac{\partial P_b}{\partial x}\right|_{x=d}, \quad P_e(d,t) = P_b(d,t), \tag{122b}
$$

$$
D_{H_e} \left.\frac{\partial H_e}{\partial x}\right|_{x=d} = D_{H_b} \left.\frac{\partial H_b}{\partial x}\right|_{x=d}, \quad H_e(d,t) = H_b(d,t). \tag{122c}
$$

These conditions mean that the fluxes of the substrate, the product and hydrogen peroxide through the stagnant external diffusion layer are equal to the corresponding fluxes entering the surface of the enzyme layer. The partitions of the substrate, the product and hydrogen peroxide in the enzyme layer versus the bulk are assumed to be equal [26, 36].

6.1.3 Response of Optical Biosensor

The light absorbance was assumed as the response of the optical biosensor. The optical signal arises due to the product absorbance in the enzyme and diffusion layers. The optical biosensor was assumed to be placed in the flow or inside of a very high volume of mixed solution. The product molecules which escape the enzyme and diffusion layers do not contribute to the signal. The absorbance $A(t)$ at time t may be obtained as follows:

$$A(t) = \varepsilon_P l_{ef} \bar{P}, \quad l_{ef} = d + \delta, \tag{123}$$

where ε_P is the molar extinction coefficient of the product, $\bar{P}$ – the concentration of the product averaged through the enzyme and diffusion layers, l_{ef} – the effective thickness of the enzyme and diffusion layers [263]. For organic compounds ε_P varies between 10^2 and 10^4 m^2/mol.

The system (117)–(122) approaches a steady state as $t \to \infty$,

$$A_s = \lim_{t \to \infty} A(t), \tag{124}$$

where A_s is the steady state absorbance.

The sensitivity of the optical biosensor is defined as the gradient of the steady state absorbance with respect to the substrate concentration. The absorbance varies in orders of magnitude with the concentration of the substrate to be analyzed [161]. Therefore the dimensionless expression of the sensitivity is preferable,

$$B_O(S_0) = \frac{S_0}{A_s(S_0)} \times \frac{dA_s(S_0)}{dS_0}, \tag{125}$$

where B_O stands for the dimensionless sensitivity of the optical biosensor, $A_s(S_0)$ is the steady state absorbance calculated at the substrate concentration S_0 in the bulk solution.

6.1.4 Response of Fluorescence Biosensor

The reaction product may be fluorescent and it may be the fluorescence that is measured [53, 161]. By extension of the Beer–Lambert law, the response of a fluorescence $F_I(t)$ at time t can be obtained as an inversely exponential function of the average concentration of the product [17, 123, 263],

$$F_I(t) = I_0 \phi \left(1 - 10^{-\varepsilon_P l_{ef} \bar{P}}\right), \tag{126}$$

where I_0 stands for the intensity of excitation light, φ is a quantum yield of fluorophore, ε_P is the molar extinction coefficient of the product, l_{ef} – the effective thickness of the enzyme and diffusion layers, and $\bar{P}$ – the concentration of the product averaged through the enzyme and diffusion layers [263]. The quantum yield of the product fluorescence φ, defined as the ratio of the number of photons emitted to the number of photons absorbed, practically varies between 0.001 and 1.

When the absorbance which was expressed as $\varepsilon_P l_{ef} \bar{P}$ is less than 0.1, the fluorescence $F_I(t)$ is almost linearly proportional to the averaged concentration of the product $\bar{P}$ [263],

$$F_I(t) = 2.303 I_0 \varphi \varepsilon_P l_{ef} \bar{P} \,. \tag{127}$$

Since the fluorescence $F_I(t)$ is directly proportional to the intensity I_0 of the excitation light, the measured fluorescence $F(t)$ is usually reported in the relative fluorescence units (RFU),

$$F(t) = F_I(t)/I_0 \,. \tag{128}$$

The steady state fluorescence F_∞ is defined as follows:

$$F_\infty = \lim_{t\to\infty} F(t) \,. \tag{129}$$

The dimensionless sensitivity B_F of the fluorescence biosensor is defined by

$$B_F(S_0) = \frac{S_0}{F_\infty(S_0)} \times \frac{\mathrm{d}F_\infty(S_0)}{\mathrm{d}S_0} \,, \tag{130}$$

where $F_\infty(S_0)$ is the steady state fluorescence calculated at the substrate concentration S_0 in the bulk solution.

Since the optical absorbance is directly proportional to the concentration of the reaction product (see (123)), the fluorescence can be calculated from the corresponding absorbance. Because of this, only the dynamics of species concentrations and of the absorbance is analysed below.

6.2 Numerical Simulation

The initial boundary value problem (88), (91)–(95) can be solved numerically using the finite difference technique [56, 222]. Since the governing equations, the initial as well as the boundary conditions are of the same type as in a two-compartment model and as in the modeling of a biosensor with a porous membrane discussed in the previous sections, the model equations can be approximated with the difference equations by using the same technique.

The following constant-concentration conditions can be derived from Eqs. 117–122:

$$E(x,t) + C(x,t) = E_0, \quad 0 \le x \le d, \quad t > 0, \tag{131}$$

$$\begin{aligned} S_e(x,t) + P_e(x,t) = S_0, &\quad 0 \le x \le d, \quad t \to \infty, \\ S_b(x,t) + P_b(x,t) = S_0, &\quad d \le x \le d+\delta, \quad t \to \infty, \end{aligned} \tag{132}$$

$$H_e(x,t) + P_e(x,t) = H_0, \quad 0 \le x \le d, \quad t \to \infty,$$
$$H_b(x,t) + P_b(x,t) = H_0, \quad d \le x \le d+\delta, \quad t \to \infty. \tag{133}$$

These conditions can be used in testing the numerical solution of the model.

For further representation of the averaged concentrations of the substrate ($\bar{S}_{eb}$), the product ($\bar{P}_{eb}$) and the hydrogen peroxide ($\bar{H}_{eb}$) through the enzyme and diffusion layers, the following designations are introduced:

$$\bar{U}_{eb}(t) = \frac{1}{d+\delta}\left(\int_0^d U_e(x,t)\mathrm{d}x + \int_d^{d+\delta} U_b(x,t)\mathrm{d}x\right), \quad U \in \{S, P, H\}. \tag{134}$$

The concentrations of the substrate ($\bar{S}_e$), the product ($\bar{P}_e$), the hydrogen peroxide ($\bar{H}_e$), the enzyme ($\bar{E}_e$) and the compound I ($\bar{C}_e$) can be averaged only through the enzyme layer and are given by

$$\bar{U}_e(t) = \frac{1}{d}\int_0^d U_e(x,t)\mathrm{d}x, \quad U \in \{S, P, H\},$$
$$\bar{U}_e(t) = \frac{1}{d}\int_0^d U(x,t)\mathrm{d}x, \quad U \in \{E, C\}. \tag{135}$$

The concentrations S, P and H of the substrate, the product and the hydrogen peroxide, respectively, can be defined for the entire interval $x \in [0, d+\delta]$ as follows ($t \ge 0$):

$$U(x,t) = \begin{cases} U_e(x,t), & x \in [0,d], \\ U_b(x,t), & x \in (d, d+\delta], \end{cases} \qquad U = S, P, H. \tag{136}$$

The concentration functions $S(x,t)$, $P(x,t)$ and $H(x,t)$ are continuous in the entire interval $x \in [0, d+\delta]$ for $t \ge 0$.

6.3 Simulated Biosensor Action

The following values of the model parameters were employed in all of the numerical experiments:

$$D_{S_e} = D_{P_e} = D_{H_e} = 300\,\mu\mathrm{m}^2/\mathrm{s}, \quad D_{S_b} = D_{P_b} = D_{H_b} = 600\,\mu\mathrm{m}^2/\mathrm{s},$$
$$d = 1\,\mu\mathrm{m}, \quad \delta = 400d = 400\,\mu\mathrm{m},$$
$$k_1 = 7.1 \times 10^6\,1/\mathrm{Ms}, \quad k_2 = 2 \times 10^7\,1/\mathrm{Ms}, \quad \varepsilon_P = 10^3\,\mathrm{m}^2/\mathrm{mol}. \tag{137}$$

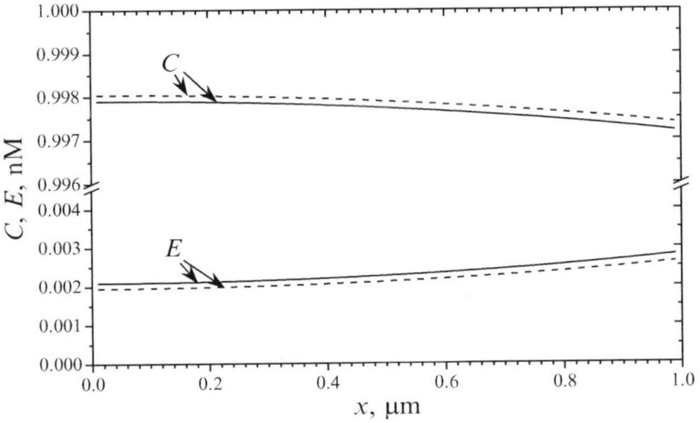

Fig. 24 The steady state ($T = 305$ s, *solid lines*) and the half of it ($T_{0.5} = 116$ s, *dashed lines*) concentration profiles of compound I (C) and peroxidase (E) in the enzyme layer at $S_0 = 100\,\mu$M, $E_0 = 1$ nM, $H_0 = 1$ mM

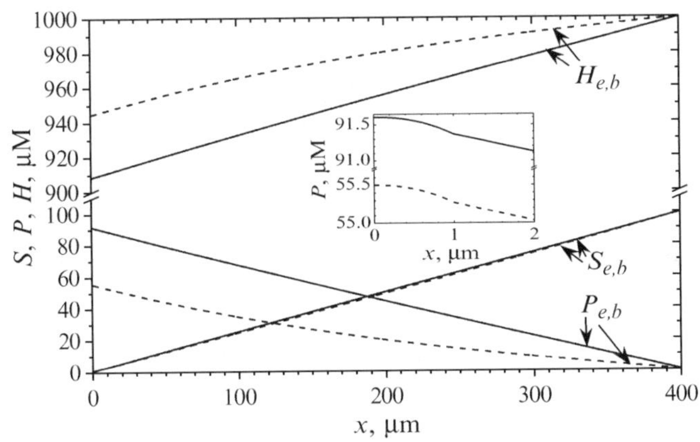

Fig. 25 The concentration profiles of the substrate (S), the product (P) and the hydrogen peroxide (H) in the enzyme and the diffusion layers. The parameters and notation are the same as in Fig. 24

Figures 24 and 25 show the concentration profiles of the substrate, the product, hydrogen peroxide, compound I and enzyme peroxidase in the enzyme and diffusion layers. These concentration profiles were obtained when the steady state and when the half of it were reached.

As one can see in Figs. 24 and 25, the constraints (133) on the concentrations are ensured. When the biosensor operation starts, the initial ($t = 0$) concentration of the enzyme (E) equals E_0 and the compound I (C) starts at zero concentration. Figure 24 shows, that the final (at steady state conditions) concentration E of the enzyme is less than 0.3% of the initial concentration E_0 while the concentration C of the compound I is equal approximately to the initial concentration E_0 of the enzyme. These concentrations quickly become invariable. The dynamics of the substrate concentration is also quit fast. The final (steady state) concentrations of these three compounds differ only slightly from the concentrations obtained at the half time of the steady state. The concentrations of the hydrogen peroxide (H) and of the product (P) change notably slower.

Although the dependence of the product concentration is linear as seen in Fig. 25, the linear dependence is not followed in the enzyme layer (1 μm). This is highlighted in the inset of Fig. 25. The non-linear dependence could be explained by the enzymatic reaction occurring in the enzyme layer.

The dynamics of the concentrations of the compounds is also presented in Figs. 26 and 27. Figure 27 shows the concentrations averaged through the enzyme

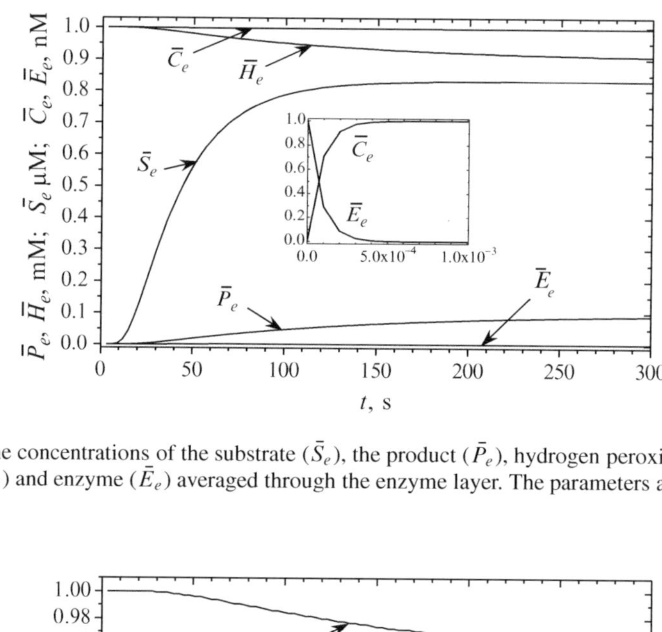

Fig. 26 The concentrations of the substrate ($\bar{S}_e$), the product ($\bar{P}_e$), hydrogen peroxide ($\bar{H}_e$), compound I ($\bar{C}_e$) and enzyme ($\bar{E}_e$) averaged through the enzyme layer. The parameters are the same as in Fig. 24

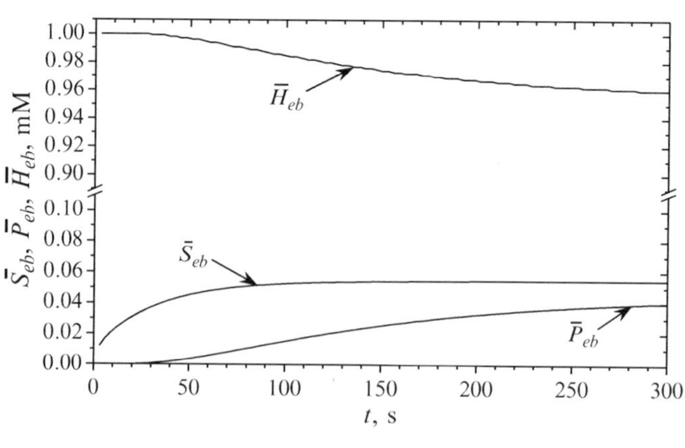

Fig. 27 The concentrations of the substrate ($\bar{S}_{eb}$), the product ($\bar{P}_{eb}$) and hydrogen peroxide ($\bar{H}_{eb}$) averaged through the enzyme and diffusion layers. The parameters are the same as in Fig. 24

layer while Fig. 27 shows the concentrations averaged through both compartments, the enzyme layer and the diffusion layer. After a certain time the equilibrium approaches and the concentrations become invariable.

During the biosensor action the substrate diffuses into the enzyme layer and this results in a decrease of the enzyme as well as of the hydrogen peroxide and in an increase of the compound I as well as of the product concentrations. The inset in Fig. 27 shows a very high concentration dynamics of the enzyme ($\bar{E}$) as well as of the compound I ($\bar{C}$) in the beginning of the reacting process. In about 1 ms these concentrations become approximately constant. The concentration dynamics of all other compounds is significantly lower.

6.4 Impact of the Substrate Concentration

To investigate the dependence of the biosensor response on the outer substrate concentration S_0, the ratio of the substrate concentration S_0 to the concentration H_0 of the hydrogen peroxide, combining with the rates of the corresponding reactions (116a) and (116b), is considered,

$$\Sigma = \frac{k_2 S_0}{k_1 H_0}. \tag{138}$$

The dependence of the absorbance and the sensitivity of the biosensor on the dimensionless ratio Σ is depicted in the Figs. 28 and 29, respectively.

One can see in Fig. 28 a linear range of the calibration curve up to $\Sigma \approx 5 \times 10^3$ ($S_0 \approx 200\,\mu$M). The dependence of the absorbance A_s on the ratio Σ is noticeably affected by the hydrogen peroxide (H_0). The absorbance is directly

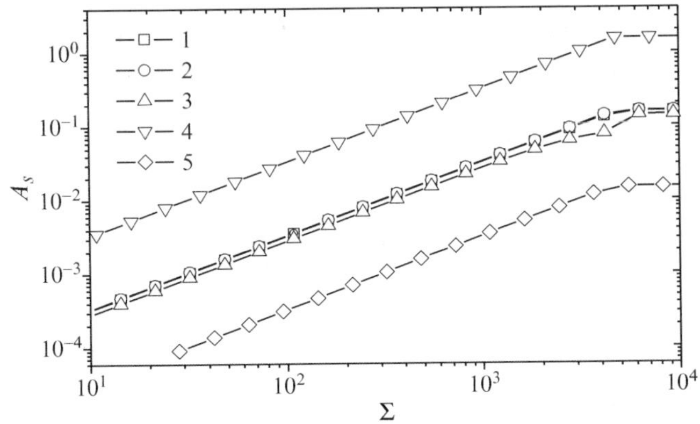

Fig. 28 The dependence of the absorbance A_s on the dimensionless ratio Σ changing the substrate concentration S_0 at three initial concentrations of the hydrogen peroxide (H_0): 0.1 (5), 1 (1, 2, 3), 10 (4) mM and three initial concentrations of the enzyme (E_0): 0.1 (3), 1 (1, 4, 5), 10 (2) nM

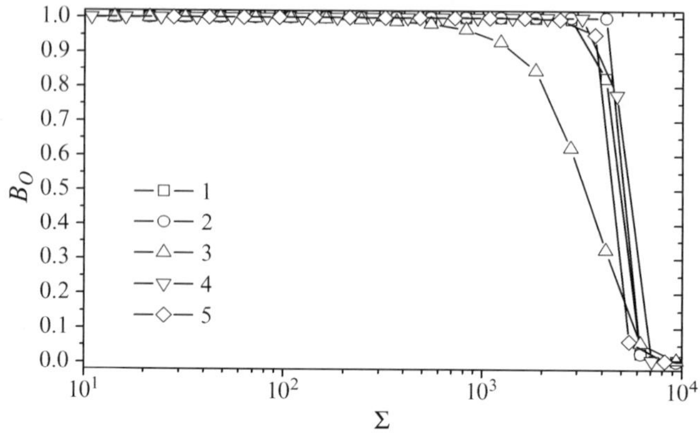

Fig. 29 The dependence of the dimensionless sensitivity B_O on the dimensionless ratio Σ. The parameters and the notations are the same as in Fig. 28

proportional to the concentration H_0 of the hydrogen peroxide. A tenfold increase in the concentration H_0 increases the absorbance approximately tenfold (curve 4). The corresponding decrease in H_0 decreases the A_R tenfold (curve 5). A variation in the initial concentration E_0 of enzyme effects the absorbance slightly (curves 2 and 3).

Figure 29 shows that the biosensor sensitivity notably decreases with a decrease in the concentrations E_0 on the enzyme (curve 3). The concentrations of the enzyme and the hydrogen peroxide determine the concentration of the compound I (reaction (116a)), which interacts with the substrate to form the reaction product (reaction (116b)). The decrease in the enzyme concentration E_0 decreases the rate of the product formation, while the increase in the substrate concentration S_0 increases the reaction rate up to the saturation [229, 258]. A lower concentration E_0 of the enzyme corresponds to a lower substrate concentration S_0 at which the enzyme is saturated with the substrate. Figure 29 shows this effect as the decreasing sensitivity of the biosensor with a decrease in the enzyme concentration E_0.

Modeling Biosensors of Complex Geometry

Contents

1 Biosensor Based on Heterogeneous Microreactor 204
1.1 Structure of Modeling Biosensor 204
1.2 Mathematical Model 206
1.3 Numerical Simulation 211
1.4 Effect of the Tortuosity of the Microreactor Matrix 214
1.5 Effect of the Porosity of the Microreactor Matrix 214
1.6 Concluding Remarks 216
2 Biosensor Based on Array of Microreactors 217
2.1 Principal Structure of Biosensor 217
2.2 Mathematical Model 219
2.3 Numerical Simulation 222
2.4 Effect of the Electrode Coverage with Enzyme 225
2.5 Concluding Remarks 227
3 Plate-Gap Biosensor 228
3.1 Principal Structure of Biosensor 228
3.2 Mathematical Model 229
3.3 Governing Equations 230
3.4 Initial Conditions 231
3.5 Boundary and Matching Conditions 231
3.6 Biosensor Response 232
3.7 Numerical Simulation 233
3.8 Effect of the Gaps Geometry 236
3.9 Concluding Remarks 236
4 Biosensors with Selective and Perforated Membranes 237
4.1 Principal Structure of Biosensor 238
4.2 Mathematical Model 239
4.3 Numerical Simulation 242
4.4 Effect of the Perforation Topology 245
4.5 Concluding Remarks 246

Usually, when modeling a biosensor as a flat electrode having one or several layers sandwich-likely applied onto the electrode surface, a mathematical model of the biosensor is formulated in a one-dimensional-in-space domain. This chapter deals with the modeling of biosensors for which two-dimensional-in-space domains are used when describing mathematically the biosensor action.

R. Baronas et al., *Mathematical Modeling of Biosensors*, Springer Series on Chemical Sensors and Biosensors 9, DOI 10.1007/978-90-481-3243-0_9,

1 Biosensor Based on Heterogeneous Microreactor

A remarkable variety of the carbon paste electrodes (CPEs) belonging to a special group of heterogeneous electrodes has been used for biosensors construction [59, 85, 96, 112, 179, 193]. A carbon paste (CP) consists of a mixture of graphite powder and an organic binder, e.g. mineral oil, which is immiscible with water. The electrodes prepared from the CP show exceptionally low background current, a wide operating potential window, a convenient modification, renewability, miniaturization and low cost [142, 266]. Because of these fascinating properties, The CPEs are currently in extensive use in electroanalysis [54, 182, 265]. A number of biosensors based on the CPE have been constructed for determination of glucose [6, 7, 105, 139, 140, 141, 155, 159, 186, 189, 190, 203, 221, 225, 268, 277].

A miniaturization of analytical systems is regarded as a trend in biosensors because of a large number of advantages [251, 252]. The reduction of the device size, reagents and sample consumption are among the most important advantages. An integration of such systems with enzymatic microreactors proved to be a very suitable approach to the biosensor miniaturization [86, 175, 214, 259]. The application field of microreactors becomes wider every year [170, 251, 260, 264, 279].

This section considers amperometric biosensors based on carbon paste electrodes encrusted with a single microreactor (MR). The constructed biosensor explores an idea to separate the enzyme and the electron transfer components in a microreactor, the silica particle, and use the well-established carbon paste electrode [142]. The microreactor contained an enzyme (glucose oxidase), a mediator, an electron acceptor together with a polymer. A biosensor was assembled by encrusting the CPE with a single MR. The biosensor showed a linear dynamic response up to 50 mM of glucose, a high stability and selectivity [142].

1.1 Structure of Modeling Biosensor

The MR was prepared from the CPC-silica carrier (CPC) and was loaded with the glucose oxidase (GO), mediator (M) and acceptor (A). The volume of the MR was about 0.16 mm^3. The MR was placed onto a freshly cut CPE and pressed into the carbon paste [142]. The principal structure of the biosensor is shown in Fig. 1. The biosensor operation starts after placing the buffer solution (50 mm^3) of glucose onto the surface of the electrode.

During the biosensor action, the glucose (substrate) diffuses into the microreactor (MR), where the glucose reacts with the GO and the reduced GO is formed. Then the reduced GO acts with the oxidized mediator (M^+) followed by the mediator reoxidation with the acceptor (A) [142],

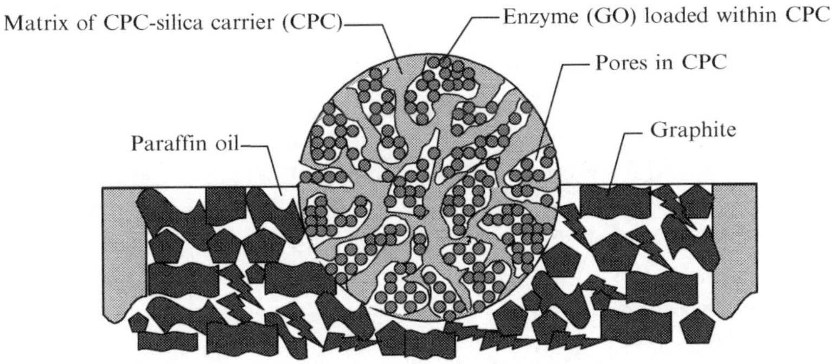

Fig. 1 The principal structure of a biosensor based on CPE and microreactor. The figure is not to scale. The diameter of the microreactor is 0.67 mm and of the carbon paste electrode is 2.3 mm

$$\beta\text{-D-Glucose} + \text{GO}_{ox} \longrightarrow \delta\text{-Glucolactone} + \text{GO}_{red} \tag{1a}$$
$$\text{GO}_{red} + 2\,\text{M}^{+} \longrightarrow \text{GO}_{ox} + 2\,\text{M} + 2\,\text{H}^{+} \tag{1b}$$
$$\text{M} + \text{A} \longrightarrow \text{M}^{+} + \text{A}^{-} \tag{1c}$$

where GO_{ox} and GO_{red} are oxidized and reduced glucose oxidase, respectively.

Due to the large excess of the acceptor used and the fast mediator reoxidation, the simplified scheme of the biosensors action includes the glucose oxidase catalyzed glucose oxidation as well as the reaction of the reduced glucose oxidase with the mediator, the concentration of which does not change due to larger excess of the acceptor,

$$\beta\text{-D-Glucose} + 2\,\text{A} \longrightarrow \delta\text{-Glucolactone} + 2\,\text{A}^{-} \tag{2}$$

The reduced acceptor (A^{-}) diffuses out from the MR and is oxidized on the carbon paste electrode (CPE) producing the current,

$$\text{A}^{-} \longrightarrow \text{A} + \text{e} \tag{3}$$

Following this scheme the biosensor current is a function of the concentration gradient of the acceptor on the CPE. In mathematical modeling of the biosensor it was assumed that the GO, the mediator and the acceptor are homogeneously distributed within the MR. The modeling of the heterogeneous enzymatic process is associated with the solving diffusion equations containing a non-linear term related to the enzymatic reaction. In the simplest case this term is expressed by the Michaelis–Menten function.

In terms of the substrate (S) and the reaction product (P), the simplified scheme of the biosensor action is expressed as

$$\text{S} \xrightarrow{\text{GO}} \text{P} \tag{4}$$

where the substrate (S) binds to the enzyme (GO) and is converted to the product (P).

1.2 Mathematical Model

The biosensor action includes a heterogeneous enzymatic process and the diffusion. The mathematical model involves a system of non-linear differential equations with the inclusion of the enzymatic reaction and the diffusion of glucose (substrate) and acceptor (product). The model to be considered consists of three regions: the enzyme (GO) region where the enzyme reaction as well as the mass transport by diffusion takes place, the impermeable CPC-silica carrier (CPC), and the diffusion limiting region where only the mass transport by diffusion takes place.

Let Ω_b be the open region of the bulk solution containing some substrate, and Ω_{mr} – the open region of the entire microreactor. Since the microreactor was constructed from the CPC and was loaded with GO, the region Ω_{mr} of the entire MR consists of two subregions: Ω_{cpc} – the CPC-carrier and Ω_{go} – the GO, $\Omega_{mr} = \Omega_{cpc} \cup \Omega_{go}$ (see Fig. 1). Let $\bar{\Omega}$ denote a closure of the corresponding open region Ω, and Γ denote a boundary of the corresponding domain Ω, $\bar{\Omega} = \Omega \cup \Gamma$, $\Gamma = \bar{\Omega} \setminus \Omega$.

1.2.1 Governing Equations

The biosensor operation includes the heterogeneous enzymatic process (reaction) and the diffusion. The stimulus of the reaction is the MR, but the reaction performs only in the domain Ω_{go} of the MR which was filled with the GO.

Assuming the homogeneous distribution of the GO in the CPC pores, the mass transport and the reaction kinetics in the enzyme region Ω_{go} can be described by the following system of the reaction–diffusion equations:

$$\begin{aligned}
\frac{\partial S_{go}}{\partial t} &= D_{S_{go}} \Delta S_{go} - \frac{V_{max} S_{go}}{K_M + S_{go}}, \\
\frac{\partial P_{go}}{\partial t} &= D_{P_{go}} \Delta P_{go} + \frac{V_{max} S_{go}}{K_M + S_{go}}, \quad \mathbf{x} \in \Omega_{go}, \quad t > 0,
\end{aligned} \tag{5}$$

where the vector $\mathbf{x}$ stands for space, t is time, Δ is the Laplace operator, $S_{go}(\mathbf{x}, t)$ and $P_{go}(\mathbf{x}, t)$ are the concentrations of the substrate and reaction product, respectively, in the region loaded with the GO, $D_{S_{go}}$ and $D_{P_{go}}$ are the diffusion coefficients, V_{max} is the maximal enzymatic rate and K_M is the Michaelis constant.

However, due to the technology of the MR preparation, the number of cells loaded with the GO is very large, and the geometrical shape of the cells cannot be precisely defined. Nevertheless, it was ascertained that the average size of a cell was much less than the size of the entire MR. Because of the inaccurate geometry of the domain Ω_{go}, there is no hope of solving Eq. 5 neither analytically nor numerically.

The governing equations (5) were reduced by applying the homogenization process [18]. Let $V(\Omega)$ denote the volume of the domain Ω, and γ – the volume fraction of the GO in the MR, i.e. $\gamma = V(\Omega_{go})/V(\Omega_{mr})$. The fraction γ can also be regarded as the porosity of the MR. It was easy to calculate the porosity γ experimentally.

Assuming the MR as a periodic media, the governing equations (5) were reformulated for the entire MR [18, 271],

$$\begin{aligned}\frac{\partial S_{mr}}{\partial t} &= D_{S_{mr}} \Delta S_{mr} - \gamma \frac{V_{max} S_{mr}}{K_M + S_{mr}},\\ \frac{\partial P_{mr}}{\partial t} &= D_{P_{mr}} \Delta P_{mr} + \gamma \frac{V_{max} S_{mr}}{K_M + S_{mr}}, \quad \mathbf{x} \in \Omega_{mr}, \quad t > 0,\end{aligned} \tag{6}$$

where $S_{mr}(\mathbf{x}, t)$ and $P_{mr}(\mathbf{x}, t)$ are the averaged concentrations of the substrate and the reaction product in the MR, respectively, $D_{S_{mr}}$ and $D_{P_{mr}}$ are the effective diffusion coefficients averaged for the entire MR, and γ is the porosity of the MR.

Outside the MR only the mass transport by diffusion of the substrate and the product takes place. Due to restricted amount of the buffer solution placed onto the surface of the electrode, the external mass transport obeys a finite diffusion regime

$$\begin{aligned}\frac{\partial S_b}{\partial t} &= D_{S_b} \Delta S_b,\\ \frac{\partial P_b}{\partial t} &= D_{P_b} \Delta P_b, \quad \mathbf{x} \in \Omega_b, \quad t > 0,\end{aligned} \tag{7}$$

where $S_b(\mathbf{x}, t)$ and $P_b(\mathbf{x}, t)$ are the concentrations of the substrate and the reaction product in the buffer solution, respectively.

Due to the symmetry of the homogenized biosensor, only a half of the cross-section of the biosensor can be considered. Figure 2 shows the profile of the homogenized biosensor in the Descartes as well as in the spherical coordinates. In Fig. 2, Ω_{mr} denotes the MR, Ω_b represents the buffer solution, r_1 is the radius of the homogenized MR, r_2 is the radius of the CPE, r_3 is the radius of the biosensor (outer boundary of the bulk),

$$\begin{aligned}\Omega_{mr} &= (0, r_1) \times (0, \pi),\\ \Omega_b &= (r_1, r_3) \times (0, \pi/2),\\ \Gamma_{mr} &= \{r_1\} \times [0, \pi/2],\\ \Gamma_b &= \{r_3\} \times [0, \pi/2],\\ \Gamma_{cpe,mr} &= \{r_1\} \times [\pi/2, \pi],\\ \Gamma_{cpe,b} &= [r_1, r_2] \times \{\pi/2\},\\ \Gamma_p &= [r_2, r_3] \times \{\pi/2\},\end{aligned} \tag{8}$$

where Γ_{mr} corresponds to the interface between the MR and the bulk solution, $\Gamma_{cpe,mr}$ is the interface between the CPE and the MR, $\Gamma_{cpe,b}$ is the interface between the CPE and the bulk, $\Gamma_{cpe,mr} \cup \Gamma_{cpe,b}$ corresponds to the entire surface of the CPE, Γ_p corresponds to the surface of the plate, Γ_b is the outer boundary of the bulk solution.

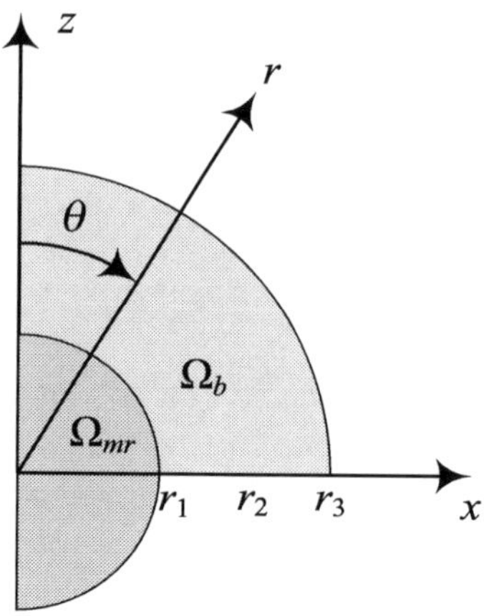

Fig. 2 The profile of the modeling homogenized biosensor based on the CPE and the MR. The figure is not to scale

In the governing equations (6), (7) adjusted for the two-dimensional domains (8), the substrate and the product concentrations are functions of r, θ and t, $S_{mr}(r, \theta, t)$, $P_{mr}(r, \theta, t)$, $S_b(r, \theta, t)$, $P_b(r, \theta, t)$. Let us remind that the Laplace operator Δ in two dimensions in spherical coordinates r and θ is described as follows [73]:

$$\Delta U = \frac{1}{r^2}\frac{\partial}{\partial r}\left(r^2\frac{\partial U}{\partial r}\right) + \frac{1}{r^2 \sin\theta}\frac{\partial}{\partial \theta}\left(\sin\theta\frac{\partial U}{\partial \theta}\right). \tag{9}$$

1.2.2 Effective Diffusion Coefficient

The values of the effective diffusion coefficients $D_{S_{mr}}$ and $D_{P_{mr}}$ in the MR depend on the corresponding diffusion coefficients in the GO, the porosity γ of the MR and the geometry of pores [18, 271]. The diffusion coefficients $D_{S_{mr}}$ and $D_{P_{mr}}$ of the substrate and the product in the MR are lower than those in the GO.

In the case when the model material is a two-phase composite, the effective diffusion coefficient D_{eff} in a periodic media usually satisfies the following condition [18]:

$$\frac{2D_1D_2}{D_1 + D_2} \leq D_{eff} \leq \frac{D_1 + D_2}{2}. \tag{10}$$

where D_i is the diffusion coefficient of the species in the phase i, $i = 1, 2$. When one of the aggregates is impermeable ($D_2 = 0$) this estimation reduces to

$$0 \leq D_{eff} \leq \frac{D_1}{2}. \tag{11}$$

There are several models to evaluate the effective diffusion coefficient more accurately for a porous material. The coefficient D_{eff} is considered as a function of D_1, D_2 and the porosity γ [83, 111], and is estimated by

$$\frac{D_1 D_2}{\gamma D_2 + (1-\gamma)D_1} \leq D_{eff} \leq \gamma D_1 + (1-\gamma)D_2. \tag{12}$$

When the constituent is impermeable ($D_2 = 0$), D_{eff} is estimated by

$$0 \leq D_{eff} \leq \gamma D_1. \tag{13}$$

Assuming a regular spherical matrix configuration of a porous material ($D_2 \neq 0$) [95, 121, 283], value of D_{eff} is given by

$$D_{eff} = D_1 \frac{2D_1\gamma + D_2(3-2\gamma)}{D_1(3-\gamma) + D_2\gamma}. \tag{14}$$

When $D_2 = 0$, Eq. 14 reduces to the following estimation of D_{eff}:

$$D_{eff} = D_1\gamma\frac{2}{3-\gamma}. \tag{15}$$

In a more common case of an irregular matrix of a porous material, the estimation (15) is replaced by

$$D_{eff} = D_1\gamma\beta. \tag{16}$$

where β is the tortuosity of the pore space [83, 95].

Although the geometry of pores in the MR was really irregular (Fig. 1), assuming zero diffusivities of the substrate and the product in the CPC and the regular matrix of the CPC, from (15) an estimation for $D_{S_{mr}}$ and $D_{P_{mr}}$ is obtained,

$$\begin{gathered} D_{S_{mr}} = D_{S_{go}}\gamma\beta, \quad D_{P_{mr}} = D_{P_{go}}\gamma\beta, \\ \beta = \frac{2}{3-\gamma}, \end{gathered} \tag{17}$$

where β corresponds to the tortuosity (irregularity) of the CPC matrix. A very similar approach to the effective diffusion coefficient was applied in the modeling of glucose diffusion through an isolated pancreatic islet of the Langerhans [48].

1.2.3 Initial and Boundary Conditions

The biosensor operation starts after placing the buffer solution of the substrate (glucose) onto the surface of the electrode and the MR. The simulation of the biosensor operation starts when some substrate appears in the bulk solution ($t = 0$),

$$\begin{aligned} S_{mr}(r,\theta,0) &= 0, \quad (r,\theta) \in \bar{\Omega}_{mr} \setminus \Gamma_{mr}, \\ S_{mr}(r,\theta,0) &= S_0, \quad (r,\theta) \in \Gamma_{mr}, \\ S_b(r,\theta,0) &= S_0, \quad (r,\theta) \in \bar{\Omega}_b, \\ P_{mr}(r,\theta,0) &= 0, \quad (r,\theta) \in \bar{\Omega}_{mr}, \\ P_b(r,\theta,0) &= 0, \quad (r,\theta) \in \bar{\Omega}_b, \end{aligned} \tag{18}$$

where S_0 is the concentration of the substrate in the bulk solution.

Assuming the amperometry, the substrate is an electro-inactive substance, while the product is an electro-active substance. The electrode potential is chosen to keep zero concentration of the product at the electrode surface ($t > 0$),

$$\begin{aligned} &P_{mr}(r,\theta,t) = 0, \quad (r,\theta) \in \Gamma_{cpe,mr}, \\ &D_{S_{mr}} \frac{\partial S_{mr}}{\partial r}\bigg|_{\Gamma_{cpe,mr}} = 0, \\ &P_b(r,\theta,t) = 0, \quad (r,\theta) \in \Gamma_{cpe,b}, \\ &D_{S_b} \frac{\partial S_b}{\partial \theta}\bigg|_{\Gamma_{cpe,b}} = 0. \end{aligned} \tag{19}$$

Due to the symmetry of the biosensor, the non-leakage boundary conditions on the borders $\theta = 0$ and $\theta = \pi$ are defined,

$$\begin{aligned} &D_{S_{mr}} \frac{\partial S_{mr}}{\partial \theta}\bigg|_{\theta=0} = D_{S_{mr}} \frac{\partial S_{mr}}{\partial \theta}\bigg|_{\theta=\pi} = 0, \quad r \in [0, r_1], \\ &D_{P_{mr}} \frac{\partial P_{mr}}{\partial \theta}\bigg|_{\theta=0} = D_{P_{mr}} \frac{\partial P_{mr}}{\partial \theta}\bigg|_{\theta=\pi} = 0, \quad r \in [0, r_1], \\ &D_{S_b} \frac{\partial S_b}{\partial \theta}\bigg|_{\theta=0} = 0, \quad r \in [r_1, r_3], \\ &D_{P_b} \frac{\partial P_b}{\partial \theta}\bigg|_{\theta=0} = 0, \quad r \in [r_1, r_3]. \end{aligned} \tag{20}$$

Under the circumstances where both external and internal diffusion gradients are found, the flux of the substrate and the product through the stagnant layer must equal the flux entering the surface of the MR ($t > 0$),

$$\begin{aligned} &D_{S_{mr}} \frac{\partial S_{mr}}{\partial r}\bigg|_{\Gamma_{mr}} = D_{S_b} \frac{\partial S_b}{\partial r}\bigg|_{\Gamma_{mr}}, \quad S_{mr}\Big|_{\Gamma_{mr}} = S_b\Big|_{\Gamma_{mr}}, \\ &D_{P_{mr}} \frac{\partial P_{mr}}{\partial r}\bigg|_{\Gamma_{mr}} = D_{P_b} \frac{\partial P_b}{\partial r}\bigg|_{\Gamma_{mr}}, \quad P_{mr}\Big|_{\Gamma_{mr}} = P_b\Big|_{\Gamma_{mr}}. \end{aligned} \tag{21}$$

The external diffusion layer of the thickness $r_3 - r_1$ remains unchanged with time. Due to a limited volume of the bulk, at the boundary of the bulk the non-leakage condition for both species, the substrate and the product, is used ($t > 0$),

$$D_{S_b} \frac{\partial S_b}{\partial r}\bigg|_{\Gamma_b} = 0, \qquad D_{P_b} \frac{\partial P_b}{\partial r}\bigg|_{\Gamma_b} = 0. \tag{22}$$

1.2.4 Biosensor Response

The total current $i(t)$ at time t of the amperometric biosensor is proportional to the gradient of the product concentration at the electrode surface $\Gamma_{cpe,mr} \cup \Gamma_{cpe,b}$,

$$\begin{aligned} i(t) &= n_e F \int_0^{2\pi} \left(\int_{\pi/2}^{\pi} D_{P_{mr}} \frac{\partial P_{mr}}{\partial r}\bigg|_{r=r_1} r_1^2 sin\theta \, \mathrm{d}\theta \right. \\ &\quad \left. + \int_{r_1}^{r_2} D_{P_b} \frac{\partial P_b}{\partial \theta}\bigg|_{\theta=\pi/2} r \, \mathrm{d}r \right) \mathrm{d}\varphi \\ &= 2\pi n_e F \left(r_1^2 D_{P_{mr}} \int_{\pi 2}^{\pi} \frac{\partial P_{mr}}{\partial r}\bigg|_{r=r_1} sin\theta \, \mathrm{d}\theta + D_{P_b} \int_{r_1}^{r_2} \frac{\partial P_b}{\partial \theta}\bigg|_{\theta=\pi/2} r \, \mathrm{d}r \right), \end{aligned} \tag{23}$$

where n_e is the number of electrons, and F is the Faraday constant.

The reaction–diffusion process (6), (7), (18)–(22) approaches a steady state as $t \to \infty$.

$$I = \lim_{t \to \infty} i(t), \tag{24}$$

where I is the density of the stationary current.

1.3 Numerical Simulation

The initial boundary value problem (6), (7), (18)–(22) can be solved numerically using the finite difference technique [21, 22, 56, 222]. Although the mathematical model has been formulated in a two-dimensional space, the governing, the initial as well as the boundary conditions are of the same type as in a one-dimensional space and can be approximated very similarly.

The following values of the model parameters were employed in the modeling of a practical biosensor [142, 160, 270]:

$$
\begin{aligned}
&D_{S_b} = D_{P_b} = 673\,\mu\text{m}^2/\text{s}, \quad D_{S_{go}} = D_{P_{go}} = 0.5D_{S_b} = 0.5D_{P_b},\\
&r_1 = 0.34\,\text{mm}, \quad r_2 = 1.15\,\text{mm}, \quad r_3 = 2.29\,\text{mm},\\
&\gamma = 0.5, \quad \beta = 0.8,\\
&K_M = 83\,\text{mM}, \quad V_{max} = 44\,\text{mM/s}, \quad S_0 = 25\,\text{mM}, \quad n_e = 2.
\end{aligned}
\tag{25}
$$

Figure 3 shows the concentrations of the substrate (S) and the product (P) in the enzyme and diffusion regions at the steady state conditions (at time $t = 334$ s).

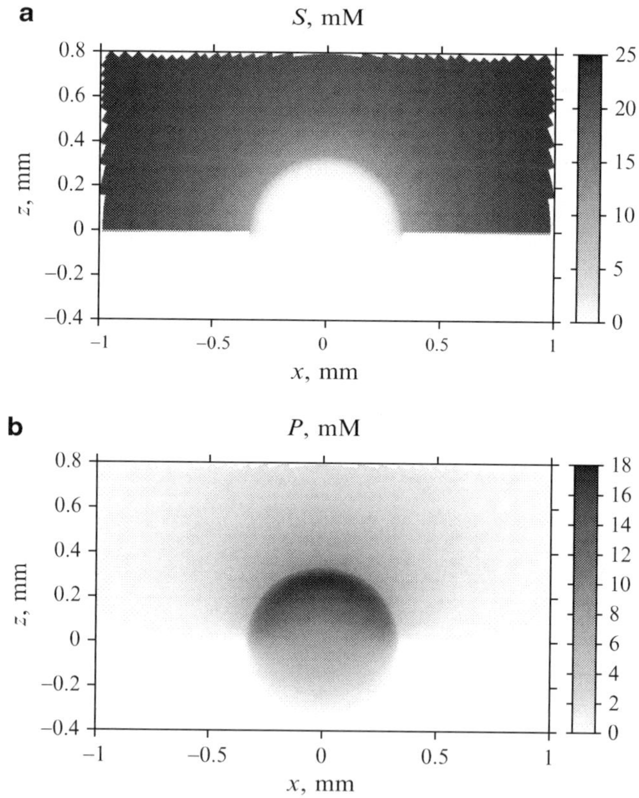

Fig. 3 The stationary concentrations of the substrate (**a**) and the product (**b**) in the MR and the surrounding region, obtained at $t = 334$ s. The values of the model parameters are defined in (25)

The concentrations S and P for the entire domain $\overline{\Omega}_{mr} \cup \overline{\Omega}_b$ can be introduced as follows ($t \geq 0$):

$$S(r,\theta,t) = \begin{cases} S_{mr}(r,\theta,t), & (r,\theta) \in \overline{\Omega}_{mr}, \\ S_b(r,\theta,t), & (r,\theta) \in \overline{\Omega}_b \setminus \overline{\Omega}_{mr}, \end{cases}$$
$$P(r,\theta,t) = \begin{cases} P_{mr}(r,\theta,t), & (r,\theta) \in \overline{\Omega}_{mr}, \\ P_b(r,\theta,t), & (r,\theta) \in \overline{\Omega}_b \setminus \overline{\Omega}_{mr}. \end{cases} \quad (26)$$

Although the dynamics of the concentrations was simulated in the spherical coordinates, Fig. 3 presents the profiles in the Descartes coordinates. The process was simulated in a domain shown in Fig. 2. To have a more comprehensive view, the mirror-image along the z-axis is also shown in Fig. 3 where the point $(0,0)$ corresponds to the center of the MR. Only a part of the entire domain is presented in the figure because a very small variation of the concentrations appeared far from the centre of the MR.

Since the porosity γ and the tortuosity β are undersell grounded parameters of the model, and the geometry of the CPC-matrix of the MR can not be defined precisely (Fig. 1), it is important to evaluate the sensitivity of the biosensor response to the changes in these parameters. Figure 4 shows the dynamics of the biosensor current calculated at different values of the porosity γ and the tortuosity β.

One can see in Fig. 4 that both parameters, γ and β, defining the internal structure of the CPC-matrix of the MR noticeably effect the stationary current as well as the time moment of occurrence of the stationary current. Below the effect of the porosity γ as well as of the tortuosity β on the biosensor response is investigated in detail.

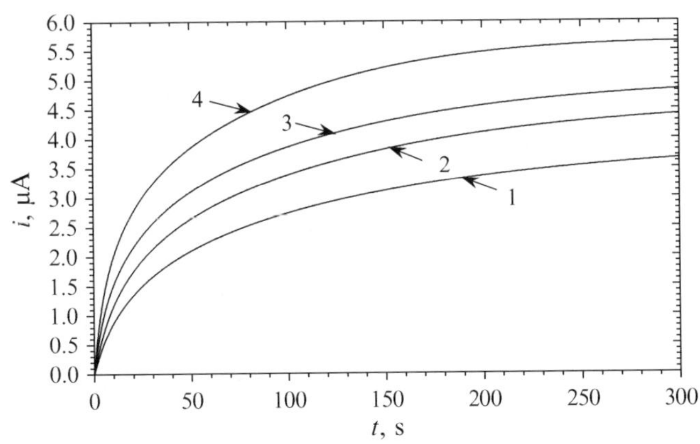

Fig. 4 The dynamics of the biosensor current i at two values of the porosity γ: 0.5 (1, 2), 0.8 (3, 4), and two values of tortuosity β: 0.5 (1, 3), 0.8 (2, 4). The other parameters are the same as in Fig. 4

1.4 Effect of the Tortuosity of the Microreactor Matrix

Since the geometry of the CPC-matrix of the MR can not be defined precisely (Fig. 1), it is important to know how the tortuosity β effects the biosensor response. The dependence of the steady state biosensor current on the tortuosity β of the CPC-matrix is investigated at different concentrations of the substrate. Since the steady state current is very sensitive to the concentration S_0 of the substrate, the current is normalized to evaluate the effect of the tortuosity β on the biosensor response. The normalized steady state biosensor current I_β is expressed by the steady state current at the tortuosity β divided by the steady state current assuming a unitary tortuosity β of the CPC-matrix,

$$I_\beta(S_0, \beta) = \frac{I(S_0, \beta)}{I(S_0, 1)}, \qquad 0 < \beta \leq 1, \tag{27}$$

where $I(S_0, \beta)$ is the density of the steady state current as defined by (24) calculated at given substrate concentration S_0 and the tortuosity β of the CPC-matrix of the MR. The upper limiting value of the tortuosity ($\beta = 1$) corresponds to an unrealistic CPC-matrix containing no curvature. The results of calculations are depicted in Fig. 5. The results were obtained at four values of the substrate concentration S_0 changing the tortuosity β in a wide range and keeping constant values of all other parameters as defined in (25).

As one can see in Fig. 5a the stationary current (I_β as well as I) increases with an increase in the tortuosity β. The effect of the tortuosity β on the steady state current practically does not depend on the substrate concentration S_0. All calculated values of I_β were fitted with a positive allometric function

$$I_\beta(\beta) = \beta^{0.35}. \tag{28}$$

The allometric function (28) is presented in Fig. 5a by a straight solid line.

Figure 5b shows a relatively stable sensitivity B_S of the biosensor at high values of the tortuosity, $\beta > 0.5$. At lower values of the tortuosity, $\beta < 0.5$, the sensitivity notably decreases with a decrease in the tortuosity β. As usual, the biosensor sensitivity is higher at a low substrate concentration S_0 rather than at a higher one.

1.5 Effect of the Porosity of the Microreactor Matrix

The porosity γ of the MR influences the effective diffusion coefficients and the rate of the enzymatic reaction (see (17) and (6)). To investigate the dependence of the steady state biosensor current on the porosity γ of the MR the biosensor response was simulated at different values of γ and of the substrate concentration S_0. The normalised stationary current I_γ was expressed as the steady state current calculated at the porosity γ and divided by the stationary current assuming a unitary porosity,

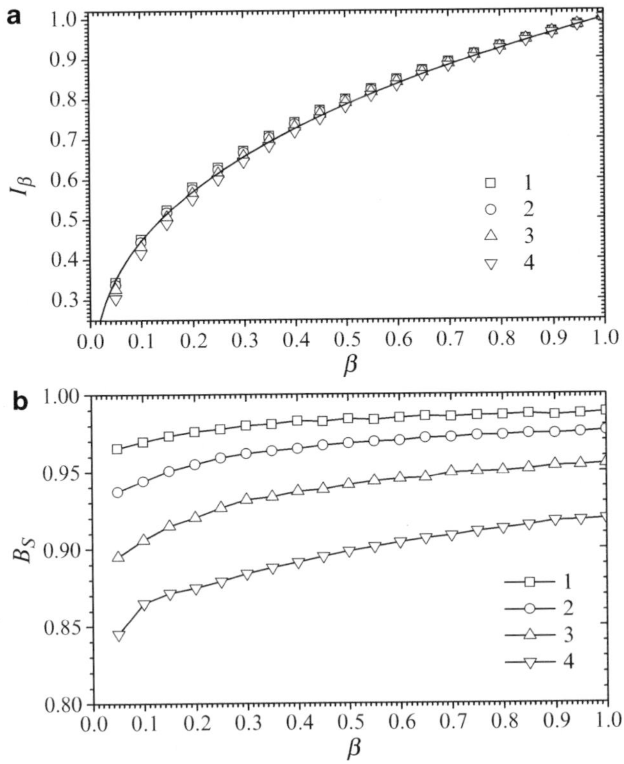

Fig. 5 The normalised steady state current I_β (**a**) and the sensitivity B_S (**b**) versus the tortuosity β at four substrate concentrations (S_0): 25 (1), 50 (2), 100 (3) and 200 (4) mM. The values of the other parameters were defined in (25)

$$I_\gamma(S_0, \gamma) = \frac{I(S_0, \gamma)}{I(S_0, 1)}, \quad 0 < \gamma \leq 1, \tag{29}$$

where $I(S_0, \gamma)$ is the stationary current as defined by (24) calculated at given substrate concentration S_0 and the porosity γ of the MR. The upper limiting value of the porosity ($\gamma = 1$) corresponds to an unrealistic case of the MR containing no CPC-silica carrier, while the lower limiting value ($\gamma = 0$) corresponds to another unrealistic case of the MR constructed exclusively from the CPC-silica carrier. The calculated stationary currents as well as the sensitivities at four values of the substrate concentration S_0 are depicted in Fig. 6.

As one can see in Fig. 6 the effect of the porosity γ on the stationary current (I_β and I) as well as on the biosensor sensitivity B_S is very similar to that of the tortuosity β shown in Fig. 5. The calculated values of the normalised stationary current I_γ were fitted with the following exponential function

$$I_\gamma(\gamma) = \beta^{0.63}. \tag{30}$$

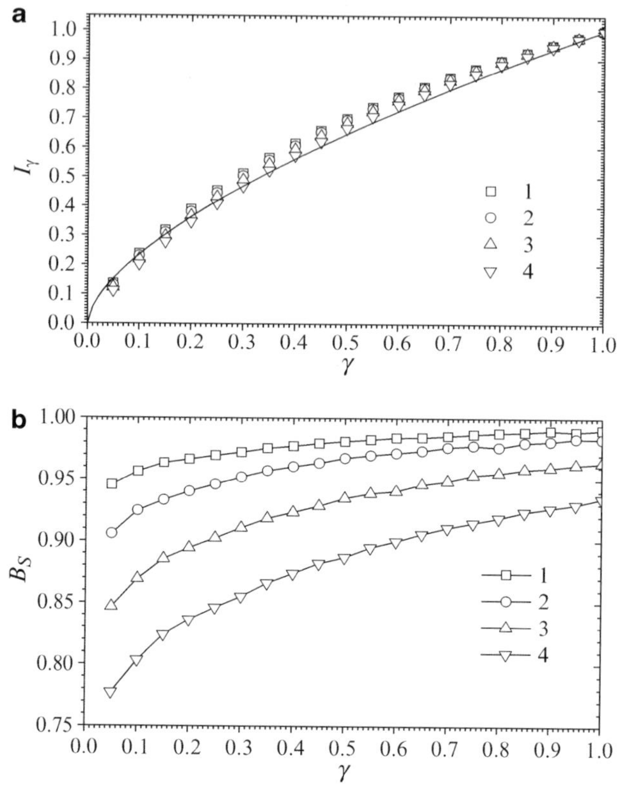

Fig. 6 The normalised steady state current I_γ (**a**) and the sensitivity B_S (**b**) versus the microreactor porosity γ. The parameters and notations are the same as in Fig. 5

1.6 Concluding Remarks

When modeling an amperometric biosensor based on a carbon paste electrode encrusted with a single MR prepared from the CPC-silica carrier and loaded with a glucose oxidase, the homogenization process can be applied to model the MR as a homogeneous media [18, 271]. Assuming the microreactor as a porous media, the MR can be modeled as a reaction–diffusion system (6) with an effective diffusion coefficient directly proportional to the porosity and tortuosity of the CPC-silica carrier. Due to the homogenization process the rate of the enzymatic reaction has to be corrected with the porosity of the MR.

The stationary biosensor current increases in an allometric fashion with an increase in the porosity γ as well as in the tortuosity β (Figs. 5a and 6a). The biosensor sensitivity increases with an increase in the porosity and the tortuosity (Figs. 5b and 6b).

2 Biosensor Based on Array of Microreactors

In some applications of biosensors, enzymes are archival and only available in every limited quantity or are the products of combinatorial synthesis procedures and thus they are only produced in microgram to milligram quantities. These include point-of-care testing [79], high throughput drug discovery [77], detection of biological warfare agents [20], astrobiology [239] and others. Such applications of biosensors require high-density arrays of microvolume reaction vessels. Because of this, miniaturization of biosensors is a very important trend in biotechnology [251, 252]. The application of arrays of microreactors is one way of the miniaturization [207, 281].

This section discusses mathematical modeling of a biosensor system based on an array of enzyme microreactors immobilized on a single electrode [33, 36]. The model involves three regions: an array of microreactors where the enzymatic reaction as well as the mass transport by diffusion takes place, a diffusion limiting region where only the diffusion takes place, and a convective region, where the analyte concentration is maintained constant. The enzyme microreactors were modeled by identical particles (right cylinders) and by strips (right longitudinal quadrangular prisms) distributed uniformly on the electrode surface.

Using the numerical simulation the influence of the geometry of the microreactors as well as of the diffusion region on the biosensor response has been investigated.

2.1 Principal Structure of Biosensor

Two shapes of the enzyme microreactors immobilized on a single electrode are investigated. In the case of the first kind of the biosensor geometry, the microreactors were modeled by identical enzyme filled right cylinders. Figure 7 shows a biosensor system, where the enzyme cylinders of radius a and height c are arranged in a rigid hexagonal array. The distance between the centres of two adjacent cylinders equals $2b$.

The mass transport during the biosensor action obeys a finite diffusion regime. The principal structure of the electrode and the profile of the biosensor at the perpendicular plane are depicted in Fig. 8 where d stands for the thickness of the external diffusion layer.

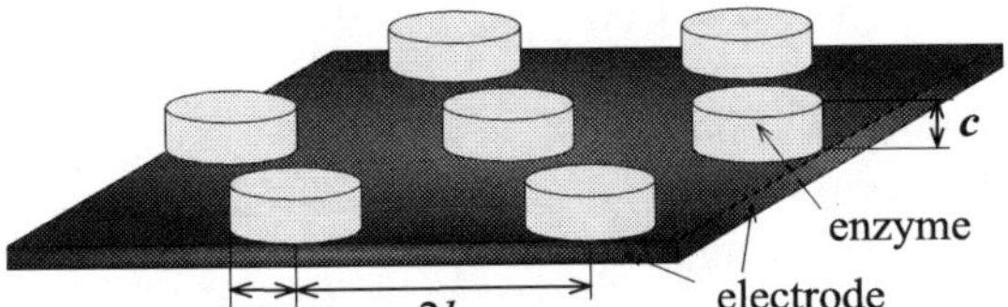

Fig. 7 The principal structure of an array of the enzyme microreactors immobilized on a single electrode. Microreactors are modeled by cylinders. The figure is not to scale

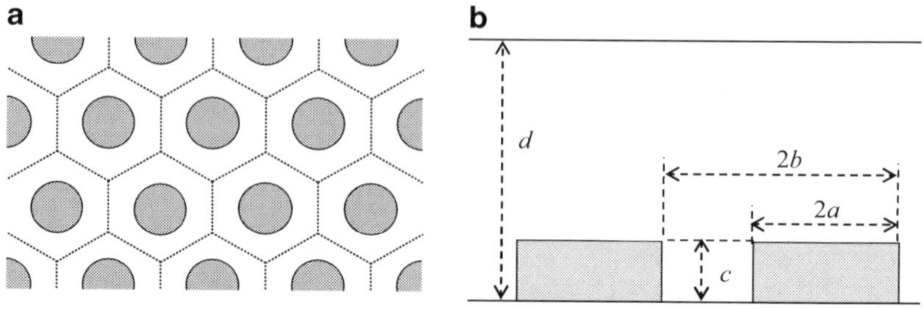

Fig. 8 The principal structure of the enzyme electrode (**a**) and the profile at the perpendicular plane (**b**)

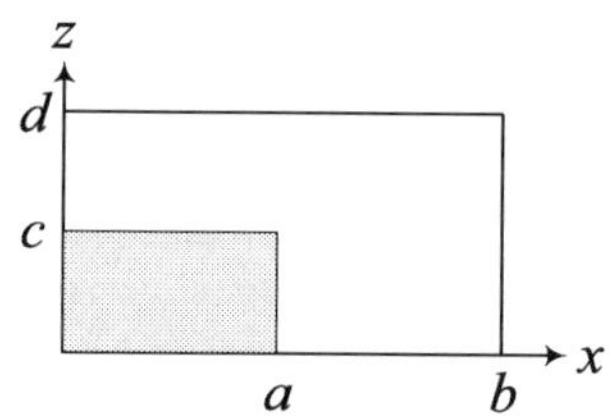

Fig. 9 The modeling domain as the profile of a unit cell at the perpendicular to the electrode plane

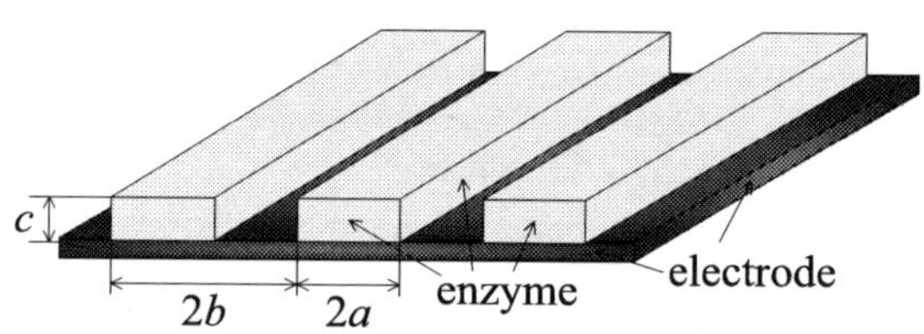

Fig. 10 The principal structure of an array of enzyme microreactors immobilized on a single electrode. The microreactors are modeled by strips. The figure is not to scale

Assuming the uniform distribution of the enzyme microreactors on the electrode surface, the biosensor may be divided into equal hexagonal prisms with regular hexagonal bases. For simplicity, it is reasonable to consider a circle of radius b whose area equals that of the hexagon and to regard one of the cylinders as a unit cell. Due to the symmetry of the unit cell, only a half of the transverse section of the unit cell can be considered. The profile of the biosensor at the perpendicular to the electrode plane is depicted in Fig. 9. A very similar approach has been used in modeling of partially blocked electrodes [23,75,97,272] and in modeling of surface roughness of the enzyme membrane [28].

In the second case of the biosensor geometry, the microreactors were modeled by identical strips distributed uniformly on the electrode surface. Figure 10 shows the principal structure of a biosensor, where enzyme microreactors are right quadrangular prisms of base $2a$ by c distributed uniformly so, that the distance between adjacent prisms equals $2(b - a)$. Due to the uniform distribution of the enzyme strips, it is reasonable to consider only a unit consisting of a single strip together with region between two adjacent strips. Because of the symmetry and the relatively great length of the unit cell only the transverse section of a half of the unit can be considered. Figure 9 also represents the profile of that kind of microreactors.

2.2 *Mathematical Model*

Consider a scheme where a substrate (S) is enzymatically converted to a product (P),

$$S \xrightarrow{E} P \tag{31}$$

Two different kinds of the geometry of the enzyme microreactors, the cylinders and the strips, have been discussed. Although the profile at y-plane (Fig. 8) is the same for both kinds of the microreactors, the corresponding mathematical models have to be formulated differently. In the case of the cylinders (Fig. 7), a two-dimensional-in-space (2-D) model in the cylindrical coordinates is formulated, while in the second case (Fig. 10) 2-D model is formulated in the Cartesian ones.

In the profile (figure 9), parameter b stands for the half width (radius) of the entire unit cell, while a is the half width (radius) of the enzyme microreactor, c is the height of the microreactor. The fourth parameter d is the thickness of the diffusion layer. The diffusion region surrounding the microreactors is assumed as the Nernst diffusion layer [73,158,199,267]. Away from it, the solution is in motion and uniform in concentration. The thickness d of the diffusion layer remains unchanged with time.

Let Ω and Ω_0 be open regions corresponding to the entire domain to be considered and to the enzyme region, respectively, Γ_0 – the boundary between the buffer solution and the enzyme region, and Γ_N – the upper boundary of the entire cell,

$$\begin{aligned} \Omega &= (0,b) \times (0,d), \\ \Omega_0 &= (0,a) \times (0,c), \\ \Gamma_0 &= \{a\} \times [0,c] \cup [0,a] \times \{c\}, \\ \Gamma_N &= [0,b] \times \{d\}. \end{aligned} \tag{32}$$

Let $\overline{\Omega}$ and $\overline{\Omega}_0$ denote the corresponding closed regions.

2.2.1 Governing Equations

In the enzyme region, coupling the enzyme-catalyzed reaction with the two-dimensional-in-space mass transport by diffusion described by Fick's law leads to the system of the reaction–diffusion equations ($t > 0$),

$$\frac{\partial S_e}{\partial t} = D_{S_e}\Delta S_e - \frac{V_{max}S_e}{K_M + S_e},$$
$$\frac{\partial P_e}{\partial t} = D_{P_e}\Delta P_e + \frac{V_{max}S_e}{K_M + S_e}, \quad (x,z) \in \Omega_0, \tag{33}$$

where z stands for the distance from the electrode, x is the distance from the z-axis, t is time, Δ is the Laplace operator, $S_e(x,z,t)$ and $P_e(x,z,t)$ are the concentrations of the substrate and reaction product, respectively, D_{S_e} and D_{P_e} are the diffusion coefficients, V_{max} is the maximal enzymatic rate and K_M is the Michaelis constant.

The expression of the Laplace operator depends on the system of coordinates. Let us recall that the Laplace operator Δ in two dimensions in cylindrical coordinates x and z is defined as follows [73]:

$$\Delta U = \frac{1}{x}\frac{\partial}{\partial x}\left(x\frac{\partial U}{\partial x}\right) + \frac{\partial^2 U}{\partial z^2}. \tag{34}$$

The operator Δ in the Descartes coordinates x and z is defined by

$$\Delta U = \frac{\partial^2 U}{\partial x^2} + \frac{\partial^2 U}{\partial z^2}. \tag{35}$$

Outside the enzyme region only the mass transport by diffusion of the substrate and the product takes place,

$$\frac{\partial S_b}{\partial t} = D_{S_b}\Delta S_b,$$
$$\frac{\partial P_b}{\partial t} = D_{P_b}\Delta P_b, \quad (x,z) \in \Omega \setminus \overline{\Omega}_0, \tag{36}$$

where $S_b(x,z,t)$ and $P_b(x,z,t)$ are the concentrations of the substrate and the reaction product, respectively, in the buffer solution.

2.2.2 Initial and Boundary Conditions

In the domain presented in Fig. 9, $z = 0$ represents the electrode surface. The simulation of the biosensor operation starts when some substrate appears in the bulk solution ($t = 0$),

$$S_e(x,z,0) = 0, \quad P_e(x,z,0) = 0, \quad (x,z) \in \overline{\Omega}_0,$$
$$S_b(x,z,0) = 0, \quad P_b(x,z,0) = 0, \quad (x,z) \in \overline{\Omega} \setminus (\overline{\Omega}_0 \cup \Gamma_N), \tag{37}$$
$$S_b(x,z,0) = S_0, \quad P_b(x,z,0) = 0, \quad (x,z) \in \Gamma_N,$$

where S_0 is the concentration of the substrate to be analyzed.

The following boundary conditions express the symmetry of the biosensor:

$$
\begin{aligned}
&D_{S_e}\frac{\partial S_e}{\partial x}\bigg|_{x=0} = D_{P_e}\frac{\partial P_e}{\partial x}\bigg|_{x=0} = 0, \quad z \in [0, c],\\
&D_{S_b}\frac{\partial S_b}{\partial x}\bigg|_{x=0} = D_{P_b}\frac{\partial P_b}{\partial x}\bigg|_{x=0} = 0, \quad z \in [c, d],\\
&D_{S_b}\frac{\partial S_b}{\partial x}\bigg|_{x=b} = D_{P_b}\frac{\partial P_b}{\partial x}\bigg|_{x=b} = 0, \quad z \in [0, d].
\end{aligned}
\tag{38}
$$

Assuming the amperometry, the product is an electro-active substance. The electrode potential is chosen to keep zero concentration of the product at the electrode surface. The substrate does not react at the electrode surface. These characteristics are expressed in the boundary conditions ($t > 0$) given by

$$
\begin{aligned}
&P_e(x, 0, t) = 0, \quad D_{S_e}\frac{\partial S_e}{\partial z}\bigg|_{z=0} = 0, \quad x \in [0, a],\\
&P_b(x, 0, t) = 0, \quad D_{S_b}\frac{\partial S_b}{\partial z}\bigg|_{z=0} = 0, \quad x \in [a, b].
\end{aligned}
\tag{39}
$$

According to the Nernst approach, away from the diffusion layer, the buffer solution is uniform in concentration,

$$
S_b\big|_{\Gamma_N} = S_0, \quad P_b\big|_{\Gamma_N} = 0. \tag{40}
$$

On the boundary Γ_0 the matching conditions are defined ($t > 0$)

$$
\begin{aligned}
&D_{S_e}\frac{\partial S_e}{\partial \mathrm{n}}\bigg|_{\Gamma_0} = D_{S_b}\frac{\partial S_b}{\partial \mathrm{n}}\bigg|_{\Gamma_0}, \quad S_e\big|_{\Gamma_0} = S_b\big|_{\Gamma_0},\\
&D_{P_e}\frac{\partial P_e}{\partial \mathrm{n}}\bigg|_{\Gamma_0} = D_{P_b}\frac{\partial P_b}{\partial \mathrm{n}}\bigg|_{\Gamma_0}, \quad P_e\big|_{\Gamma_0} = P_b\big|_{\Gamma_0},
\end{aligned}
\tag{41}
$$

where n stands for the normal direction.

In a very special case when $a = b$, the model (33), (36)–(41) describes an operation of a monolayer enzyme biosensor discussed in chapter 7.

2.2.3 Biosensor Response

The generated current of the amperometric biosensor is proportional to the gradient of the product concentration at the electrode surface, i.e. on the border $z = 0$. Due to the direct proportionality, the total current is normalized with the area of the electrode surface. In the case of the modeling enzyme reactors by right cylinders, the current density $i(t)$ has to be calculated from the model formulated in the cylindrical coordinates,

$$i(t) = \frac{n_e F}{\pi b^2} \int_0^{2\pi} \left(\int_0^a D_{P_e} \frac{\partial P_e}{\partial z}\bigg|_{z=0} x \, \mathrm{d}x + \int_a^b D_{P_b} \frac{\partial P_b}{\partial z}\bigg|_{z=0} x \, \mathrm{d}x \right) \mathrm{d}\varphi$$

$$= \frac{2n_e F}{b^2} \left(D_{P_e} \int_0^a \frac{\partial P_e}{\partial z}\bigg|_{z=0} x \, \mathrm{d}x + D_{P_b} \int_a^b \frac{\partial P_b}{\partial z}\bigg|_{z=0} x \, \mathrm{d}x \right), \tag{42}$$

where n_e is the number of electrons involved in a charge transfer, F is the Faraday constant, and φ is the third cylindrical coordinate.

In the case of the Cartesian coordinates, the density $i(t)$ of the biosensor current is expressed as follows:

$$i(t) = \frac{n_e F}{b} \left(\int_0^a D_{P_e} \frac{\partial P_e}{\partial z}\bigg|_{z=0} \mathrm{d}x + \int_a^b D_{P_b} \frac{\partial P_b}{\partial z}\bigg|_{z=0} \mathrm{d}x \right). \tag{43}$$

The density I of the steady state biosensor current is calculated alike for both kinds of the reactor geometry,

$$I = \lim_{t \to \infty} i(t). \tag{44}$$

2.3 Numerical Simulation

The initial boundary value problem (33), (36), (37)–(41) can be successfully solved using the finite difference technique [33, 36].

In the case of $a = b$, this model describes the operation of a mono-layer enzyme biosensor. This feature of the model can be applied for validation of the mathematical and the numerical models. Accepting $a = b$, the solution of the problem (33), (36), (37)–(41) must coincide with the corresponding analytical solution known for the two-compartment model of the amperometric biosensor discussed in Sect. 2 of third chapter of Part II. Being invariant to both considered systems (the cylindrical and the Descartes) of coordinates is another useful feature of mathematical model (33), (36), (37) - (41) which could be used when evaluating the correctness of the corresponding numerical model.

The upper layer of the thickness $\delta = d - c$ from the enzyme region was assumed as the Nernst diffusion layer. The thickness δ of the Nernst layer depends upon the nature and stirring of the buffer solution (see Sect. 2.6 of third chapter of Part II). In practice, the zero thickness of the Nernst layer can not be achieved. In the case when the buffer solution to be analyzed is stirred by magnetic stirrer, the thickness δ may be minimized up to 20 μm by increasing stirring intensity [134, 267]. That thickness of the Nernst layer, $\delta = d - c = 20\,\mu\text{m}$, was used in the simulation of the biosensor action changing some other parameters,

$$\begin{aligned} &D_{S_e} = D_{P_e} = 300\,\mu\text{m}^2/\text{s}, \quad D_{S_b} = D_{P_b} = 2D_{S_e} = 2D_{P_e}, \\ &a = 40\,\mu\text{m}, \quad b = 100\,\mu\text{m}, \quad c = 100\,\mu\text{m}, \quad d = 120\,\mu\text{m}, \\ &K_M = 100\,\mu\text{M}, \quad V_{max} = 100\,\mu\text{M/s}, \quad S_0 = 100\,\mu\text{M}, \quad n_e = 2. \end{aligned} \tag{45}$$

Some results of the numerical simulations are depicted in Figs. 11 and 12. The enzyme microreactors were modeled by cylinders (Figs. 11a and 12) and by strips (Fig. 11b).

One can see in Fig. 11 that the radius (half-width) a of the enzyme microreactors significantly affects the density i of the biosensor current as well as the response time. The biosensor current was simulated at the following five values

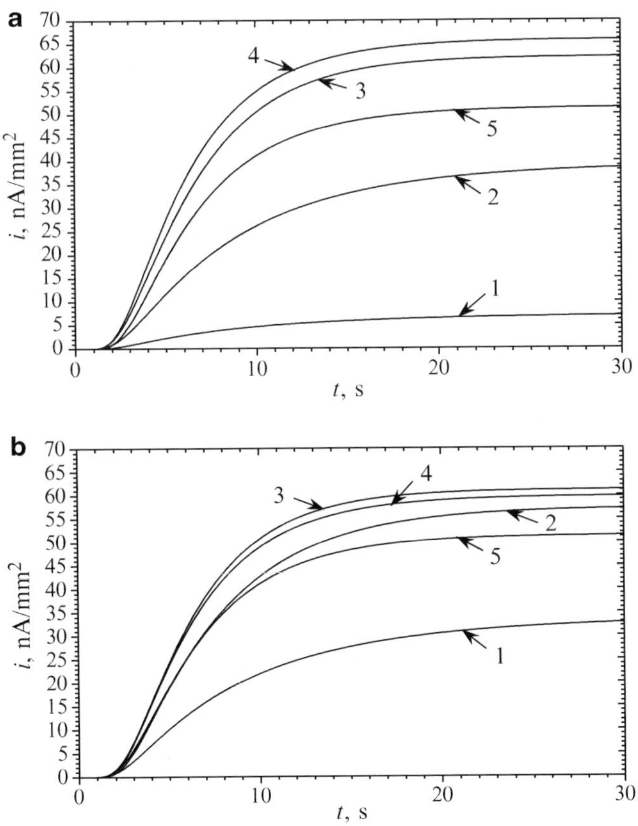

Fig. 11 The dynamics of the biosensor current at different values of the half-width a of the microreactors in the cases of cylinders (**a**) and strips (**b**), $a = 0.2b$ (1), $0.4b$ (2), $0.6b$ (3), $0.8b$ (4), b (5). The values of all the other parameters are defined in (45)

of a: 20 μm (curve 1 in Fig. 11), 40 (2), 60 (3), 80 (4) and 100 (5). The case of $a = 100\,\mu\text{m} = b$ corresponds to a flat biosensor with continuous enzyme membrane of thickness c. The calculated biosensor current notably differs for different shapes of the microreactors unchanging values of all the model parameters. For example, at $a = 20\,\mu\text{m} = 0.2b$ (curves 1 in Fig. 11), the density of the steady state current ($I = 7.1\,\text{nA}/\text{mm}^2$ at $t = 41.1\,\text{s}$, Fig. 11a) in the case of the cylinders is about 4.7 times less than the corresponding current calculated in the case of strips ($I = 33.3\,\text{nA}/\text{mm}^2$ at $t = 38.3\,\text{s}$, Fig. 11b). Figure 11 also shows a non-monotonic dependence of the stationary current on the base area of microreactors. The next section discusses this effect in detail.

Figure 12 shows the concentrations of substrate (S) and the product (P) in the enzyme and diffusion regions at time $t = 34.8\,\text{s}$ when the steady state was reached. The enzyme reactors were modeled by right cylinders. The concentration S of the

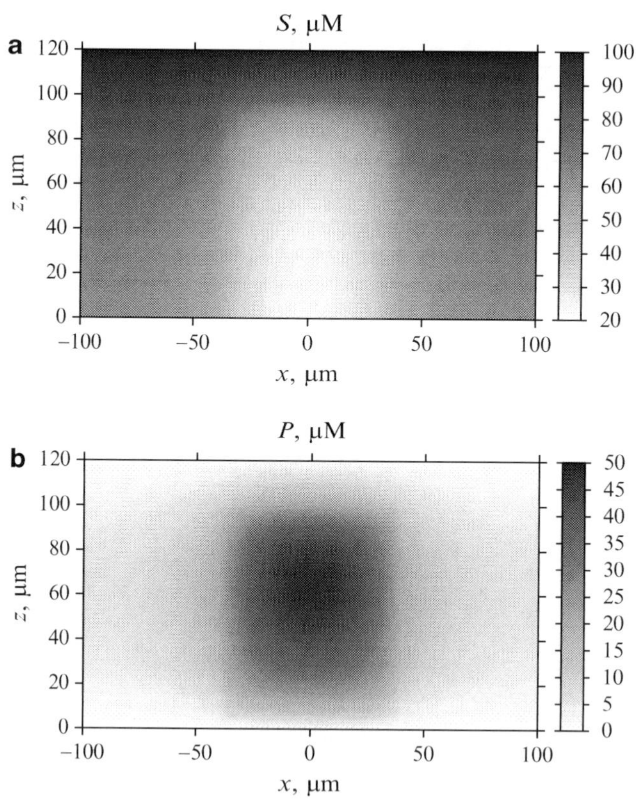

Fig. 12 The stationary concentrations of the substrate (**a**) and the product (**b**) in the enzyme and the surrounding regions, obtained at $t = 34.8\,\text{s}$. The values of the model parameters are defined in (45)

substrate and the concentration P of the reaction product for the entire domain $\overline{\Omega}$ were introduced as follows ($t \geq 0$):

$$S(x,z,t) = \begin{cases} S_e(x,z,t), & (x,z) \in \overline{\Omega}_0, \\ S_b(x,z,t), & (x,z) \in \overline{\Omega} \setminus \overline{\Omega}_0, \end{cases}$$
$$P(x,z,t) = \begin{cases} P_e(x,z,t), & (x,z) \in \overline{\Omega}_0, \\ P_b(x,z,t), & (x,z) \in \overline{\Omega} \setminus \overline{\Omega}_0. \end{cases} \tag{46}$$

At any time, $t \geq 0$, both concentration functions, $S(x,z,t)$ and $P(x,z,t)$ are continuous at all $(x,z) \in \overline{\Omega}$.

Although the dynamics of the concentrations was simulated in the spherical coordinates, Fig. 12 presents the profiles in the Descartes coordinates. To have a more comprehensive view, the mirror-image along the z-axis is also shown in Fig. 12 where the point $(0,0)$ corresponds to the center of the base of an enzyme cylinder.

2.4 Effect of the Electrode Coverage with Enzyme

Figure 11 shows a non-monotonic dependence of the stationary current on the radius (half width) of the base of microreactors. To investigate that effect in detail the density I of the steady state current has to be calculated at different values of the radius (half-width) b of the entire unit changing the radius (half-width) a of the microreactor. To compare the response of the biosensor based on an array of the microreactors ($a < b$) with the response of the corresponding flat (membrane) biosensor ($a = b$), the normalized steady state current is introduced as a function of the degree θ of the electrode surface coverage,

$$I_\theta(b,\theta) = \frac{I(b,\theta)}{I(b,1)}, \quad 0 < \theta \leq 1, \tag{47}$$

where $I_\theta(b,\theta)$ is the density of the steady state biosensor current calculated as defined in (44) at the radius (half-width) b and the degree θ of the electrode coverage, $0 < \theta \leq 1$. The dimensionless degree θ was expressed as the area of the bases of all enzyme microreactors divided by the area of the whole electrode surface. The case of $\theta = 1$ corresponds to a flat biosensor, i.e. an electrode fully covered by an enzyme mono-layer. In the case when microreactors are modeled by cylinders (Fig. 7), $\theta = a^2/b^2$, and $\theta = a/b$ in the case of strips (Fig. 10).

Figure 13 shows the normalized steady state current I_θ (Fig. 13a) and the biosensor sensitivity B_S (Fig. 13b) versus the degree θ of the electrode coverage at four values of b. The values of all other parameters were the same as defined in (45). Since in the case of strips, the behaviour of the stationary current versus the degree θ is very similar to that in the case of cylinders, Fig. 13 shows the calculation results

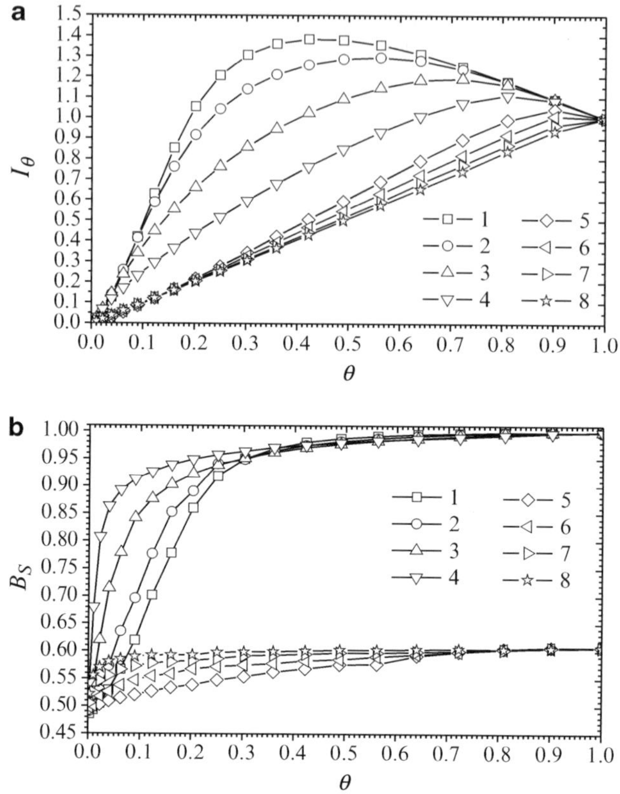

Fig. 13 The normalized steady state current I_θ (**a**) and the sensitivity B_S (**b**) versus the electrode coverage θ, $b = 50$ (1, 5), 100 (2, 6), 200 (3, 7), 400 (4, 8) μm, $c = 100$ (1–4), 10 (5–8) μm. The values of the other parameters are defined in (45)

obtained only for the microreactors modeled by right cylinders. The calculations have been performed at four values of b (50 (curves 1 and 5), 100 (2, 6), 200 (3, 7), 400 (4, 8) μm) and two values of c (100 (1–4), 10 (5–8) μm) changing the radius a from $0.05b$ up to b. The electrode coverage θ changes from 0.0025 to 1 when a increases from $0.05b$ to b.

As one can see in Fig. 13a, in the case of $c = 100$ μm (curves 1–4), the steady state current is a non-monotonous function of the degree θ of the electrode coverage. At 10 times smaller value of c (curves 5–8) the normalized current I_θ is approximately a linearly increasing function of θ. Additional calculations confirmed this property also for the Descartes coordinates [36]. In the case of $c = 100$ μm and $b = 0.5c$ (curve 1 in Fig. 13a) the relative difference between the steady state current at $\theta = 0.5$ and another one at $\theta = 1$ exceeds 38%. Let us notice that the volume of enzyme microreactors is directly proportional to the degree θ, when the height c of the microreactors is kept constant. Although, the biosensor based on an array of microreactors ($\theta < 1$) is of less enzyme volume than the corresponding flat biosensor ($\theta = 1$), the array biosensor can generate even higher steady state current than the flat one.

The variation of the half-width b of the entire unit keeping θ and c constant does not change the volume of the microreactors. Since I_θ varies with b (Fig. 13a), the biosensor response depends also on the shape of the enzyme microreactors but not only on their volume. The smaller value of b corresponds to the denser distribution of the enzyme microreactors on the electrode surface. According to Fig. 13a, the denser the microreactors are distributed the higher steady state current is generated.

Assuming the continuous enzyme layer ($a = b$, $\theta = 1$) of the thickness $c = 100\,\mu\text{m}$, the diffusion module σ^2 at values (45) is equal approximately to 33.3, i.e. the biosensor response is under diffusion control. While at ten times smaller value (0.01 mm) of c, σ^2 is equal approximately to 0.33 and consequently the enzyme kinetics controls the biosensor response. The steady state biosensor current is a notably non-monotonous function of the degree θ of the electrode coverage when $\sigma^2 > 1$, i.e. when the biosensor response is significantly under diffusion control. Additional calculations confirmed this property. When comparing curves 1–4 with the corresponding curves 5–8, one can see that the biosensor sensitivity is notable higher when the response is under diffusion control (curves 1–4) rather than under the enzyme kinetics (curves 5–8).

Figure 13b shows that a moderate decrease in the electrode coverage θ practically does not change the biosensor sensitivity B_S. In the case of $c = 100\,\mu\text{m}$, the sensitivity B_S decreases from only 1 down to 0.98 changing the coverage θ from 1 down to 0.5 (curves 1–4), i.e. reducing the volume of the enzyme two times. Although at $c = 10\,\mu\text{m}$ the overall sensitivity is significantly less than at $c = 100\,\mu\text{m}$, the shape of curves B_S is very similar in both these cases. Selecting the geometry of the microreactors one can minimize the volume of an enzyme practically without loosing the sensitivity.

2.5 Concluding Remarks

In the case when the biosensor response is under the diffusion control, a biosensor based on an array of microreactors is able to generate a greater steady state current than a corresponding flat mono-layer biosensor the thickness of which being the same as the height of the microreactors (Fig. 13a). The denser the microreactors are distributed on the electrode surface the higher steady state current is generated. This feature of array biosensors can be applied to design novel highly sensitive biosensors when the minimization of the enzyme volume is of crucial importance. Selecting the geometry of microreactors one can minimize the volume of enzyme without loosing the sensitivity (Fig. 13b).

3 Plate-Gap Biosensor

A working electrode is often considered as a complex device consisting of the conducting electrode (metal, carbon, or carbon paste) coated with a biochemical film [85,90,100,229]. Such definitions suggest a planar structure of a working electrode widely investigated in mathematical models of biosensors [99, 235]. These approaches to the electrodes omit a specific but widely used in practice class of non-planar biosensors that are based on a bulk modification of entire electrode material, e.g. enzyme modified porous carbon electrodes [96]. The dimensionality of such electrodes is seldom taken into account [3, 22, 37, 117, 118, 240].

This section considers computational modeling of an enzyme modified screen printed porous electrode. The porous electrode based biosensors seem particularly promising for the detection of glucose, galactose, ethanol, phenol and some other substrates [89, 153, 241]. Carbon paste porous electrodes have been theoretically investigated by applying a plate-gap model [114, 118]. The ultimate goal of the modeling of plate-gap biosensors was verification of the proposed model by comparison of theoretical and experimental responses for biosensors based on the PQQ-dependent glucose dehydrogenases [118, 154].

3.1 Principal Structure of Biosensor

The pores of the carbon paste electrode are assumed to be the enzyme deposited. The electrode is covered with an inert porous membrane [153, 154]. The enzyme activity is gradually dispersed in the volume of porous electrode, and the distances between the enzymatic reaction sites and the conducting walls of porous electrode are as short as an average radius of pores. According to this physical model, the enzyme activity is uniformly dispersed in the gap between two parallel conducting plates. The modeled physical system, in general, mimics the main features of the porous electrode. Firstly, the uniform dispersion of the enzyme activity is affirmed according to the definition of the modeled physical system. Secondly, the gap width dependent characteristic distances between the enzymatic reaction sites and the conducting plates of the modeled system can be admitted to be similar to the average radius of pores in the porous electrode. In addition, the substrate or product molecules in the modeled plate gap electrode may diffuse distantly in the directions, which are parallel to the surface of the electrode, i.e. as it is in the three dimensional network of a porous electrode. Figure 14 shows the principal structure of a biosensor, where enzyme-filled gaps are modeled by right quadrangular prisms distributed uniformly.

The thickness of the outer membrane as well as the depth of the gaps of the electrode is assumed to be much less than its length. The porous membrane is assumed to be of a uniform thickness. Because of the symmetry and the relatively great length of the gaps only the transverse section of the biosensor can be considered. Figure 15a shows the profile of a biosensor where the rectangular section

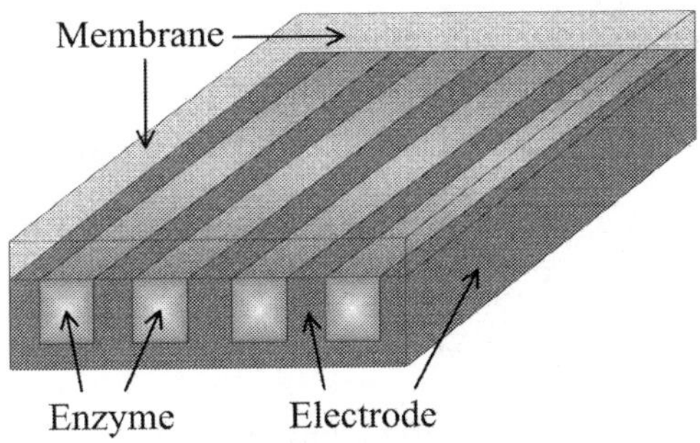

Fig. 14 The principal structure of a plate-gap biosensor. The figure is not to scale

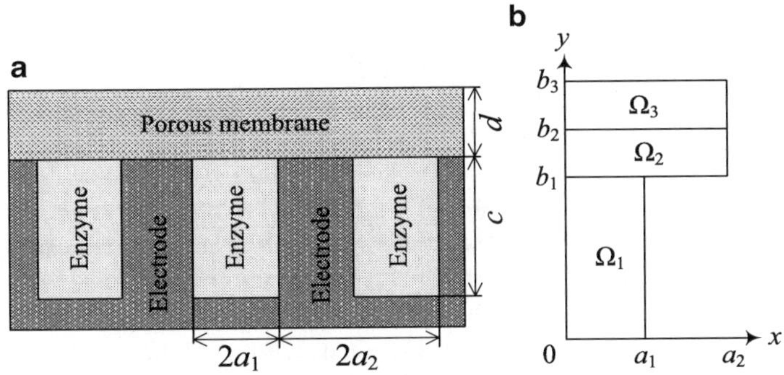

Fig. 15 The profile (**a**) and the unit cell (**b**) of a plate-gap biosensor

of the enzyme filled gaps is $2a_1$ by c, the distance between adjacent prisms equals $2(a_2 - a_1)$, a_1 is the half width of the gaps, c is the gap depth, and d is the thickness of the outer porous membrane.

Due to the uniform distribution of the gaps, it stands to reason to consider only a unit consisting of a single gap together with the region between two adjacent gaps. Figure 15b shows the profile of a unit cell to be considered in mathematical modeling of the biochemical behaviour of the plate-gap biosensor represented schematically in Fig. 14.

In Fig. 15b, Ω_1 represents the enzyme-filled gaps, Ω_2 corresponds to the porous membrane, and Ω_3 stands for the external diffusion layer, $c = b_1$ is the depth of the gaps, $d = b_2 - b_1$ is the thickness of the porous membrane and $\delta = b_3 - b_2$ is the thickness of the external diffusion layer. A very similar approach has been used in modeling of partially blocked electrodes [23, 75].

3.2 Mathematical Model

The mathematical model of a plate-gap biosensor with an outer porous membrane (Fig. 15) can be formulated in a two dimensional domain consisting mainly of three regions: the enzyme region Ω_1, the region Ω_2 corresponding to the porous membrane, and the region Ω_3 of the external diffusion,

$$\begin{aligned}\Omega_1 &= (0, a_1) \times (0, b_1),\\ \Omega_2 &= (0, a_2) \times (b_1, b_2),\\ \Omega_3 &= (0, a_2) \times (b_2, b_3).\end{aligned} \tag{48}$$

Let $\overline{\Omega}_i$ denote the closed region corresponding to Ω_i, $i = 1, 2, 3$.

In the enzyme region Ω_1 the mass transport by diffusion and the enzyme-catalyzed reaction are considered,

$$\mathrm{S} \xrightarrow{\mathrm{E}} \mathrm{P} \tag{49}$$

where a substrate (S) is enzymatically converted to a product (P),

Assuming the outer porous membrane as the periodic media, the homogenization process has been applied to the domain Ω_2 corresponding of the membrane [18]. After this, the porous membrane was modeled as a diffusion layer with an effective diffusion coefficient. Thus, in the region Ω_2 of the outer membrane only the mass transport by diffusion takes place. In the homogeneous external region Ω_3 also only the mass transport by diffusion of the substrate as well as of the product takes place. According to the Nernst approach, the thickness $\delta = b_3 - b_2$ of the diffusion layer was assumed to be constant during the biosensor action.

3.3 Governing Equations

The action of the biosensor, presented schematically in Figs. 14 and 15 can be described by the following system of reaction–diffusion equations ($t > 0$) [37, 117, 235]:

$$\begin{aligned}\frac{\partial S_1}{\partial t} &= D_{S_1} \Delta S_1 - \frac{V_{max} S_1}{K_M + S_1},\\ \frac{\partial P_1}{\partial t} &= D_{P_1} \Delta P_1 + \frac{V_{max} S_1}{K_M + S_1}, \quad (x, y) \in \Omega_1,\end{aligned} \tag{50}$$

$$\begin{aligned}\frac{\partial S_j}{\partial t} &= D_{S_j} \Delta S_j,\\ \frac{\partial P_j}{\partial t} &= D_{P_j} \Delta P_j, \quad (x, y) \in \Omega_j, \quad j = 2, 3,\end{aligned} \tag{51}$$

where Δ is the Laplacian, $S_i(x, y, t)$ is the concentration of the substrate in the region Ω_i, $P_i(x, y, t)$ is the concentration of the reaction product in Ω_i, $i = 1, 2, 3$, V_{max} is the maximal enzymatic rate, and K_M is the Michaelis constant.

3.4 Initial Conditions

Let Γ_1 be the electrode surface, and Γ_2 – the porous membrane/bulk solution boundary,

$$\begin{aligned} \Gamma_1 &= ([0, a_1] \times \{0\}) \cup (\{a_1\} \times [0, b_1]) \cup ([a_1, a_2] \times \{b_1\}), \\ \Gamma_2 &= [0, a_2] \times \{b_3\}. \end{aligned} \tag{52}$$

The biosensor operation starts when the substrate of concentration S_0 appears in the bulk solution. This is used in the initial conditions ($t = 0$)

$$\begin{aligned} &S_k(x, y, 0) = 0, \quad (x, y) \in \overline{\Omega}_k, \quad k = 1, 2, \\ &S_3(x, y, 0) = 0, \quad (x, y) \in \overline{\Omega}_3 \setminus \Gamma_2, \\ &S_3(x, y, 0) = S_0, \ (x, y) \in \Gamma_2, \\ &P_i(x, y, 0) = 0, \quad (x, y) \in \overline{\Omega}_i, \quad i = 1, 2, 3. \end{aligned} \tag{53}$$

3.5 Boundary and Matching Conditions

Assuming $b_0 = 0$, the following boundary conditions express the symmetry of the biosensor ($t > 0$):

$$D_{P_i} \frac{\partial P_i}{\partial x}\bigg|_{x=0} = D_{S_i} \frac{\partial S_i}{\partial x}\bigg|_{x=0} = 0, \quad y \in [b_{i-1}, b_i], \quad i = 1, 2, 3, \tag{54}$$

$$D_{P_j} \frac{\partial P_j}{\partial x}\bigg|_{x=a_2} = D_{S_j} \frac{\partial S_j}{\partial x}\bigg|_{x=a_2} = 0, \quad y \in [b_{j-1}, b_j], \quad j = 2, 3. \tag{55}$$

The substrate is an electro-inactive substance, while the product is an electro-active substance. The electrode potential is chosen to keep zero concentration of the product at the electrode surface ($t > 0$),

$$D_{S_k} \frac{\partial S_k}{\partial n}\bigg|_{\Gamma_1} = 0, \quad P_k = 0, \quad (x, y) \in \Gamma_1, \quad k = 1, 2, \tag{56}$$

where n stands for the normal direction.

Assuming the bulk solution to be well-stirred and in a powerful motion, the diffusion layer ($b_2 < y < b_3$) may be treated as the Nernst diffusion layer [73, 158, 199,

267]. According to the Nernst approach, a layer of thickness $\delta = b_3 - b_2$ remains unchanged with time. Away from it, the bulk solution is in motion and is uniform in concentration ($t > 0$),

$$\begin{aligned} S_3(x, b_3, t) &= S_0, \\ P_3(x, b_3, t) &= 0, \quad x \in [0, a_2]. \end{aligned} \tag{57}$$

On the boundary between adjacent regions Ω_k and Ω_{k+1} the matching conditions are defined ($t > 0$),

$$\begin{aligned} &D_{S_k} \left. \frac{\partial S_k}{\partial y} \right|_{y=b_k} = D_{S_{k+1}} \left. \frac{\partial S_{k+1}}{\partial y} \right|_{y=b_k}, \\ &S_k(x, b_k, t) = S_{k+1}(x, b_k, t), \quad (x, y) \in \overline{\Omega}_k \cap \overline{\Omega}_{k+1}, \qquad k = 1, 2. \end{aligned} \tag{58}$$

$$\begin{aligned} &D_{P_k} \left. \frac{\partial P_k}{\partial y} \right|_{y=b_k} = D_{P_{k+1}} \left. \frac{\partial P_{k+1}}{\partial y} \right|_{y=b_k}, \\ &P_k(x, b_k, t) = P_{k+1}(x, b_k, t), \quad (x, y) \in \overline{\Omega}_k \cap \overline{\Omega}_{k+1}, \qquad k = 1, 2. \end{aligned} \tag{59}$$

The governing equations (50), (51) together with the initial (53), the boundary (54)–(57) and the matching (58), (59) conditions form together a initial boundary value problem.

3.6 *Biosensor Response*

The measured current is accepted as a response of a biosensor in an actual experiment. The current depends upon the flux of the reaction product at the electrode surface, i.e. on the border Γ_1. The density $i(t)$ of the current at time t can be obtained explicitly from the Faraday and the Fick laws

$$\begin{aligned} i(t) = \frac{n_e F}{a_2} \Bigg(& D_{P_1} \int_0^{a_1} \left. \frac{\partial P_1}{\partial y} \right|_{y=0} \mathrm{d}x + D_{P_1} \int_0^{b_1} \left. \frac{\partial P_1}{\partial x} \right|_{x=a_1} \mathrm{d}y \\ & + D_{P_2} \int_{a_1}^{a_2} \left. \frac{\partial P_2}{\partial y} \right|_{y=b_1} \mathrm{d}x \Bigg), \end{aligned} \tag{60}$$

where n_e is the number of electrons involved in a charge transfer and F is the Faraday constant.

The system (50)–(59) approaches a steady state when $t \to \infty$,

$$I = \lim_{t \to \infty} i(t), \tag{61}$$

where I is the density of the steady state current of the plate-gap biosensor.

3.7 Numerical Simulation

The finite difference technique can be successfully applied to solve numerically the initial boundary value problem (50), (51), (53)–(59) [37, 114, 117, 118].

To find an efficient numerical solution of the problem a bilinear discrete grid in all directions: x, y and t was introduced [26, 37]. Using alternating direction method a semi-implicit linear finite difference scheme has been built as a result of the difference approximation [222]. The resulting system of linear algebraic equations was solved rather efficiently because of the tridiagonality of the matrix of the system.

Due to high gradients of the concentrations of both species, the substrate and the product, an accurate and stable numerical solution was achieved only at a very small step size in y direction at the boundaries $y = 0$ and $y = b_3$. Because of the concavity of an angle at point (a_1, b_1) it was necessary to use a very small step size in both space directions, x and y, also at the boundaries $x = a_1$, $y = b_1$. Due to the matching conditions between the adjacent regions with different diffusivities, a small step size near the boundary $y = b_2$ was also used. In the direction x, an exponentially increasing step size was used to both sides from a_1: to a_2 and down to 0. In the direction y, an exponentially increasing step size was used from 0 to $b_1/2$, from b_3 down to $(b_2 + b_3)/2$, from b_j down to $(b_j + b_{j-1})/2$ and from b_j to $(b_j + b_{j+1})/2$, $j = 1, 2$, where $b_0 = 0$.

The step size in the direction of time was restricted due to the nonlinear reaction term in (50), boundary conditions and the geometry of the domain. In order to achieve an accurate and stable solution of the problem, at the beginning of the reaction–diffusion process the restrictive condition was required. Since the biosensor action obeys the steady state assumption when $t \to \infty$, it was reasonable to apply an increasing step size in the time direction. The final step size was in a few orders of magnitude higher than the first one. The numerical simulator has been programmed in JAVA language [191].

Assuming $a_1 = a_2 \gg b_1$ and the zero thickness either of the porous membrane ($d = 0$, $b_2 = b_1$) or of the external diffusion layer ($\delta = 0$, $b_3 = b_2$), the mathematical model (50), (51), (53)–(59) approaches the two-compartment model [235] (see Sect. 2 of third chapter of Part II). At relatively low ($S_0 \ll K_M$) as well as at very high ($S_0 \gg K_M$) concentrations of the substrate, the two-compartment mathematical model can be solved analytically [235].

The adequacy of the mathematical model (50), (51), (53)–(59) of the plate-gap biosensor as well as of the numerical solution of the problem was evaluated using well-known analytical solutions for the two-compartment model [235]. Accepting $a_1 = a_2 = 20b_1$, the density of the steady state biosensor current was calculated at different values of the model parameters: the maximal enzymatic rate V_{max}, the substrate concentration S_0 ($S_0 \ll K_M$ as well as $S_0 \gg K_M$), the gap depth $c = b_1$, the thickness d of the outer membrane (accepting $\delta = 0$) and the thickness δ of the external diffusion layer (accepting $d = 0$). In all these cases the relative difference between the numerical and analytical solutions was less than 1%.

The upper layer of the thickness $\delta = b_3 - b_2$ was assumed to be the Nernst diffusion layer. Assuming the intensively stirred buffer solution, the thickness δ of $2\,\mu$m was used in the numerical simulation of the biosensor action changing some other parameters. The following values of the model parameters were constant in all the numerical experiments discussed below:

$$\begin{aligned} &D_{S_1} = D_{P_1} = 300\,\mu\text{m}^2/\text{s}, \quad D_{S_2} = D_{P_2} = 2D_{S_1} = 2D_{P_1}, \\ &K_M = 100\,\mu\text{M}, \quad V_{max} = 100\,\mu\text{M/s}, \quad S_0 = 100\,\mu\text{M}, \quad n_e = 2. \end{aligned} \tag{62}$$

Some results of the numerical simulation are depicted in Figs. 16 and 17.

One can see in Fig. 16 that the biosensor current is very sensitive to changes in the depth $c = b_1$ and in the relative width of the gaps. The threefold increase in gap depth c (from 2 to 6 μm) increases the steady state current about 2.7 times (from 6.1 to 16.4 nA/mm^2, curves 1 and 4). At $a_2 = 1.5\,\mu$m, the increase in b_1 from 2 upto 6 μm increases the total area of the electrode surface about 2.14 times, $(1.5 + 6)/(1.5 + 2) \approx 2.14$.

As it is possible to notices in Fig. 16, the two pairs of curves (2, 3) and (5, 6) are very close to each other. When comparing the parameters of the simulation corresponding to curves 2 and 3, one can see the significant (two times) difference in absolute width a_1. However, the relative width a_1/a_2 of the gaps is the same. A very similar situation is observed when comparing another pair (5 and 6) of curves. Probably, the biosensor response depends mainly on the relative width a_1/a_2 of the gaps, and the biosensor current is practically invariant to the absolute width a_1. The next section discusses this effect in greater detail.

Figure 17 shows the concentrations of the substrate (S) and the product (P) in the enzyme and the diffusion regions at time $t = 3.2$ s when the steady state was reached. The concentration S of the substrate and the concentration P of the reaction product in the entire modeling domain were introduced as follows ($t \geq 0$):

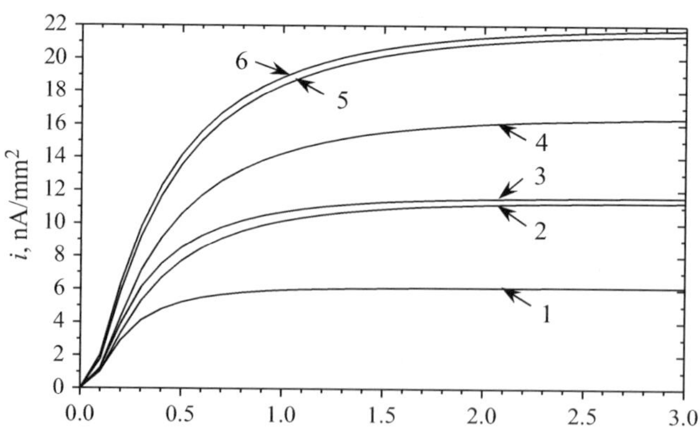

Fig. 16 The dynamics of the density i of the biosensor current, a_1: 0.5 (1, 3, 4), 1 (2, 6), 2 (5) a_2: 1.5 (1, 3, 4, 6) , 3.0 (2, 5) b_1: 2 (1), 4 (2, 3, 5, 6), 6 (4), $b_2 = b_1 + 2$, $b_3 = b_2 + 2$ (μm). The values of all other parameters are defined in (62)

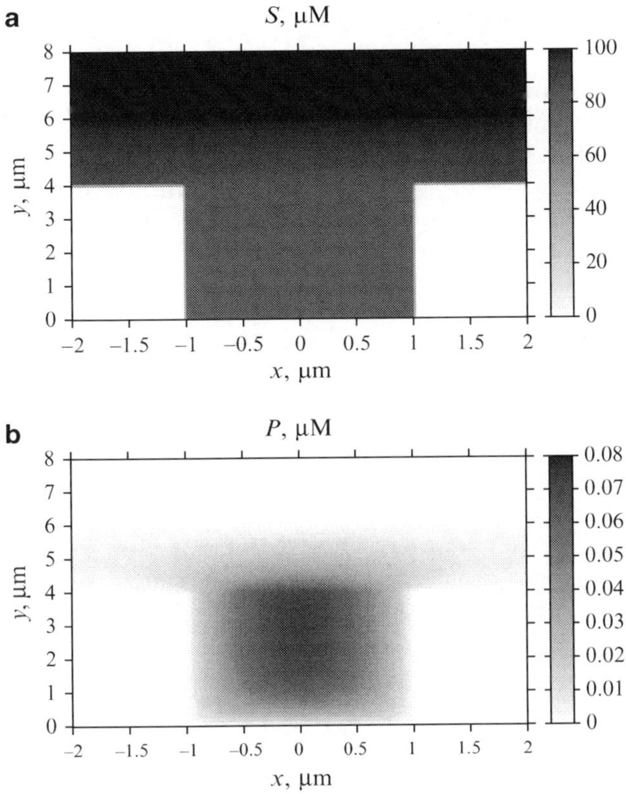

Fig. 17 The steady state concentrations of the substrate (**a**) and the product (**b**) in the gaps, the outer membrane and the surrounding regions obtained at $t = 3.2$ s, $a_1 = 1$, $a_2 = a_1 + 1$, $b_1 = 2$, $b_2 = b_1 + 2$, $b_3 = b_2 + 2$ (μm). The values of the model parameters are defined in (62)

$$
\begin{aligned}
S(x,y,t) &= \begin{cases} S_1(x,y,t), & (x,y) \in \overline{\Omega}_1, \\ S_2(x,y,t), & (x,y) \in \overline{\Omega}_2 \setminus \overline{\Omega}_1, \\ S_3(x,y,t), & (x,y) \in \overline{\Omega}_3 \setminus \overline{\Omega}_2, \end{cases} \\
P(x,y,t) &= \begin{cases} P_1(x,y,t), & (x,y) \in \overline{\Omega}_1, \\ P_2(x,y,t), & (x,y) \in \overline{\Omega}_2 \setminus \overline{\Omega}_1, \\ P_3(x,y,t), & (x,y) \in \overline{\Omega}_3 \setminus \overline{\Omega}_2. \end{cases}
\end{aligned}
\tag{63}
$$

At any time $t \geq 0$, both concentration functions, $S(x,y,t)$ and $P(x,y,t)$, are continuous at all $(x,z) \in \overline{\Omega}_1 \cup \overline{\Omega}_2 \cup \overline{\Omega}_3$.

To have a more comprehensive view, the mirror-image along the y-axis is also shown in Fig. 17 where the point $(0,0)$ corresponds to the center of the base of the enzyme-filled gap.

3.8 Effect of the Gaps Geometry

To investigate the effect of the geometry of gaps on the biosensor response the density of the steady state current has to be calculated at different values of the gap depth c and the half-width a_2 of the cell changing the half-width a_1 of the gaps. To compare the response of a plate-gap biosensor with the response of the corresponding flat (membrane) biosensor, the normalized dimensionless steady state current was defined as a function of the degree ρ of the rimosity of the plate-gap electrode,

$$I_\rho(\rho) = \frac{I(\rho)}{I_F}, \quad \rho = \frac{a_1}{a_2}, \quad 0 < \rho < 1, \tag{64}$$

where $I_\rho(\rho)$ is the density of the steady state biosensor current calculated as defined in (61) at the relative width a_1/a_2 of gaps of the electrode. The dimensionless degree ρ was expressed as the total area of the bases of all gaps divided by the base area of the whole electrode. I_F corresponds to the density of the steady state current of the flat biosensor having the enzyme mono-layer immobilized on a flat electrode. In general, $I_\rho(1) \neq I_F$, however, $I_\rho(1) \to I_F$ when $a_1 = a_2 \to \infty$.

Figure 18 shows the dependence of the steady state current I_ρ and of the sensitivity B_S of the plate-gap biosensor on the rimosity ρ (relative width) of the plate-gap electrode at different values of the gap depth c and the half-width a_2 of the cell.

Figure 18a shows that the steady state current of the plate-gap biosensor decreases with the decrease in the rimosity ρ of the electrode. The model parameter a_2 stands for the density of the gaps distribution. As one can see in Fig. 18 there is no notable difference between the curves 1–3 obtained at a constant gap depth $c = b_1$ and different values of a_2. The biosensor response is practically invariant to the topology of the electrode gaps.

It is possible to notice in Fig. 18b that the narrowing of the gaps slightly decreases the biosensor sensitivity B_S, and thereby truncates the calibration curve of the plate-gap biosensor. This is definitely a disadvantage of the plate-gap biosensors. However, significantly less amount of the enzyme used in a plate-gap biosensor in comparison with the corresponding flat biosensor redresses this disadvantage with a vengeance. The increase in the rimosity ρ increases linearly the total volume of the enzyme used in the plate gap biosensors.

3.9 Concluding Remarks

Reducing the relative width of gaps reduces the effective steady state current (Fig. 18a) and can slightly decrease the biosensor sensitivity B_S (Fig. 18b). However, reducing the volume of the enzyme redresses this disadvantage with a vengeance.

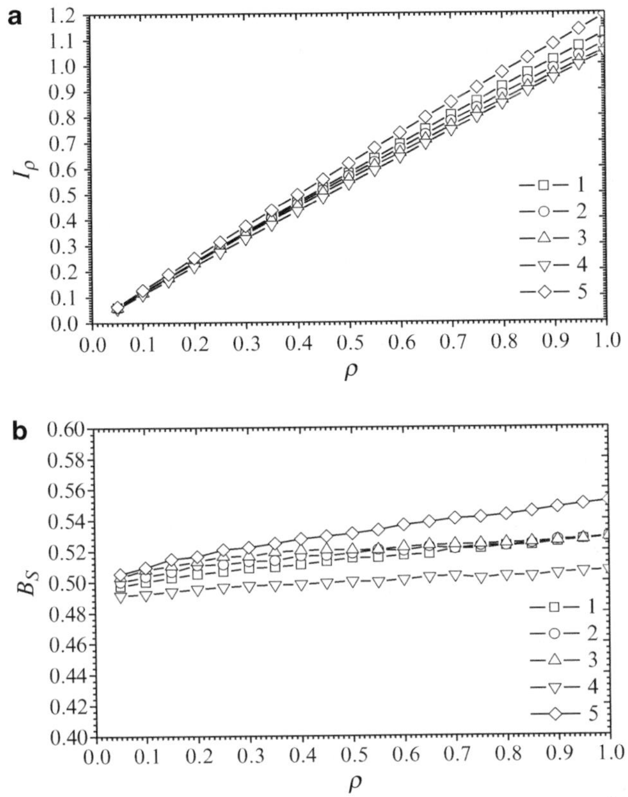

Fig. 18 The normalized steady state current I_ρ (**a**) and the sensitivity B_S (**b**) versus the rimosity ρ of the plate-gap electrode, a_2: 2 (1), 4 (2, 4, 5), 6 (3) μm, b_1: 2 (4), 4 (1–3), 6 (5) μm. The other parameters and notations are the same as in Fig. 16

4 Biosensors with Selective and Perforated Membranes

A practical biosensor usually contains a multi-layer enzyme membrane [15, 16, 47, 91, 236]. The electrode acting as a transducer of the biosensor is often covered by a selective membrane, following a layer of the immobilized enzyme and an outer membrane. The outer membrane can be prepared from a dialysis membrane or, in modern biosensors, from the perforated nucleopore type membrane [8]. The selective membrane is permeable only for selected species, usually, the product of the biochemical reaction. The selective membrane is used to increase the biosensors selectivity [50, 205, 226].

To improve the productivity and efficiency of the biosensors design as well as to optimize the biosensors configuration a model of the real biosensors should be build [3, 9, 71, 87, 247, 276]. The modeling of the biosensors with the perforated membranes has been performed by Schulmeister and Pfeiffer [236]. Their model did not take into account the geometry of the holes in the membranes and included the diffusion coefficients having a limited physical sense [236]. This section focuses on the modeling of the practical amperometric biosensor containing the selective and the perforated membranes. The perforated membrane is analyzed with a special emphasis to the geometry of the membrane perforation [34, 35, 41, 115].

4.1 Principal Structure of Biosensor

The thickness of the selective membrane as well as of the perforated membrane of the biosensor is assumed to be much less than its length and width. In the biosensor, the selective membrane is of a uniform thickness. The holes in the perforated membrane are modeled by right cylinders. The holes are of uniform diameter and spacing, forming a hexagonal pattern. Figure 19 presents the biosensor schematically.

Due to the uniform distribution of the holes, the entire biosensor may be divided into equal hexagonal prisms with regular hexagonal bases. For simplicity, it is reasonable to consider a circle whose area equals that of the hexagon and to regard one of the cylinders as a unit cell of the biosensor. Figure 20a shows the profile of the biosensor schematically represented in Fig. 19. Due to the symmetry of the unit cell, only a half of the transverse section of the unit cell can be considered. The unit cell to be considered is presented in Fig. 20b. A very similar approach has been taken in modeling of the analytical system based on the array of enzyme microreactors (see Sects. 2 and [33, 36]) and of the partially blocked electrodes [23, 75, 97].

In Fig. 20b, a_2 is the radius of the base of the unit cell, a_1 is the radius of the holes, b_1 stands for the thickness of the selective membrane, $b_4 - b_2$ is the thickness of the perforated membrane, $b_5 - b_4$ is the thickness of the external diffusion layer. The holes can be fully or partially filled with an enzyme.

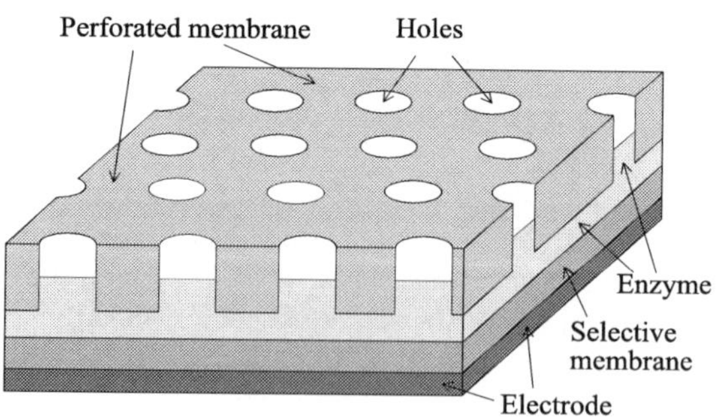

Fig. 19 The principal structure of the biosensor. The figure is not to scale

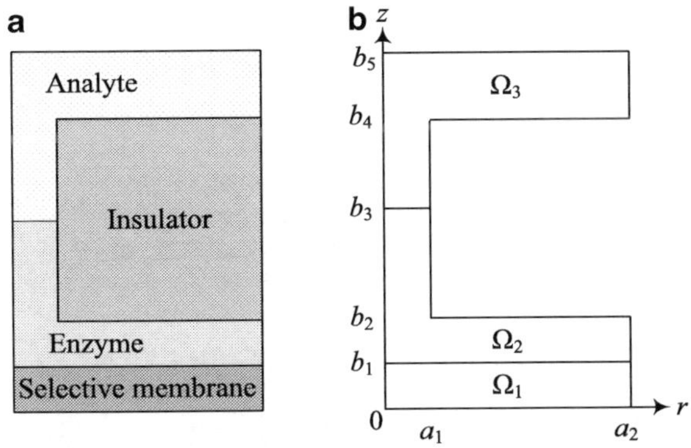

Fig. 20 The profile (**a**) and the unit cell (**b**) at y-plane. $z = 0$ corresponds to the electrode surface, a_1 is the radius of holes, a_2 is the half distance between the centres of adjacent holes. b_1, b_2-b_1 and $b_3 - b_2$ are the thicknesses of the selective membrane, the basic enzyme layer and the perforated membrane, respectively

4.2 Mathematical Model

The mathematical model of the enzyme electrode with the selective and the perforated membranes (Figs. 19 and 20) can be formulated in a two dimensional domain $\Omega_1 \cup \Omega_2 \cup \Omega_3$ in the cylindrical coordinates,

$$\begin{aligned} \Omega_1 &= (0, a_2) \times (0, b_1), \\ \Omega_2 &= ((0, a_2) \times (b_1, b_2)) \cup ((0, a_1) \times [b_2, b_3)), \\ \Omega_3 &= ((0, a_1) \times (b_3, b_4]) \cup ((0, a_2) \times (b_4, b_5)). \end{aligned} \tag{65}$$

Let $\overline{\Omega}_i$ denote the closed region corresponding to the open region Ω_i, $S_i(r, z, t)$ be the concentration of the substrate in the region $\overline{\Omega}_i$, $P_i(r, z, t)$ – the concentration of the reaction product in $\overline{\Omega}_i$, $(r, z) \in \overline{\Omega}_i$, $t \geq 0$, $i = 1, 2, 3$.

4.2.1 Governing Equations

In the selective membrane (region Ω_1) only the mass transport by diffusion of the reaction product takes place,

$$\frac{\partial P_1}{\partial t} = D_{P_1} \Delta P_1, \quad (r, z) \in \Omega_1, \tag{66}$$

where Δ is the Laplace operator in the cylindrical coordinates,

In the enzyme region Ω_2 the mass transport by diffusion and the enzyme-catalyzed reaction are considered,

$$\mathrm{S} \xrightarrow{\mathrm{E}} \mathrm{P} \tag{67}$$

Coupling the enzyme-catalyzed reaction with the two-dimensional-in-space mass transport by diffusion leads to the system of the reaction–diffusion equations ($t > 0$),

$$\begin{aligned} \frac{\partial S_2}{\partial t} &= D_{S_2}\Delta S_2 - \frac{V_{max}S_2}{K_M + S_2}, \\ \frac{\partial P_2}{\partial t} &= D_{P_2}\Delta P_2 + \frac{V_{max}S_2}{K_M + S_2}, \quad (r,z) \in \Omega_2, \end{aligned} \tag{68}$$

where V_{max} is the maximal enzymatic rate, and K_M is the Michaelis constant.

The region $\overline{\Omega}_3$ stands for the external diffusion,

$$\begin{aligned} \frac{\partial S_3}{\partial t} &= D_{S_3}\Delta S_3, \\ \frac{\partial P_3}{\partial t} &= D_{P_3}\Delta P_3, \quad (r,z) \in \Omega_3. \end{aligned} \tag{69}$$

4.2.2 Initial Conditions

Let Γ be the upper boundary of $\overline{\Omega}_3$,

$$\Gamma = [0, a_2] \times \{b_5\}. \tag{70}$$

The enzyme electrode operation starts when the substrate of the concentration S_0 appears in the bulk solution, i.e. on the boundary Γ. This is defined in the initial conditions ($t = 0$)

$$\begin{aligned} &S_2(r,z,0) = 0, \quad (r,z) \in \overline{\Omega}_2, \\ &S_3(r,z,0) = 0, \quad (r,z) \in \overline{\Omega}_3 \setminus \Gamma, \\ &S_3(r,z,0) = S_0, \quad (r,z) \in \Gamma, \\ &P_i(r,z,0) = 0, \quad (r,z) \in \overline{\Omega}_i, \quad i = 1,2,3. \end{aligned} \tag{71}$$

Assuming well-stirred and being in the powerful motion bulk solution, the upper layer of the constant thickness $b_5 - b_4$ can be treated as the Nernst diffusion layer [56, 235]. Away from this layer, the solution is in motion and is of the uniform concentration. This is described in the following boundary condition ($t > 0$):

$$S_3(r,z,0) = S_0, \quad P_3(r,z,0) = 0, \quad (r,z) \in \Gamma. \tag{72}$$

4.2.3 Boundary and Matching Conditions

Assuming the amperometry and the electric activity of the product leads to the condition

$$P_1(r,0,t) = 0, \quad r \in [0,a_2]. \tag{73}$$

On the boundary between the selective membrane and the enzyme layer the matching conditions for the product and the non-leakage condition for the substrate are defined ($t > 0$),

$$\begin{aligned} &D_{P_1}\frac{\partial P_1}{\partial z}\Big|_{z=b_1} = D_{P_2}\frac{\partial P_2}{\partial z}\Big|_{z=b_1}, \quad P_1(r,b_1,t) = P_2(r,b_1,t), \\ &D_{S_2}\frac{\partial S_2}{\partial z}\Big|_{z=b_1} = 0, \quad r \in [0,a_2]. \end{aligned} \tag{74}$$

On the boundary between adjacent regions $\overline{\Omega}_2$ and $\overline{\Omega}_3$ the matching conditions are defined ($t > 0, \quad r \in [0,a_1]$),

$$\begin{aligned} &D_{S_2}\frac{\partial S_2}{\partial z}\Big|_{z=b_3} = D_{S_3}\frac{\partial S_3}{\partial z}\Big|_{z=b_3}, \quad S_2(r,b_3,t) = S_3(r,b_3,t), \\ &D_{P_2}\frac{\partial P_2}{\partial z}\Big|_{z=b_3} = D_{P_3}\frac{\partial P_3}{\partial z}\Big|_{z=b_3}, \quad P_2(r,b_3,t) = P_3(r,b_3,t). \end{aligned} \tag{75}$$

Non-leakage conditions are used for the boundaries of the symmetry of the unit cell and for the insulator boundary,

$$D_{P_1}\frac{\partial P_1}{\partial r}\Big|_{r=0} = D_{P_1}\frac{\partial P_1}{\partial r}\Big|_{r=a_2} = 0, \quad z \in [0,b_1], \tag{76}$$

$$D_{S_2}\frac{\partial S_2}{\partial r}\Big|_{r=0} = D_{P_2}\frac{\partial P_2}{\partial r}\Big|_{r=0} = 0, \quad z \in [b_1,b_3], \tag{77}$$

$$D_{S_2}\frac{\partial S_2}{\partial r}\Big|_{r=a_2} = D_{P_2}\frac{\partial P_2}{\partial r}\Big|_{r=a_2} = 0, \quad z \in [b_1,b_2], \tag{78}$$

$$D_{S_2}\frac{\partial S_2}{\partial r}\Big|_{r=a_1} = D_{P_2}\frac{\partial P_2}{\partial r}\Big|_{r=a_1} = 0, \quad z \in (b_2,b_3], \tag{79}$$

$$D_{S_3}\frac{\partial S_3}{\partial r}\Big|_{r=0} = D_{P_3}\frac{\partial P_3}{\partial r}\Big|_{r=0} = 0, \quad z \in [b_3,b_5], \tag{80}$$

$$D_{S_3}\frac{\partial S_3}{\partial r}\Big|_{r=a_1} = D_{P_3}\frac{\partial P_3}{\partial r}\Big|_{r=a_1} = 0, \quad z \in [b_3,b_4), \tag{81}$$

$$D_{S_3}\frac{\partial S_3}{\partial r}\Big|_{r=a_2} = D_{P_3}\frac{\partial P_3}{\partial r}\Big|_{r=a_2} = 0, \quad z \in [b_4,b_5], \tag{82}$$

$$D_{S_2}\frac{\partial S_2}{\partial z}\Big|_{z=b_2} = D_{P_2}\frac{\partial P_2}{\partial z}\Big|_{z=b_2} = 0, \quad r \in (a_1,a_2], \tag{83}$$

$$D_{S_3}\frac{\partial S_3}{\partial z}\bigg|_{z=b_4} = D_{P_3}\frac{\partial P_3}{\partial z}\bigg|_{z=b_4} = 0, \quad r \in (a_1, a_2]. \tag{84}$$

4.2.4 Biosensor Response

The density $i(t)$ of the anodic current at time t can be obtained explicitly from the Faraday and the Fick laws,

$$\begin{aligned} i(t) &= n_e F D_{P_1}\frac{1}{\pi a_2^2}\int_0^{2\pi}\int_0^{a_2}\frac{\partial P_1}{\partial z}\bigg|_{z=0} r\mathrm{d}r\mathrm{d}\varphi \\ &= n_e F D_{P_1}\frac{2}{a_2^2}\int_0^{a_2}\frac{\partial P_1}{\partial z}\bigg|_{z=0} r\,\mathrm{d}r, \end{aligned} \tag{85}$$

where φ is the third cylindrical coordinate, n_e is the number of electrons involved in a charge transfer at the electrode surface, and F is the Faraday constant.

The system approaches the steady state when $t \to \infty$,

$$I = \lim_{t\to\infty} i(t), \tag{86}$$

where I is the density of the steady state current.

4.3 *Numerical Simulation*

Because of the nonlinearity of the governing equations and a rather complex geometry of the domain, the initial boundary value problem was solved numerically using the finite difference technique [49, 56, 222]. To find a numerical solution of the problem a non-uniform discrete grid in all the directions, r, z and t, was introduced [35, 41, 115]. Using the alternating direction method an implicit linear finite difference scheme has been built as a result of the difference approximation [222]. The resulting system of the linear algebraic equations was solved rather efficiently because of the tridiagonality of the matrix of the system.

Due to the high gradients of the concentrations of the species, an accurate and stable numerical solution was only achieved at a very small step size in z direction at the boundaries $z = 0$ and $z = b_5$. Because of the concavity of angles at points (a_1, b_2) and (a_1, b_4) it was also necessary to use a very small step size at the boundaries $r = a_1$, $z = b_2$ and $z = b_4$ in both space directions: r and z.

Due to the matching conditions between the adjacent regions with very different diffusivities, a small step size was also used at the boundaries $z = b_1$ and $z = b_3$. In the direction r, an exponentially increasing step size was used to both sides from a_1: to a_2 and down to 0. In the direction z, an exponentially increasing step size was used from 0 to $b_1/2$, from b_5 down to $(b_4 + b_5)/2$, from b_j down to $(b_j + b_{j-1})/2$ and from b_j to $(b_j + b_{j+1})/2$, $j = 1, 2, 3, 4$, where $b_0 = 0$.

The step size in the time direction was restricted due to the nonlinear reaction term in (68), the boundary conditions and the domain geometry. Since the biosensor action obeys the steady state assumption when $t \to \infty$, it was reasonable to apply an increasing step size in the time direction. The final step size in time was in a few orders of magnitude higher than the fist one. The numerical simulator has been programmed in JAVA language [191].

The upper layer of the thickness $\delta = b_5 - b_4$ was assumed to be the Nernst diffusion layer. Assuming intensively stirred the buffer solution, the thickness δ of 2 μm was used in the numerical simulation of the biosensor action changing some other parameters. The following values of the model parameters were constant in all the numerical experiments discussed below:

$$\begin{aligned} &D_{P_1} = 1\,\mu\text{m}^2/\text{s}, \quad D_{S_2} = D_{P_2} = 300\,\mu\text{m}^2/\text{s},\\ &D_{S_3} = D_{P_3} = 2D_{S_2} = 2D_{P_2}, \quad n_e = 2,\\ &K_M = 100\,\mu\text{M}, \quad V_{max} = 100\,\mu\text{M/s}, \quad S_0 = 100\,\mu\text{M}. \end{aligned} \tag{87}$$

Some results of the simulation are depicted in Figs. 21 and 22.

One can see in Fig. 21 that the biosensor current is very sensitive to changes in the radius of the holes and in the degree of the holes filling with the enzyme. The density of the steady state current changes non-monotonously from 2.9 (curve 1) to 1.9 (curve 3) nA/mm^2 when the holes radius a_1 changes from 0.1 to 0.4 μm (curves 1–3).

On the other hand, three curves (1, 4, 5) show the current dynamics at the same relative radius of the holes when the ratio a_1/a_2 equals 0.1. Although, in these three cases the radius a_2 differs in several times (1, 2 and 4 μm, respectively),

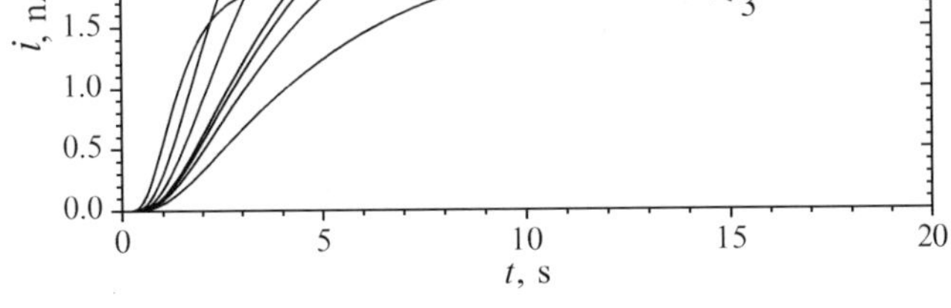

Fig. 21 The dynamics of the density i of the biosensor current, a_1: 0.1 (1, 4, 5–7), 0.2 (2), 0.4 (3), a_2: 1 (1–3, 6, 7), 2 (4) , 4(5), b_3: b_2 (6), $(b_2 + b_4)/2$ (1–5), b_4 (7), $b_1 = 2$, $b_2 = b_1 + 2$, $b_4 = b_2 + 10$, $b_5 = b_4 + 2$ (μm). The values of all other parameters are defined in (87)

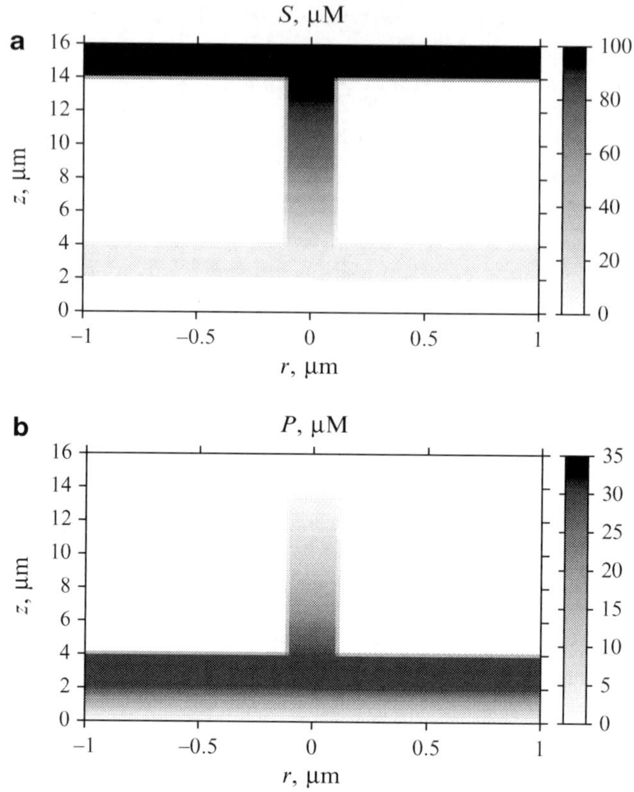

Fig. 22 The steady state concentrations of the substrate (**a**) and the product (**b**) in the selective membrane, enzyme region and buffer solution at $t = 26$ s, $a_1 = 0.1, a_2 = 1, b_1 = 2, b_2 = b_1 + 2, b_3 = b_2 + 5, b_4 = b_3 + 5, b_5 = b_4 + 2$ (μm). The values of the model parameters are defined in (87)

the density of the steady state current only differs in about 10% (changes from 2.9 to 3.3 nA/mm^2). The next section discusses the effect of the relative radius of the holes on the biosensor response in detail.

Figure 22 shows the steady state concentrations of the substrate (S) and the product (P) in the enzyme and the diffusion regions at time $t = 26$ s. The concentrations were introduced as follows ($t \geq 0$):

$$
\begin{aligned}
S(r,z,t) &= \begin{cases} 0, & (r,z) \in \Omega_1, \\ S_2(r,z,t), & (r,z) \in \overline{\Omega}_2, \\ S_3(r,z,t), & (r,z) \in \overline{\Omega}_3 \setminus \overline{\Omega}_2, \end{cases} \\
P(r,z,t) &= \begin{cases} P_1(r,z,t), & (r,z) \in \overline{\Omega}_1, \\ P_2(r,z,t), & (r,z) \in \overline{\Omega}_2 \setminus \overline{\Omega}_1, \\ P_3(r,z,t), & (r,z) \in \overline{\Omega}_3 \setminus \overline{\Omega}_2. \end{cases}
\end{aligned}
\tag{88}
$$

At any time, $t \geq 0$, both concentration functions, $S(r,z,t)$ and $P(r,z,t)$ are continuous at all $(r,z) \in \overline{\Omega}_1 \cup \overline{\Omega}_2 \cup \overline{\Omega}_3$.

To have a more comprehensive view, the mirror-image along the z-axis is also shown in Fig. 22.

4.4 *Effect of the Perforation Topology*

Using computer simulation the dependence of the response of the enzyme electrode on the geometry of the membrane perforation has been investigated. For this purpose the density I of the steady state current was calculated at different values of the radius a_2 of the entire unit and several degrees of the holes filling with the enzyme. In the calculations the radius a_1 of holes was changed in a wide range. To compare the response of the biosensor ($a_1 < a_2$) with the response of the corresponding flat biosensor ($a_1 = a_2$), the steady state current was expressed as a function of the degree α of the membrane openness and was normalized with the steady state current of the corresponding flat biosensor having no perforated membrane,

$$I_\alpha = \frac{I(\alpha)}{I(1)}, \quad \alpha = \frac{a_1^2}{a_2^2}, \quad 0 < \alpha \leq 1, \tag{89}$$

where $I(\alpha)$ is the density of the steady state current calculated at the concrete value of α. The degree α expresses the level of the membrane perforation. The theoretically extreme case of $\alpha = 0$ corresponds to the fully impermeable membrane having no holes. In the opposite case of $\alpha = 1$ ($a_1 = a_2$), the perforated membrane becomes so opened that it disappears at all.

Figure 23 shows the effect of the openness α on the steady state current and on the biosensor sensitivity. One can see in Fig. 23a that the steady state current is a non-monotonous function of the degree α of the perforated membrane openness. The steady state current of the biosensor having a perforated membrane onto the top of the enzyme layer can generate the steady state current which is significantly higher than the current if the enzyme electrode is without the perforated membrane ($a_1 = a_2, \alpha = 1$). This effect is especially meaningful at small values of the degree α ($\alpha \approx 0.05$).

Figure 23 also shows that the biosensor response is practically invariant to the absolute value of the holes radius a_1. No notable difference is observed between the cases when the values of the radius a_1 differ but the relative radius a_1/a_2 remains stagnant (curves 1–3). Such difference was already noticed in Fig. 21. According to Fig. 23, the difference is observed only in the cases of very small values of α.

Figure 23b shows that the biosensor sensitivity B_S can be notably increased by decreasing the holes radius a_1. The membrane containing the holes of a relatively small radius creates the diffusion limitation to the substrate, i.e. lowers the substrate concentration in the enzymatic layer and thereby prolongs the calibration curve of the biosensor [152, 166, 229, 258].

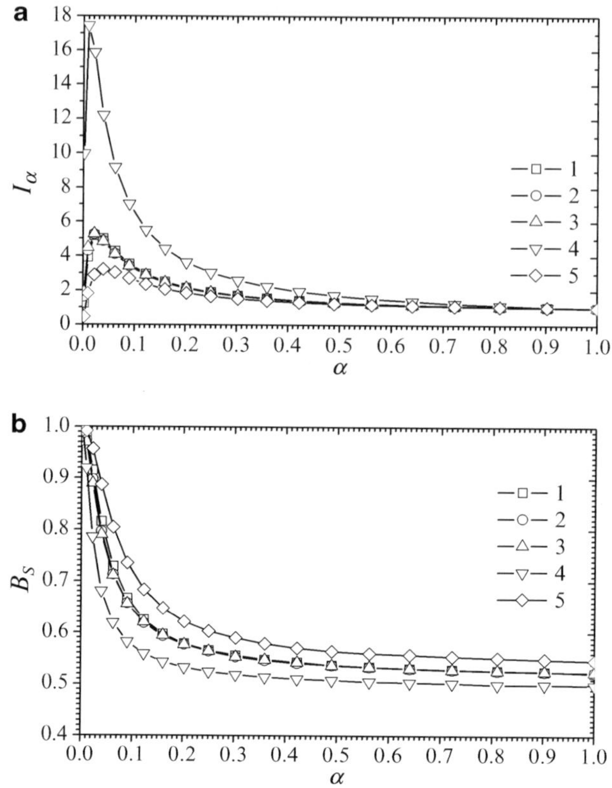

Fig. 23 The normalized current I_α (**a**) ant the sensitivity B_S (**b**) versus the degree α of the perforated membrane openness, a_2: 1 (1, 4, 5), 2 (2), 4 (3), b_3: b_2 (4), $(b_2 + b_4)/2$ (1–3), b_4 (5) (μm). The values of all other parameters are the same as in Fig. 21

4.5 Concluding Remarks

The numerical simulation of the response of the amperometric biosensor with the selective and the perforated membranes shows that the steady state current is a non-monotonous function of the relative radius of the holes of the perforated membrane (Figs. 21 and 23a).

The biosensor with the perforated membrane can produce the steady state current which is notably higher than the current of the corresponding biosensor without the membrane (Fig. 23a). The sensitivity of the biosensors can be significantly increased by choosing an appropriate topology of the membrane perforation (Fig. 23b).

Part III
Numerical Methods for Reaction–Diffusion Equations

Abstract This is the part three of the book Mathematical Modeling of Biosensors. The part is dedicated to the basic concepts of the theory of the finite difference schemes for the digital solving of linear diffusion equations which are basis for biosensors modeling. The part introduces the relevant facts about the basic concepts of the theory of the finite difference schemes for the linear diffusion equations. The most popular effective difference schemes for the nonlinear reaction-diffusion equations are presented. The concepts of a primary nature and being sufficient for the solution of the problems of the biosensor are discussed. The difference schemes are extensively applied to the solution of biosensor problems.

Keywords Finite difference scheme · PDE · Reaction-diffusion equation

The Difference Schemes for the Diffusion Equation

Contents

1 The Grids ... 250
1.1 An Equidistant Grid in the Straight ... 251
1.2 A Non-equidistant Grid in a Straight ... 251
1.3 The Equidistant Grid in a Plane ... 252
1.4 The Non-equidistant Grid in a Plane ... 253
1.5 The Grid in a Multidimensional Case ... 253
2 The Approximation of the Function Derivatives ... 254
2.1 The Derivative of the First Order ... 254
2.2 The Approximation of the Second Order Derivative ... 256
2.3 The Approximation of the Second Order Derivative on a Non-equidistant Grid ... 256
3 The Explicit Difference Scheme ... 258
3.1 The Calculation of a Solution ... 261
3.2 The Convergence and the Stability ... 262
4 The Implicit Difference Scheme ... 264
4.1 The Convergence and the Stability ... 266
4.2 The Calculation of a Solution ... 266
5 The Elimination Method for the System of Linear Equations ... 268
5.1 Stability of the Elimination Method ... 271
6 The Crank–Nicolson Difference Scheme ... 271
6.1 The Calculation of a Solution ... 273
6.2 The Convergence and the Stability ... 274
7 The Difference Scheme with the Weights ... 274
7.1 The Convergence and the Stability ... 276
8 The Crank–Nicolson Difference Scheme on Non-equidistance Grid ... 276
8.1 The Calculation of a Solution ... 278
8.2 The Convergence and the Stability ... 279
9 The Explicit Difference Scheme in the Cylindrical Coordinates ... 280
9.1 The Calculation of a Solution ... 281
9.2 The Convergence and Stability ... 282
10 The Crank–Nicolson Difference Scheme in the Cylindrical Coordinates ... 282
10.1 The Calculation of a Solution ... 283
10.2 The Convergence and Stability ... 285
11 The Discontinuous Diffusion Coefficient ... 285
12 The Explicit Difference Scheme ... 287
12.1 The Calculation of a Solution ... 288
13 The Crank–Nicolson Difference Scheme ... 288
13.1 The Calculation of a Solution ... 289

R. Baronas et al., *Mathematical Modeling of Biosensors*, Springer Series on Chemical Sensors and Biosensors 9, DOI 10.1007/978-90-481-3243-0_10,

Contemporary numerical methods for solving problems of the mathematical chemistry are gaining increasing popularity. The aim of this chapter is to introduce the reader with the relevant facts about the basic concepts of the theory of the difference schemes for the linear diffusion equations. The linear diffusion equations play an important and crucial role in most models of a biosensor theory. The selection of the difference methods for the solution of these equations is motivated by the following two arguments:

- The geometry (size) of a biosensor really does not change during the measurements;
- The simplicity and efficiency of the difference method.

The most popular simple and together effective difference schemes are presented here. These difference schemes are extensively applied to the solution of a biosensor problems in the next chapter. This method is being frequently used in solving applied problems not only by professional mathematicians, but also by laymen. The concepts presented below are of a primary nature and are sufficient for the solution of the problems of the biosensor. In this book the notations of [222] are mainly applied. The many aspects of the numerical methods for the solution of the partial differential equations are presented in [5, 12, 187, 222].

1 The Grids

We will consider the one-dimensional in a space diffusion equation with the constant coefficients and the difference schemes in order to develop various methods of designing appropriate schemes for nonlinear and multidimensional problems of the biosensor mathematical theory. The composition of the difference equation approximating the differential equation of interest amounts to performing the following operations:

- To substitute the domain of discrete variation of an argument for the domain of continuous variation;
- To impose the difference analogs of the boundary conditions and initial data.

Obviously, in the numerical solution of the mathematical problems it is unrealistic to reproduce a difference solution for all the values of the argument varying in a certain domain. The traditional way of covering this is to select some finite set of points in this domain and look for an approximate solution only at those points. Any such set of points is called a grid and the isolated points are termed the grid nodes. Any such function defined at the nodal points is called a grid function. For this reason the time of solution, the properties of the difference solution and the proximity to an exact solution depend essentially on the proper choice of the grid. We will give several simple example of the grids.

1.1 An Equidistant Grid in the Straight

We now consider a set of the functions $u(x)$ in the segment [0,d]. The segment [0,d] of the length d is divided into n equal intervals. The spacing between the adjacent nodes $x_i - x_{i-1} = h = d/n$ is termed a grid step or simply a step. The splitting points $x = x_i = ih$ constitute what is called the set of grid nodes $\Omega_h = \{x_i = ih, i = 1, \ldots, n-1\}$ and generate one possible grid on this segment [0, d] (see Fig. 1).

When the boundary point $x_0 = 0$ and $x_n = d$ are put together with the grid Ω_h, it will be denoted by $\bar{\Omega}_h = \{x_i = ih, i = 0, 1, \ldots n\}$. On the segment $[0, d]$ we are working with a new function $u_h(x_i)$ of the discrete argument instead of a given function $u(x)$ of the continuous argument. The values of such a function are calculated at the grid nodes x_i and the function itself depends on the step h as on the parameter.

Definition 1. The function $u(x_i)$ which is defined on the grid $\bar{\Omega}_h$ nodes x_i is called a grid function.

Consequantly, the function $u_h(x_i)$ is a grid function.

1.2 A Non-equidistant Grid in a Straight

The next example concentrates on the segment $0 \leq x \leq d$ with n subintervals without concern for how the points $0 < x_1 < x_2 < \ldots < x_{n-1} < d$ will be chosen. The nodes $\{x_i, i = 0, \ldots, n, x_0 = 0, x_n = d\}$ constitute what is called a non-equidistant grid $\hat{\bar{\Omega}}_h$ on $[0, d]$ (see Fig. 2). The distance between the adjacent nodes, being a grid step, equals $h_i = x_i - x_{i-1}$ and depends on the subscript i. The steps should satisfy the normalization condition $\sum_{i=1}^{n} h_i = d$. Any distance of this type falls within the category of the grid functions. The step h_i is taken smaller in the region where the variation of a function (function derivative) is major and the step h_i is taken bigger in region where the variation of the function is smaller. Usually this is done in order to reduce an error of a solution.

Fig. 1 An equidistant grid on a segment

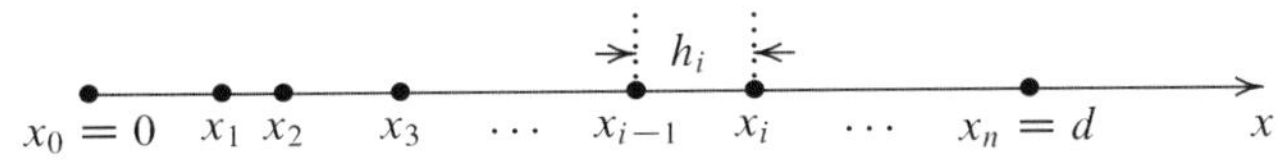

Fig. 2 A non-equidistant grid on the segment

1.3 The Equidistant Grid in a Plane

We now consider a set of the functions $u(x,t)$ of two arguments in the rectangle $\bar{Q} = \{0 \le x \le d, 0 \le t \le T\}$. Because in the mathematical theory of the biosensor the processes are non-stationary it is important to separate the time and the spatial coordinates. In future we will denote the time by t and the spatial coordinates by x, y and z. Now we will split up the segment $[0, d]$ on the x-axis and the segment $[0, T]$ on the t-axis into n and m parts with the steps $h = d/n$ and $\tau = T/m$, respectively. After that, we draw the straight lines parallel to the appropriate axes through the splitting points. The points of the intersection of those lines are adopted, as usual, as the nodes $(x_i, t_j) = (ih, j\tau)$, $i = 0, 1, \ldots, n$, $j = 0, 1, \ldots, m$, which constitute a widespread grid $\bar{Q}_{h\tau} = \{(x_i, t_j) \in \bar{Q}\}$ with steps h and τ along the directions Ox and Ot, respectively (see Fig. 3). The nodes lying on one and the same straight line (horizontal or vertical), the distance between which is equal to the grid step (h or τ), are called the adjacent grid nodes. The nodes (x_i, t_j) inside the region $\bar{Q}$ are called inner nodes. The nodes on the border $x = 0$, $x = d$, $t \in [0, T]$ are known as the boundary nodes, the nodes on $t = 0$, $x \in [0, d]$ are known as the initial nodes. The nodes

$$(x_i, t_j),\ i = 1, 2, \ldots, n-1,\ j = 1, 2, \ldots, m$$

form the set of the inner points. An equidistant grid in a plane is usually applied when the variation of the function $u(x,t)$ is almost the same on the rectangle $\bar{Q}$. If the variation of a function is different on the rectangle $\bar{Q}$ then a non-equidistant grid in a plane is applied.

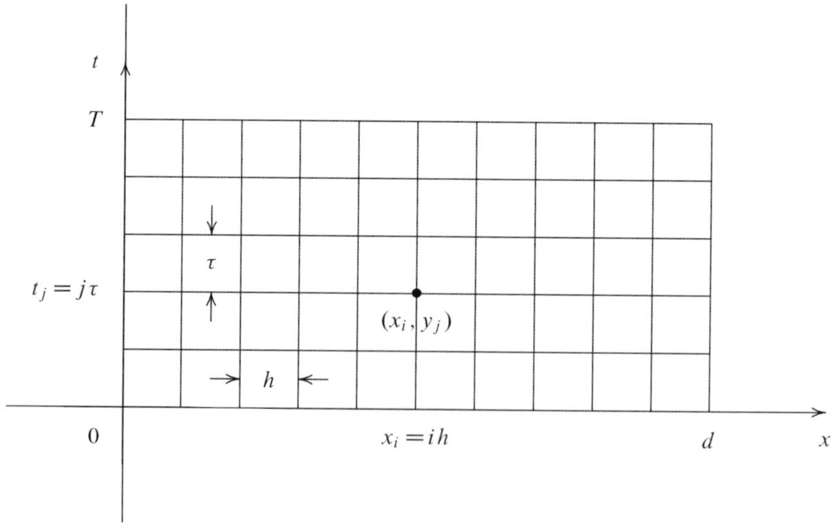

Fig. 3 An equidistant grid on the plane

1.4 The Non-equidistant Grid in a Plane

The non-equidistant grid in a plane can be dealt with in a similar way. We now consider a set of functions $u(x, t)$ of two arguments in the rectangle $\bar{Q} = \{0 \leq x \leq d, 0 \leq t \leq T\}$. The nodes $\{(x_i, t_j), i = 0, \ldots, n,\ x_0 = 0, x_n = d,\ j = 0, 1, \ldots, m,\ t_0 = 0, t_m = T\}$ constitute what is called a non-equidistant grid $\bar{Q}_{h\tau} = \{(x_i, t_j) \in \bar{Q}\}$. The distance between the adjacent nodes, being a grid step, equals $h_i = x_i - x_{i-1}$ along the directions Ox and $\tau_j = t_j - t_{j-1}$ along the directions Ot respectively. The grid steps depend on the subscripts i and j. The steps should satisfy the normalization conditions $\sum_{i=1}^{n} h_i = d$ and $\sum_{j=1}^{m} \tau_j = T$.

As in the previous case any distance between the adjacent grid points falls within the category of the grid functions. The steps h_i or τ_i are taken smaller in the region where the variation of a function (the partial derivatives of the function) is major and the steps h_i or τ_i are taken bigger in the region where the variation of the function is smaller.

1.5 The Grid in a Multidimensional Case

The grid in a multidimensional case is being introduced similarly. We consider the set of the functions $u(x, y, z, t)$ in region $\bar{Q}$, where

$$\bar{Q} = \bar{\Omega} \times [0, T],\ \bar{\Omega} = \{(x, y, z) : 0 \leq x \leq d_x,\ 0 \leq y \leq d_y,\ 0 \leq z \leq d_z\}.$$

In a uniform grid case we take the steps of the grid $h_x,\ h_y,\ h_z$ on the axes $x,\ y,\ z$ and τ on the axis t. The points of the grid are

$$(x, y, z, t) = (x_i, y_k, z_l, t_j),\ i = 0, 1, \ldots, n_x,\ k = 0, 1, \ldots, n_y,\ l = 0, 1, \ldots, n_z,\quad j = 0, 1, \ldots, m.$$

The steps satisfy the normalization conditions

$$h_x n_x = d_x,\ h_y n_y = d_y,\ h_z n_z = d_z,\ \tau m = T.$$

The surface of the parallelepiped $\bar{\Omega}$ is defined to be $\partial\Omega$. The nodes $(x_i\ , y_k, z_l, t_j)$ inside the region $\bar{Q}$ are called the inner nodes. The nodes on the border $\partial\Omega \times [0, T]$ are known as the boundary points. The nodes

$$(x_i, y_k, z_l, t_j),\ i = 1, 2, \ldots, n_x - 1,\ k = 1, 2, \ldots, n_y - 1,\ l = 1, 2, \ldots, n_z - 1,\quad j = 1, 2, \ldots, m$$

form the set of the inner points. One can introduce a non-equidistant grid in the multidimensional case. The procedure of the construction of the grid is similar to that of the two dimensional domain.

2 The Approximation of the Function Derivatives

2.1 *The Derivative of the First Order*

The difference approximation of the functions derivatives is the first step in the construction of a method for the solution of the differential problems. We next develop the approximation of the function first order derivative. The approximation of the functions derivatives is based on the definition of a function first derivative

$$u'(x) = \lim_{h \to 0} \frac{u(x+h) - u(x)}{h}.$$

The difference derivatives are obtained when the limit is rejected and some finite value of h is taken and an approximate equality is obtained

$$u'(x) \approx \frac{u(x+h) - u(x)}{h}.$$

Let us fix some point x on the Ox-axis by capturing neighboring points $x - h$ and $x + h$, were step $h > 0$. Then, we will introduce the following two expressions of the difference derivatives:
the right difference derivative

$$\frac{u(x+h) - u(x)}{h} = u_x,$$

the left difference derivative

$$\frac{u(x) - u(x-h)}{h} = u_{\bar{x}}.$$

The names of the difference derivatives are chosen according to the position of the points engaged in the definition.

The following question is of significant importance:
what is the error of one or another difference approximation and how does the difference (for example) $\Phi(x) = u'(x) - u_x(x)$ behave at a point x as $h \to 0$ refers to the error of the difference approximation to $u'(x)$ at a point x.

At first we will introduce a definition of the function order with respect to h as $h \to 0$.

Definition 2. The function $g(h)$ is an order of r with respect to h if there exists a constant c, $0 < c < \infty$, such that

$$\frac{|g(h)|}{h^r} \le c$$

as $h \to 0$ and we write $g(h) = O(h^r)$.

Now we turn back to an error of an approximation of the derivative. We next develop $u(x)$ in the series by Taylor's formula up to the derivatives of the third order

$$u(x+h) = u(x) + hu'(x) + \frac{h^2}{2}u''(x) + \frac{h^3}{6}u'''(x_+),$$

$$u(x-h) = u(x) - hu'(x) + \frac{h^2}{2}u''(x) - \frac{h^3}{6}u'''(x_-),$$

assuming $u(x)$ to be a sufficiently smooth function, where points $x_- \in [x - h, x]$, $x_+ \in [x, x+h]$. The assumption "of a sufficiently smooth function" means that

$$\max\left|u'''(x)\right| \le c < \infty$$

in some neighborhood $(x - h_0, x + h_0)$ of the point x and $h < h_0$, where the number h_0 is kept fixed. Substituting the preceding Taylor's series into the definition of the difference derivative yields

$$u_x = \frac{u(x+h) - u(x)}{h} = u'(x) + \frac{h}{2}u''(x) + O(h^2),$$

$$u_{\bar{x}} = \frac{u(x) - u(x-h)}{h} = u'(x) - \frac{h}{2}u''(x) + O(h^2).$$

So, the approximation errors $\Phi(x)$ of defined difference derivatives are

$$\Phi(x) = u_x - u' = O(h),$$

$$\Phi(x) = u_{\bar{x}} - u' = O(h).$$

Definition. The difference derivative is said to approximate the derivative of order $r > 0$ at the point x if

$$\Phi(x) = O(h^r).$$

So, we have that the order of the approximation for the right difference derivative and the left difference derivative is 1 respectively. Moreover, any linear combination such as

$$u_x^\sigma \equiv \sigma u_x + (1-\sigma)u_{\bar{x}},$$

where σ is a real number, can be adopted as the difference approximation of the derivative $u'(x)$.

2.2 The Approximation of the Second Order Derivative

The approximation of the second order derivative is based on the definition of the first derivative. In order to construct a difference approximation of the second derivative, it is necessary to rely on the three-point pattern $(x-h, x, x+h)$. In that case we introduce

$$u_{\bar{x}x} = \frac{u_x - u_{\bar{x}}}{h}.$$

It easy to prove that the relation between the second difference derivative and the values of the function $u(x-h), u(x), u(x+h)$ is

$$\frac{u(x-h) - 2u(x) + u(x+h)}{h^2} = u_{\bar{x}x}. \tag{1}$$

As in the previous case we will evaluate the error of an approximation of the second order derivative $u_{\bar{x}x}$. We next develop $u(x)$ in the series by Tailor's formula up to the derivatives of the fourth order. We get

$$u(x+h) = u(x) + hu'(x) + \frac{h^2}{2}u''(x) + \frac{h^3}{6}u'''(x) + \frac{h^4}{24}u^{(4)}(x_+),$$

$$u(x-h) = u(x) - hu'(x) + \frac{h^2}{2}u''(x) - \frac{h^3}{6}u'''(x) + \frac{h^4}{24}u^{(4)}(x_-),$$

where the points $x_- \in [x-h, x]$, $x_+ \in [x, x+h]$. This yields that approximation error

$$\Phi(x) = u_{\bar{x}x} - u'' = \frac{h^4}{12}u^{(4)}(\bar{x}),$$

where $\bar{x} \in [x-h, x+h]$. Assuming $u(x)$ to be a sufficiently smooth function we have that the difference derivative $u_{\bar{x}x}$ approximates the second derivative $u''(x)$ of the order 2, meaning $u''(x) - u_{\bar{x}x} = O(h^2)$. The assumption of a sufficiently smooth function means that

$$\max |u^{(4)}(x)| \le c < \infty$$

in some neighborhood $(x-h_0, x+h_0)$ of the point x and $h < h_0$, where the number h_0 is kept fixed.

As a result, the difference derivative $u_{\bar{x}x}$ (1) provides the approximation of the order 2 if the function $u(x)$ is a sufficiently smooth function.

2.3 The Approximation of the Second Order Derivative on a Non-equidistant Grid

This example is concerned with a non-equidistant grid on the segment $0 \le x \le d$ with n subintervals and the nodes $\{x_i,\ i = 0, \dots, n,\ x_0 = 0,\ x_n = d\}$. The distance between the adjacent nodes equals $h_i = x_{i+1} - x_i$, and depends on the subscript i. We introduce the following expressions

$$u_{\bar{x}_i} = \frac{u(x_i) - u(x_{i-1})}{h_{i-1}},$$

$$u_{x_i} = \frac{u(x_{i+1}) - u(x_i)}{h_i}.$$

The second order derivative is approximated by the value

$$u_{\bar{x}_i x_i} = \frac{u_{x_i} - u_{\bar{x}_i}}{\hat{h}_i} = \left(\frac{u(x_{i+1}) - u(x_i)}{h_i} - \frac{u(x_i) - u(x_{i-1})}{h_{i-1}}\right) / \hat{h}_i,$$

here $\hat{h}_i = (h_i + h_{i-1})/2$. For $h_{i-1} = h_i = h$ (equidistant grid) the preceding is identical with the expression (1). Now we will calculate an approximation error at the point x_i. We next develop $u(x)$ in the series by Taylor's formula

$$u(x_i + h_i) = u(x_i) + h_i u'(x_i) + \frac{h_i^2}{2} u''(x_i) + \frac{h_i^3}{6} u'''(x_i) + O(h_i^4),$$

$$u(x_i - h_{i-1}) = u(x_i) - h_{i-1} u'(x_i) + \frac{h_{i-1}^2}{2} u''(x_i) - \frac{h_{i-1}^3}{6} u'''(x_i) + O(h_{i-1}^4).$$

It leads to the following equalities

$$u_x(x_i) = u'(x_i) + \frac{h_i}{2} u''(x_i) + \frac{{h_i}^2}{6} u'''(x_i) + O(h_i^3),$$

$$u_{\bar{x}}(x_i) = u'(x_i) - \frac{h_{i-1}}{2} u''(x_i) + \frac{h_{i-1}^2}{6} u'''(x_i) + O(h_{i-1}^3),$$

$$u_{\bar{x}_i x_i} = u''(x_i) + \frac{h_i - h_{i-1}}{6\hat{h}_i} u'''(x_i) + O(\bar{h}_i^2),$$

where

$$\bar{h}_i = \max\{h_i, h_{i-1}\}.$$

With these relations in view, we derive the useful expressions for the error of the approximation

$$\Phi(x_i) = u''(x_i) - u_{\bar{x}_i x_i} = \frac{h_i - h_{i-1}}{6\hat{h}_i} u'''(x_i) + O(\bar{h}_i^2) = O(\bar{h}_i). \qquad (2)$$

Consequently, the difference derivative $u_{\bar{x}x}$ provides the approximation of the order 1 if the function $u(x)$ is sufficiently smooth function. If $h_i = h_{i-1}$ then the error of the approximation is of the order 2.

3 The Explicit Difference Scheme

The difference approximation of the diffusion equation is based on the replacement of the first and the second derivatives by the difference derivatives. In this paragraph we will consider the diffusion equation together with the initial (for $t = 0$) and boundary (for $x = 0$, $x = d$) conditions. The diffusion equation together with the initial and boundary conditions is called the boundary-value problem.

Now we will consider the boundary-value problem for the diffusion equation. We will start with the simplest one-dimensional in the space example. This problem is formulated in the following way: to find a function $u(x,t)$ which satisfies the diffusion equation

$$\frac{\partial u}{\partial t} = D\frac{\partial^2 u}{\partial x^2} \tag{3}$$

in a rectangle $Q = \{0 < x < d,\ 0 < t \leq T\}$ with the initial condition for $t = 0$

$$u(x,0) = \varphi(x),\ x \in [0,d] \tag{4}$$

and the boundary condition

$$u(0,t) = \psi_1(t),\ u(d,t) = \psi_2(t),\ t \in [0,T], \tag{5}$$

where $\varphi(x)$, $\psi_1(t)$ and $\psi_2(t)$ are the given functions, a diffusion coefficient $D > 0$. Also we assume that the solution of the problem (3)–(5) exists, is unique and sufficiently smooth.

Let us introduce an equidistant grid $\bar{Q}_{h\tau} = \{(x_i,t_j) \in \bar{Q}\}$ with the steps h and τ along the directions Ox and Ot, respectively in rectangle $\bar{Q} = \{0 \leq x \leq d, 0 \leq t \leq T\}$.

Let (x_i,t_j) be a fixed point of the grid inside the rectangle Q. Now we write the diffusion equation at this point

$$\frac{\partial u(x_i,t_j)}{\partial t} = D\frac{\partial^2 u(x_i,t_j)}{\partial x^2}. \tag{6}$$

The expressions

$$\frac{\partial u(x_i,t_j)}{\partial t} = \frac{u(x_i,t_{j+1}) - u(x_i,t_j)}{\tau} - \frac{\tau}{2}\frac{\partial^2 u(x_i,\bar{t}_j)}{\partial t^2}, \tag{7}$$

where $\bar{t}_j \in [t_j,t_{j+1}]$ and

$$\frac{\partial^2 u(x_i,t_j)}{\partial x^2} = \frac{u(x_{i+1},t_j) - 2u(x_i,t_j) + u(x_{i-1},t_j)}{h^2} - \frac{h^2}{12}\frac{\partial^4 u(\bar{x})}{\partial x^4}, \tag{8}$$

where $\bar{x} \in [x_{i-1},x_{i+1}]$ are valid.

The correctness of the two last equalities can be achieved by employing the series of Taylor's formula up to the derivatives of the second order with respect to t and up to the derivatives of the forth order with respect to x

$$u(x_i, t_{j+1}) = u(x_i, t_j + \tau) = u(x_i, t_j) + \tau \frac{\partial u(x_i, t_j)}{\partial t} + \frac{\tau^2}{2} \frac{\partial^2 u(x_i, \hat{t}_j)}{\partial t^2},$$

where $\hat{t}_j \in [t_j, t_{j+1}]$ and

$$u(x_{i+1}, t_j) = u(x_i + h, t_j) = u(x_i, t_j) + h \frac{\partial u(x_i, t_j)}{\partial x} + \frac{h^2}{2} \frac{\partial^2 u(x_i, t_j)}{\partial x^2}$$
$$+ \frac{h^3}{6} \frac{\partial^3 u(x_i, t_j)}{\partial x^3} + \frac{h^4}{24} \frac{\partial^4 u(\bar{x}_i, t_j)}{\partial x^4},$$

$$u(x_{i-1}, t_j) = u(x_i - h, t_j) = u(x_i, t_j) - h \frac{\partial u(x_i, t_j)}{\partial x} + \frac{h^2}{2} \frac{\partial^2 u(x_i, t_j)}{\partial x^2}$$
$$- \frac{h^3}{6} \frac{\partial^3 u(x_i, t_j)}{\partial x^3} + \frac{h^4}{24} \frac{\partial^4 u(\hat{x}_i, t_j)}{\partial x^4},$$

where $\bar{x} \in [x_i, x_{i+1}]$ and $\hat{x} \in [x_{i-1}, x_i]$. The proof of (7), (8) follows by using the two last Taylor's formulae. Now the partial derivatives are changed by (7), (8). Equation (6) can be written as

$$\frac{u(x_i, t_{j+1}) - u(x_i, t_j)}{\tau} = D \frac{u(x_{i+1}, t_j) - 2u(x_i, t_j) + u(x_{i-1}, t_j)}{h^2} + \Phi(x_i, t_j), \tag{9}$$

$$i = 1, 2, \ldots, n-1, \; j = 0, 1, \ldots, m-1,$$

where

$$\Phi(x_i, t_j) = \frac{\tau}{2} \frac{\partial^2 u(x_i, \bar{t}_j)}{\partial t^2} - D \frac{h^2}{12} \frac{\partial^4 u(\bar{x}, t_j)}{\partial x^4}.$$

The value $\Phi(x_i, t_j)$ is taken as an error of the approximation for the diffusion equation (6). If, for example, the partial derivatives of the function $u(x, t)$ satisfy the inequalities

$$\max |\frac{\partial^2 u(x, t)}{\partial t^2}|, \; \max |\frac{\partial^4 u(x, t)}{\partial x^4}| = c < \infty$$

in the region $\bar{Q}$, than the order of the approximation equals $O(\tau + h^2)$ as $\tau, h \to 0$. We remark that the error of an approximation for the first derivative $\partial u(x_i, t_j)/\partial t$ is $O(\tau)$ and that the error of an approximation for the second derivative $\partial^2 u(x_i, t_j)/\partial x^2$ is $O(h^2)$.

The error of the approximation $\Phi(x_i, t_j)$ can be neglected in the Eq. 9 if τ and h are small enough. In that case we have a rough equality

$$\frac{u(x_i,t_{j+1})-u(x_i,t_j)}{\tau}\approx D\frac{u(x_{i+1},t_j)-2u(x_i,t_j)+u(x_{i-1},t_j)}{h^2}.$$

Now we introduce a new function v_i^j, which is defined on the grid points (x_i,t_j) and also a new exact equality

$$\frac{v_i^{j+1}-v_i^j}{\tau}=D\frac{v_{i+1}^j-2v_i^j+v_{i-1}^j}{h^2}, \tag{10}$$

$$i=1,2,\ldots,n-1,\ j=0,1,\ldots,m-1.$$

The initial condition (4) is changed by

$$v_i^0=\varphi(x_i),\ i=0,1,\ldots,n \tag{11}$$

and the boundary condition (5)

$$v_0^j=\psi_1(t_j),\ v_n^j=\psi_2(t_j),\ j=1,2,\ldots,m. \tag{12}$$

The difference equation (10) together with the initial (11) and boundary conditions (12) form the difference scheme for the solution of boundary-value problem for the diffusion equation (3)–(5). The initial and boundary conditions are replaced exactly. The values of the solution $u(x_i,t_j)$ of the differential problem (3)–(5) are consistent with the values v_i^j of the solution of the difference scheme (10)–(12). It is easy to see that if the approximation error $\Phi(x_i,t_j)\equiv 0$ for all $i=1,2,\ldots,n-1$ and $j=0,1,\ldots,m-1$ then the difference $z_i^j=u(x_i,t_j)-v_i^j=0$. The difference z_i^j is named as an error of the solution. If the error of approximation $\Phi(x_i,t_j)\approx 0$, then one can hope that $u(x_i,t_j)\approx v_i^j$.

Further we will use the term of a pattern of the difference equation.

Definition 3. The set of the grid points which are involved in a creation of a difference equation is named a pattern of the difference equation. The diffusion equation was replaced by the difference equation using the pattern consisting of the four points (x_{i-1},t_j), (x_i,t_j),(x_{i+1},t_j), (x_i,t_{j+1}) (see Fig. 4). Three points are located on the line $t=t_j$ and one on the line $t=t_{j+1}$. Now we introduce a definition of a grid layer.

Definition 4. The set of the grid points which are located on the line $t=t_j$ is named as $j-th$ grid layer.

The $j-th$ grid layer combines points (x_i,t_j), when $i=0,1,\ldots,n$.

Now we introduce a definition of the explicit difference equation.

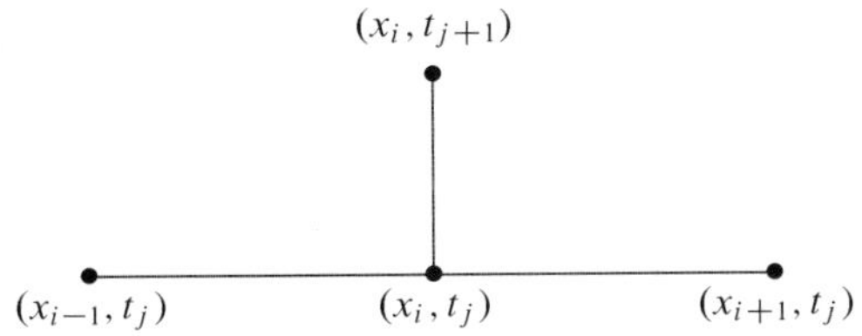

Fig. 4 The pattern of the explicit scheme

Definition 5. The difference equation is explicit if the value of the solution on the upper layer can be expressed by the values on the lower layer.

The difference equation (10) is explicit and two-layer. Indeed, from the equation we have the expression of v_i^{j+1} by $v_{i+1}^j, v_i^j, v_{i-1}^j$

$$v_i^{j+1} = v_i^j + \tau D \frac{v_{i+1}^j - 2v_i^j + v_{i-1}^j}{h^2}.$$

The questions of significant importance are:

a) what is the error of the solution $z_i^j = u(x_i, t_j) - v_i^j$ and is the difference scheme convergent, that is to say

$$\lim_{h,\tau \to 0} |u(x_i, t_j) - v_i^j| = 0?$$

b) how to solve the difference scheme?

Now we will give an answer to the second question.

3.1 The Calculation of a Solution

The explicit difference scheme is simple and has a convenient algorithm of the calculation and the programming. The calculation of the difference scheme (10)–(12) is performed in the series one grid layer after another. At first the solution on the zero layer at $t = 0$ using the initial condition (11)

$$v_i^0 = \varphi(x_i),\ i = 0, 1, \ldots, n.$$

is calculated.

Suppose, the solution v_i^j is calculated at $j - th$ layer. Thus, the values v_i^j for $i = 0, 1, \ldots, n$ are known. We proceed to solve the difference equation (10) subject to value v_i^{j+1} at $j + 1 - th$ layer. We have

$$v_i^{j+1} = v_i^j + D\frac{\tau}{h^2}(v_{i+1}^j - 2v_i^j + v_{i-1}^j),\ i = 1, 2, \ldots, n-1.$$

All values of function v_i^j are known on the right side of the equality. So one can calculate the values v_i^{j+1}, $i = 1, 2, \ldots, n-1$. The values v_0^{j+1} and v_n^{j+1} are calculated using boundary conditions (12)

$$v_0^{j+1} = \psi_1(t_{j+1}),\ v_n^{j+1} = \psi_2(t_{j+1}).$$

Further analogously we will find the solution v_i^j on the next layer and will continue up to the last layer $t = t_m$. The calculation of a solution of the explicit difference scheme is finished.

3.2 *The Convergence and the Stability*

The convergence of a difference scheme is a very important feature of the difference scheme. One can apply a difference scheme for practical purpose only if it is convergent. Now we will introduce a definition of the convergence of the explicit difference scheme (10)–(12).

Definition 6. The difference scheme is convergent if

$$\lim_{h,\tau \to 0} |u(x_i, t_j) - v_i^j| = 0$$

for $i = 0, 1, \ldots, n,\ j = 0, 1, \ldots, m$.

Here $u(x_i, t_j)$ is the solution of boundary-value problem (3)–(5) and v_i^j is the solution of difference scheme (10)–(12) at the point (x_i, t_j).

As noted above, in this book we will restrict ourselves to minimal and necessary information on a convergence of the differences scheme. The concepts presented below are of a supplementary nature and are sufficient for primary study and for the solution of the mathematical problems of the biosensor theory. For a detailed study of the theory of difference scheme we refer to [222] and the literature cited there.

The convergence of a difference scheme depends on several factors such as:

a) The smoothness of a solution of the differential problem;
b) The choice of a grid;
c) The stability of a difference scheme.

Now we will discus these factors in detail. The smoothness of a solution of the differential problem totally depends on the properties of the differential equation, the initial and boundary conditions and the region of the definition. The mathematical models, which are being investigated in this book, possess a solution and the solution generally has infinite number of the derivatives with respect to time and spatial coordinates. The question is: what are the maximal values of the derivatives in the region of investigation? This problem is quite complicated and the answer depends on concrete case. In the general case it is known that if the region is convex, than the maximal values of the derivatives are not very large. If the region is non-convex, then the values of the derivative may be very large in the neighborhoods of the interior corner. The one-dimensional in space case (the segment [0, d], (3)–(5)) is simple enough and the derivatives are not very large. The question arises: in which way do the derivatives influence the error of the difference solution? The answer lies with the estimation of the error for the solution of the difference scheme (10)–(12)

$$\max_{0\le i\le n, 0\le j\le m} |z_i^j| = \max_{0\le i\le n, 0\le j\le m} |u(x_i,t_j) - v_i^j| \le c \max_{1\le i\le n-1, 0\le j\le m-1} |\Psi(x_i,t_j)|, \tag{13}$$

here a constant c depends on the length of the segment d. The error of the approximation

$$\Psi(x_i,t^j) = \frac{\tau}{2}\frac{\partial^2 u(x_i,\bar{t}_j)}{\partial t^2} - D\frac{h^2}{12}\frac{\partial^4 u(\bar{x}_i,t_j)}{\partial x^4} \tag{14}$$

depends on the partial derivatives of the solution of the boundary-value problem (3)–(5)

$$\frac{\partial^2 u(x,t)}{\partial t^2}, \ \frac{\partial^4 u(x,t)}{\partial x^4}.$$

The estimation of the error (13) and the expression of the approximation (14) explain the importance and the meaning of the partial derivatives of the boundary-value problem (3)–(5).

The estimation of the error (13) explains the role of the steps of the grid τ and h. One can see that if $\tau,\ h \to 0$, then the error of the solution $|u(x_i,t_j) - v_i^j| \to 0$ and the difference scheme converges.

A new question is bound to arise: when does the difference scheme converge and what kind of conditions should be satisfied. The answer depends on the concrete difference scheme. Now we will give an answer in the case of an explicit difference scheme (10)–(12). The additional condition should be satisfied

$$D\frac{\tau}{h^2} \le 1/2. \tag{15}$$

It turns out that there is a restriction on the ratio between the steps of the grid τ and h. The steps can not be taken without restraint. This condition has the name, i.e. the stability condition. The stability is another important feature of a difference scheme. If the stability condition is not satisfied, then the difference scheme does not converge and can not be applied to the solution of a differential problem. Now we will explain what lies at the root of the stability. As shown above, the introduction of a difference scheme permits one to reduce the solution of a problem associated with the differential equation to an algebraic problem. In this situation the boundary condition and the initial conditions, all of which are called in the sequel the input data, are specified with a certain error. This procedure causes rounding errors to be inevitable in the process of numerical solution. Because of this, it seems natural to require that small errors in specifying the input data should not increase during the process of the execution and the result in wrong reasoning. The schemes in which initial errors grow during the course of calculations turn out to be unstable and, from the viewpoint of possible applications, cannot find response. The stability condition (15) guarantees that the error of the input data and rounding errors will not grow during the course of calculations. The explicit difference scheme (10)–(12) is called a conditionally stable difference scheme.

4 The Implicit Difference Scheme

The implicit difference scheme is often applied to the solution of the differential problems. This scheme is called the pure implicit difference scheme as well. We will start with a simple example. Thus, we will consider the similar boundary problem for the diffusion equation

$$\frac{\partial u}{\partial t} = D\frac{\partial^2 u}{\partial x^2} + f(x,t) \tag{16}$$

in a rectangle $Q = \{0 < x < d,\, 0 < t \leq T\}$ with initial condition for $t = 0$

$$u(x,0) = \varphi(x),\ x \in [0,d] \tag{17}$$

and a new type of boundary condition

$$\frac{\partial u(0,t)}{\partial x} = \gamma_0(u(0,t) - u_0),\ \frac{\partial u(d,t)}{\partial x} = \gamma_1(u(d,t) - u_1),\ t \in [0,T]. \tag{18}$$

Here $f(x,t)$, $\varphi(x)$ are the given functions, γ_0, γ_1, u_0, u_1 are the given parameters. The boundary condition of such a type is employed, for example, when there are fluxes through the boundary and the rate of the flux du/dx is proportional to the difference of the concentration $u - u_0$ inside and outside. The physical meaning of the function $f(x,t)$ is a source.

Now we will employ the difference scheme techniques for the solution of the problem (3)–(5) to (16)–(18). Let us introduce a grid $\bar{Q}_{h\tau} = \{(x_i,t_j) \in \bar{Q} = \{0 \leq x \leq d, 0 \leq t \leq T\}\}$ with steps h and τ along the directions Ox and Ot, respectively. In a similar way the initial boundary value problem for the diffusion equation can be replaced by the difference equation using the pattern consisting of four points (x_{i-1},t_{j+1}), (x_i,t_{j+1}),(x_{i+1},t_{j+1}), (x_i,t_j) (see Fig. 5). Thus, differential equation (16) converts to the difference equation

$$\frac{v_i^{j+1} - v_i^j}{\tau} = D\frac{v_{i+1}^{j+1} - 2v_i^{j+1} + v_{i-1}^{j+1}}{h^2} + f(x_i,t_j), \tag{19}$$

$$i = 1,2,\ldots,n-1,\ j = 0,1,\ldots,m-1.$$

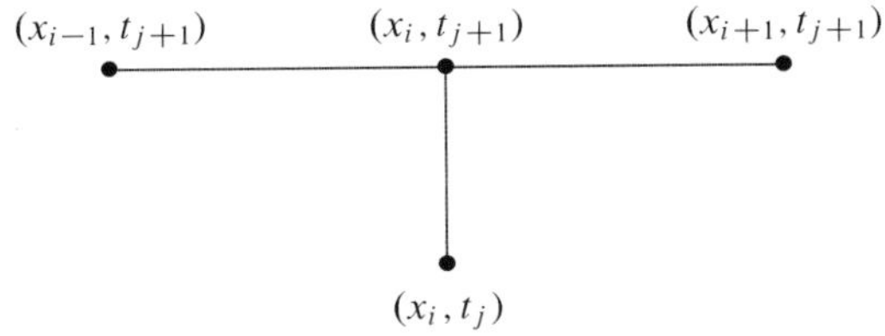

Fig. 5 The pattern of the implicit scheme

The initial and boundary conditions are changed by

$$v_i^0 = \varphi(x_i),\ i = 0, 1, \ldots, n \tag{20}$$

and

$$\frac{v_1^j - v_0^j}{h} = \gamma_0(v_0^j - u_0),\ \frac{v_n^j - v_{n-1}^j}{h} = \gamma_1(v_n^j - u_1),\ j = 1, 2, \ldots, m. \tag{21}$$

The difference equation is implicit, two-layer. It is easy to show (similar to explicit difference equation (10)) that the error of the approximation for the difference equation (19) is $O(\tau + h^2)$, the initial condition (17) is approximated exactly by (20) and the approximation error of the boundary condition (18) is $O(h)$. We will prove only the last statement. We will consider the left side boundary condition

$$\frac{\partial u(0,t)}{\partial x} = \gamma_0(u(0,t) - u_0)$$

and the corresponding difference boundary condition

$$\frac{v_1^j - v_0^j}{h} = \gamma_0(v_0^j - u_0).$$

We next develop $u(x)$ in the series by Taylor's formula up to the derivatives of the second order

$$u(x_1, t_j) = u(x_0 + h, t_j) = u(x_0, t_j) + h\frac{\partial u(x_0, t_j)}{\partial x} + \frac{h^2}{2}\frac{\partial^2 u(x_0, t_j)}{\partial x^2} + O(h^3).$$

This yields

$$\frac{u(x_1, t_j) - u(x_0, t_j)}{h} = \gamma_0(u(x_0, t_j) - u_0) + \frac{h}{2}\frac{\partial^2 u(x_0, t_j)}{\partial x^2} + O(h^2)$$

when the values v_1^j, v_0^j are replaced in the left-side difference boundary condition (21) by the values of the solution of the problem (16)–(18) $u(x_1, t_j)$, $u(x_0, t_j)$ respectively. From this equality follows the error of the approximation of the boundary condition of the order $O(h)$ if

$$\max \left| \frac{\partial^2 u(x,t)}{\partial x^2} \right| \leq c < \infty, (x,t) \in \bar{Q}.$$

In a similar way we can prove the error of the approximation for the right-side boundary condition (18) by (21).

4.1 The Convergence and the Stability

The implicit difference scheme (19)–(21) is unconditionally stable. It turns out that there is no restriction on the ratio between the steps of the grid τ and h. The steps can be taken without any restraints. It is a very important and useful feature of the implicit difference scheme.

The difference scheme (19)–(21) converges and the error of the difference solution is of the order $O(\tau + h)$.

4.2 The Calculation of a Solution

The calculation of the solution of the difference scheme (19)–(21) is performed in the series one grid layer after another. At first the solution on the zero layer at $t = 0$ using initial condition (20)

$$v_i^0 = \varphi(x_i), \ i = 0, 1, \ldots, n$$

is calculated.

Now let the solution v_i^j be calculated at $j - th$ layer. Consequently, the values v_i^j for $i = 0, 1, \ldots, n$ are known. We proceed to solve the difference equation (19) subject to value v_i^{j+1} at $j + 1 - th$ layer. The coefficients of the difference equation (19) tend to infinity if $h, \tau \to 0$. This can cause serious difficulties in the calculation. For this reason the difference equations are multiplied by some parameters in order to reduce the values of the coefficients. Let us multiply by the step of the grid τ Eq. 19. We have the difference equation for $j + 1 - th$ layer

$$\frac{\tau D}{h^2} v_{i+1}^{j+1} - \left(1 + \frac{2\tau D}{h^2}\right) v_i^{j+1} + \frac{\tau D}{h^2} v_{i-1}^{j+1} = -\tau f(x_i, t_j), \tag{22}$$

$$i = 1, 2, \ldots, n - 1.$$

Let us multiply by the step of the grid h (in order to decrease the coefficients) the initial and the boundary conditions (21). This yields

$$v_1^{j+1} - (1 + h\gamma_0)v_0^{j+1} = -h\gamma_0 u_0, \; v_n^{j+1}(1 - h\gamma_1) - v_{n-1}^{j+1} = -h\gamma_1 u_1. \quad (23)$$

The difference equation (22) together with the boundary condition (23) comprise the system of $n + 1$ linear equations with $n + 1$ unknown values

$$v_i^{j+1}, \; i = 0, 1, \ldots, n.$$

Let us introduce new notations

$$s_i = v_i^{j+1}, \; i = 0, 1, \ldots, n,$$

$$a_i = b_i = \frac{\tau D}{h^2}, \; c_i = 1 + \frac{2\tau D}{h^2}, \; \bar{f}_i = -\tau f(x_i, t_j), \; i = 1, 2, \ldots, n-1,$$

$$\beta_1 = \frac{1}{1 + h\gamma_0}, \; \beta_2 = \frac{1}{1 - h\gamma_1}, \; \mu_1 = \frac{h\gamma_0 u_0}{1 + h\gamma_0}, \; \mu_2 = \frac{h\gamma_1 u_1}{h\gamma_1 - 1}$$

if $1 - h\gamma_1 \neq 0$ and $1 + h\gamma_0 \neq 0$. In the sequel the system (22), (23) derives the form

$$a_i s_{i-1} - c_i s_i + b_i s_{i+1} = \bar{f}_i, \; i = 1, 2, \ldots, n-1, \quad (24)$$

$$s_0 = \beta_1 s_1 + \mu_1, \quad (25)$$

$$s_n = \beta_2 s_{n-1} + \mu_2. \quad (26)$$

Now we rewrite this system in the vectorial form. For this purpose we introduce a matrix C and the vectors $\mathbf{S}$, $\mathbf{F}$ as follows

$$C = \begin{pmatrix} 1 & -\beta_1 & 0 & \ldots & 0 & 0 & 0 & \ldots & 0 & 0 & 0 \\ a_1 & -c_1 & b_1 & \ldots & 0 & 0 & 0 & \ldots & 0 & 0 & 0 \\ \ldots & \ldots & \ldots & \ldots & \ldots & \ldots & \ldots & \ldots & \ldots & \ldots & \ldots \\ 0 & 0 & 0 & \ldots & a_i & -c_i & b_i & \ldots & 0 & 0 & 0 \\ \ldots & \ldots & \ldots & \ldots & \ldots & \ldots & \ldots & \ldots & \ldots & \ldots & \ldots \\ 0 & 0 & 0 & \ldots & 0 & 0 & 0 & \ldots & a_{n-1} & -c_{n-1} & b_{n-1} \\ 0 & 0 & 0 & \ldots & 0 & 0 & 0 & \ldots & 0 & -\beta_2 & 1 \end{pmatrix},$$

$$\mathbf{S} = \begin{pmatrix} s_0 \\ s_1 \\ \vdots \\ s_i \\ \vdots \\ s_{n-1} \\ s_n \end{pmatrix},$$

$$\mathbf{F} = \begin{pmatrix} \mu_1 \\ \bar{f}_1 \\ \vdots \\ \bar{f}_i \\ \vdots \\ \bar{f}_{n-1} \\ \mu_2 \end{pmatrix}.$$

With the aid of these notations we obtain

$$C\mathbf{S} = \mathbf{F}. \tag{27}$$

One can see that the first and the last rows of the matrix C consist of two non-zero coefficients and the remaining rows of three non-zero coefficients and the system of the linear equations are with three-diagonal matrix. The next step is a count of the values s_i or v_i^{j+1}. As one can easily observe, the solution of the implicit difference scheme can be reduced to a row of the systems of the linear equations of a special type. We obtain the system of linear equations for every layer, starting from the first one for $j = 1$. It turns out that it is possible to use this feature for the application of a special economical method to the solution of the system. The economical method means the time which requires a minimal execution time in giving a solution to the system of the linear equations. The flow variant of the elimination method belongs to the economical methods. It is worth emphasizing that the number of the arithmetic operations, which is necessary to solve the system, is proportional to the number of the unknown values. Thus, we can predicate, if for the solution of the boundary-value problem the implicit difference scheme is applied and consequently, if for the solution of the system of the linear equations obtained, the flow variant of the elimination method is applied then the number of the arithmetic operations, which is necessary to compute the values v_i^j, $i = 0, 1, \ldots, n$, $j = 0, 1, \ldots, m$ is proportional to the number of the unknown values $(n + 1)(m + 1)$. Now we will continue the solution of the difference scheme by developing the flow variant of the elimination method.

5 The Elimination Method for the System of Linear Equations

This section is devoted to a special elimination method (the flow variant of the elimination method) for the solution of a system of a linear equations with three diagonal matrixes. The system of the linear equations with three diagonal matrixes has been developed by constructing the difference schemes for the solution of the different differential problems. Above we obtained such system for the solution of the boundary-value problem for the diffusion equation. Now we will consider the system of linear equations

$$a_i s_{i-1} - c_i s_i + b_i s_{i+1} = f_i,\ i = 1, 2, ..., n-1, \tag{28}$$

$$s_0 = \beta_1 s_1 + \mu_1, \tag{29}$$

$$s_n = \beta_2 s_{n-1} + \mu_2. \tag{30}$$

It is easy to see that the first and the last equations have two non-zero coefficients remaining equations – three non-zero coefficients and the system of linear equations are with the three diagonal matrix.

We now turn to the solution of the system. Now we will eliminate the value s_0 in Eq. 28 for $i = 1$ with the aid of the equality (29). Thus,

$$(a_1\beta_1 - c_1)s_1 + b_1 s_2 = f_1 - a_1\mu_1.$$

We assume $a_1\beta_1 - c_1 \neq 0$. This yields

$$s_1 = d_2 s_2 + e_2, \tag{31}$$

where

$$d_2 = \frac{b_1}{c_1 - a_1\beta_1},\ e_2 = \frac{a_1\mu_1 - f_1}{c_1 - a_1\beta_1}. \tag{32}$$

We will assume the recurrence relation

$$s_{i-1} = d_i s_i + e_i$$

to be valid. The aim is to obtain a new relation

$$s_i = d_{i+1} s_{i+1} + e_{i+1} \tag{33}$$

with the indeterminate coefficients d_{i+1} and e_{i+1}. Substituting

$$s_{i-1} = d_i s_i + e_i$$

into (28) yields

$$(a_i d_i - c_i)s_i + b_i s_{i+1} = f_i - a_i e_i$$

which leads to (33) if we assume $a_i d_i - c_i \neq 0$.

Then the coefficients

$$d_{i+1} = \frac{b_i}{c_i - d_i a_i}, \tag{34}$$

$$e_{i+1} = \frac{a_i e_i - f_i}{c_i - d_i a_i},\ i = 2, 3..., n-1. \tag{35}$$

Thus, we establish the recurrence formulae for the determination of both d_i and e_i. Now we turn to the determination of s_i.

The value s_n is recovered from the system of the equations

$$s_n = \beta_2 s_{n-1} + \mu_2,$$
$$s_{n-1} = d_n s_n + e_n$$

under the constraint $1 - d_n\beta_2 \neq 0$.

Hence, the value s_n equals

$$s_n = \frac{\mu_2 + \beta_2 e_n}{1 - d_n\beta_2}. \tag{36}$$

Now we can calculate the values s_i, $i = n-1, n-2, \ldots, 1$ by employing

$$s_i = d_{i+1}s_{i+1} + e_{i+1} \tag{37}$$

for $i = n-1, n-2, \ldots, 1$ and value s_0 by employing

$$s_0 = \beta_1 s_1 + \mu_1, \tag{38}$$

By going through the matter chronologically the finding of the coefficients by the formulae (32), (34), (35) d_i, e_i is called the forward elimination path, and the finding of the values s_i by the formulae (36)–(38) is called the backward elimination path.

The algorithm presented below is called the right elimination method and is showing the gateway for the general case:

$$\overrightarrow{d_{i+1}} = \frac{b_i}{c_i - d_i a_i}, \quad \overrightarrow{e_{i+1}} = \frac{a_i e_i - f_i}{c_i - d_i a_i}, \quad i = 2, 3, \ldots, n-1,$$

$$s_n = \frac{\mu_2 + \beta_2 e_n}{1 - d_n\beta_2},$$

$$\overleftarrow{s_i} = d_{i+1}s_{i+1} + e_{i+1}, \quad i = n-1, n-2, \ldots, 1,$$

$$s_0 = \beta_1 s_1 + \mu_1.$$

Here the symbols $(\rightarrow)$ and $(\leftarrow)$ indicate the directions of the index count: either from i to $i+1$ or from $i+1$ to i.

The flow variant method can be applied to the solution of the implicit difference scheme (19)–(21) for a count at every layer starting from $j = 1$ until $j = m$. The scheme (19)–(21) was converted to the system of linear equations (24)–(26) with the three-diagonal matrixes. This system is solved by the flow variant. Consequently, the method of the solution of the implicit difference scheme (19)–(21) is presented totally.

5.1 Stability of the Elimination Method

Now we will discus the problem of the influence of the coefficients a_i, b_i, c_i, β_i errors on the solution s_i error. The stability of the method for the solution of the linear equations system means that the error of the coefficients a_i, b_i, c_i, β_i, together with the calculation round down error, practically does not influence the final result. If the method for the solution of the linear equations system is unstable, it means that the error of coefficients a_i, b_i, c_i, β_i, together with the calculation round down error, will significantly influence the final result and values s_i can be different galore from exact solution. We will give only a sufficient condition for the stability of the elimination method without proof. Common practice involves sufficient conditions

$$|c_i| \geq |a_i| + |b_i|,\ i = 1, 2, ..., n-1,$$

$$|\beta_k| \leq 1,\ k = 1, 2,\ |\beta_1| + |\beta_2| < 2.$$

The difference schemes, which are investigated in this book, generate the linear systems, which satisfy the sufficient conditions.

6 The Crank–Nicolson Difference Scheme

The Crank–Nicolson difference scheme is among the most usable difference methods. We will investigate a new example of boundary-value problem for the diffusion equation. Thus, we will consider the boundary-value problem

$$\frac{\partial u}{\partial t} = D\frac{\partial^2 u}{\partial x^2} \tag{39}$$

in a rectangle $Q = \{0 < x < d,\ 0 < t \leq T\}$ with the initial condition for $t = 0$

$$u(x, 0) = \varphi(x),\ x \in [0, d] \tag{40}$$

and a new type of the boundary condition

$$\frac{\partial u(0, t)}{\partial x} = 0,\ u(d, t) = \psi(t),\ t \in [0, T]. \tag{41}$$

Here $\varphi(x)$, $\psi(t)$ are the given functions. The boundary conditions of such a type are employed, for example, when there is no flux through the boundary $x = 0$ and the concentration $u(d, t)$ at the other boundary $x = d$ is given and equals $\psi(t)$ for every moment of the time t.

Now we will apply the technique of the difference scheme developed earlier to the solution of problem (39)–(41). Let us introduce a grid $\bar{Q}_{h\tau} = \{(x_i, t_j) \in \bar{Q}\}$ with steps h and τ along the directions Ox and Ot, respectively in rectangle

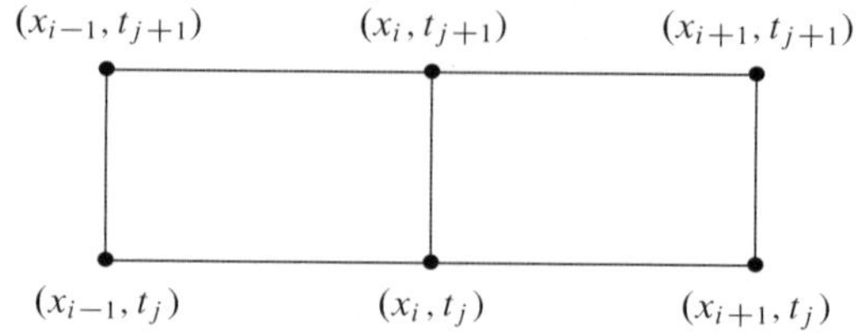

Fig. 6 The pattern of the Crank–Nicolson difference scheme

$\bar{Q} = \{0 \le x \le d, 0 \le t \le T\}$. In a similar way the boundary-value problem for the diffusion equation can be replaced by the difference equation using the pattern consisting of six grid points

$$(x_{i-1}, t_j), (x_i, t_j), (x_{i+1}, t_j), (x_{i-1}, t_{j+1}), (x_i, t_{j+1}), (x_{i+1}, t_{j+1})$$

(see Fig. 6). The diffusion equation (39) is replaced by the difference equation

$$\frac{v_i^{j+1} - v_i^j}{\tau} = \frac{D}{2}\left(\frac{v_{i+1}^{j+1} - 2v_i^{j+1} + v_{i-1}^{j+1}}{h^2} + \frac{v_{i+1}^j - 2v_i^j + v_{i-1}^j}{h^2}\right), \tag{42}$$

$$i = 1, 2, \ldots, n-1,\ j = 0, 1, \ldots, m-1.$$

The initial condition (40) is changed by

$$v_i^0 = \varphi(x_i),\ i = 0, 1, \ldots, n \tag{43}$$

and the boundary conditions (41)

$$\frac{v_1^j - v_0^j}{h} = 0,\ v_n^j = \psi(t_j),\ j = 1, 2, \ldots, m. \tag{44}$$

The difference scheme (42)–(44) is implicit and two-layer.

It easy to show that the difference equation (42) provides an approximation for the diffusion equation (39) of the order $O(\tau^2 + h^2)$ if the partial derivatives of function $u(t, x)$ satisfy inequalities

$$\max\left|\frac{\partial^3 u(x,t)}{\partial t^3}\right|,\ \max\left|\frac{\partial^4 u(x,t)}{\partial x^4}\right| = c < \infty$$

in the region $\bar{Q}$. The initial condition (40) is approximated by (43) exactly and the error of the approximation for the boundary condition (44) is $O(h)$ (analogously as for (18), (21)).

6.1 The Calculation of a Solution

The procedure of the calculation is very similar to that of the implicit difference scheme (19)–(21). The calculation of the Crank–Nicolson difference scheme (42)–(44) is performed in the series one grid layer after another. At first the solution on the zero layer at $t = 0$ using the initial condition (43)

$$v_i^0 = \varphi(x_i),\ i = 0, 1, \ldots, n$$

is calculated.

Now, suppose, that the solution v_i^j is calculated at $j - th$ layer. Thus, the values v_i^j for $i = 0, 1, \ldots, n$ are known. We proceed to solve the difference equation (42) subject to value v_i^{j+1} at $j + 1 - th$ layer. The coefficients of the difference equation (42) and the left-side boundary condition (44) tend to infinity if $h, \tau \to 0$. This can cause serious difficulties in the calculation. For this reason the difference equations are multiplied by the step of the grid τ and the boundary condition by the step of the grid h. This yields

$$v_i^{j+1} = v_i^j + \frac{\tau D}{2}\left(\frac{v_{i+1}^{j+1} - 2v_i^{j+1} + v_{i-1}^{j+1}}{h^2} + \frac{v_{i+1}^j - 2v_i^j + v_{i-1}^j}{h^2}\right), \tag{45}$$

$$i = 1, 2, \ldots, n - 1$$

and the boundary condition for $j + 1 - th$ layer

$$v_0^{j+1} = v_1^{j+1},\ v_n^{j+1} = \psi(t_{j+1}),\ j = 1, 2, \ldots, m - 1. \tag{46}$$

The difference equation together with the boundary condition establish the system of $n + 1$ linear equations with $n + 1$ unknown values

$$v_i^{j+1},\ i = 0, 1, \ldots, n.$$

Let us rewrite this system in a form, suitable for the solution by the flow variant of the elimination method. For this purpose we introduce the following notations

$$s_i = v_i^{j+1},\ i = 0, 1, \ldots, n,$$

$$a_i = b_i = \frac{\tau D}{2h^2},\ c_i = 1 + \frac{\tau D}{h^2},\ i = 1, 2, \ldots, n - 1,$$

$$\beta_1 = 1,\ \beta_2 = 0,\ \mu_1 = 0,\ \mu_2 = \psi(t_{j+1}),$$

$$f_i = -v_i^j\left(1 - \frac{\tau D}{h^2}\right) - \frac{\tau D}{2h^2}\left(v_{i+1}^j + v_{i-1}^j\right),$$

$$i = 1, 2, \ldots, n - 1.$$

With the aid of these notations the system of the linear equations (45), (46) takes the form

$$s_0 = \beta_1 s_1 + \mu_1, \tag{47}$$

$$a_i s_{i-1} - c_i s_i + b_i s_{i+1} = f_i,\ i = 1, 2, \ldots, n-1, \tag{48}$$

$$s_m = \beta_2 s_{m-1} + \mu_2. \tag{49}$$

The first and the last equations consist of two unknown values, and the remaining equations have three unknown values. As a result, the system of linear equations is with the three-diagonal matrix

$$B = \begin{pmatrix} 1 & -1 & 0 & \ldots & 0 & 0 & 0 & \ldots & 0 & 0 & 0 \\ a_1 & -c_1 & b_1 & \ldots & 0 & 0 & 0 & \ldots & 0 & 0 & 0 \\ \ldots & \ldots & \ldots & \ldots & \ldots & \ldots & \ldots & \ldots & \ldots & \ldots & \ldots \\ 0 & 0 & 0 & \ldots & a_i & -c_i & b_i & \ldots & 0 & 0 & 0 \\ \ldots & \ldots & \ldots & \ldots & \ldots & \ldots & \ldots & \ldots & \ldots & \ldots & \ldots \\ 0 & 0 & 0 & \ldots & 0 & 0 & 0 & \ldots & a_{n-1} & -c_{n-1} & b_{n-1} \\ 0 & 0 & 0 & \ldots & 0 & 0 & 0 & \ldots & 0 & 0 & 1. \end{pmatrix}$$

Now the flow variant method can be applied to the solution of the system (42)–(44) as well as for the calculations of values $s_i = v_i^{j+1}$ for $i = 0, 1, \ldots, n$. The count of the difference solution on $j-th$ layer is finished. We will continue in a similar way up to last layer $n-th$. The method of the solution of the Crank–Nicolson difference scheme is finished.

6.2 *The Convergence and the Stability*

The Crank–Nicolson difference scheme (42)–(44) is unconditionally stable. It turns out that there is no restriction on the ratio between the steps of the grid τ and h. The steps can be taken without any restraints. It is a very important and useful feature of the Crank–Nicolson difference scheme.

The difference scheme (42)–(44) converges and the error of the difference solution is of order $O(\tau^2 + h^2)$.

7 The Difference Scheme with the Weights

The difference scheme with the weights is a generalized form of the previous schemes, applied to the diffusion equation. It is a scheme with two-layer difference equation.

Here we will discuss only the question of the substitution of the diffusion equation and will not consider the question of an approximation of an initial and boundary conditions. We will use a pattern containing six points of the grid

$$(x_{i-1}, t_j), (x_i, t_j), (x_{i+1}, t_j), (x_{i-1}, t_{j+1}), (x_i, t_{j+1}), (x_{i+1}, t_{j+1}).$$

The diffusion equation

$$\frac{\partial u(x_i, t_j)}{\partial t} = D \frac{\partial^2 u(x_i, t_j)}{\partial x^2}$$

is replaced by difference equation and has the form

$$\frac{v_i^{j+1} - v_i^j}{\tau} = \sigma \frac{v_{i+1}^{j+1} - 2v_i^{j+1} + v_{i-1}^{j+1}}{h^2}$$

$$+(1-\sigma) \frac{v_{i+1}^j - 2v_i^j + v_{i-1}^j}{h^2},$$

$$i = 1, 2, \ldots, n-1, \; j = 0, 1, \ldots, m-1,$$

where σ is an arbitrary real parameter.

The difference equation converts to the explicit difference equation if $\sigma = 0$, to the pure implicit difference one if $\sigma = 1$, to implicit difference one if $\sigma \neq 0$ and to the Crank–Nicolson equation if $\sigma = 1/2$.

In a similar way one can show that the error of the approximation is $O(\tau^2 + h^2)$ for $\sigma = 1/2$ if the partial derivatives of function $u(t, x)$ satisfy the inequalities

$$\max |\frac{\partial^3 u(x,t)}{\partial t^3}|, \max |\frac{\partial^4 u(x,t)}{\partial x^4}| = c < \infty$$

and the error of the approximation is $O(\tau + h^2)$ for $\sigma \neq 1/2$ if the partial derivatives of the function $u(t, x)$ satisfy the inequalities

$$\max |\frac{\partial^2 u(x,t)}{\partial t^2}|, \max |\frac{\partial^4 u(x,t)}{\partial x^4}| = c < \infty$$

in the region of the investigation.

As in the previous cases it is necessary to add the initial and the boundary conditions and to complete the difference scheme. The procedure of the calculation is very similar to that of a pure implicit or the Crank–Nicolson difference schemes. The calculation of the difference scheme with weights is performed in series one grid layer after another. At first the solution on the zero layer at $t = 0$ using initial condition is calculated. The solution on the next layers is being calculated similarly with respect to the type of the boundary conditions.

7.1 The Convergence and the Stability

The difference scheme with the weights has the following conditions of the stability:

a) If $\sigma \geq \frac{1}{2}$, then the scheme is stable for any τ and h
b) If $0 < \sigma < \frac{1}{2}$, then the scheme is conditionally stable and the steps of the grid have to satisfy the inequality

$$\frac{D\tau}{h^2} \leq \frac{1}{2(1-2\sigma)}.$$

It turns out that there is no restriction on the ratio between the steps of the grid τ and h for

$$\sigma \geq \frac{1}{2}.$$

The steps can be taken without any restraints. It is a very important and useful feature of the difference scheme with the weights.

The difference scheme with the weights converges and the error of the difference solution depends on the errors of the approximation of the diffusion equation and the initial and boundary conditions.

8 The Crank–Nicolson Difference Scheme on Non-equidistance Grid

This section deals with to the Crank–Nicolson difference scheme on a non-equidistance grid. The non-equidistance grid plays an important role when the media is non-homogeneous or the geometry of an area of investigation is complicated. Now we will consider the boundary-value problem

$$\frac{\partial u}{\partial t} = D\frac{\partial^2 u}{\partial x^2} \tag{50}$$

in a rectangle $Q = \{0 < x < d,\ 0 < t \leq T\}$ with the initial condition for $t = 0$

$$u(x,0) = \varphi(x),\ x \in [0,d] \tag{51}$$

and the boundary condition

$$\frac{\partial u(0,t)}{\partial x} = 0,\ u(d,t) = \psi(t),\ t \in [0,T]. \tag{52}$$

Here $\varphi(x),\ \psi(t)$ are the given functions.

Now we will apply the technique of the difference scheme developed earlier for the solution of the problem (50)–(52). Let us introduce a non-equidistance grid $\bar{Q}_{h\tau} = \{(x_i, t_j) \in \bar{Q}\}$ with steps $h_i = x_i - x_{i-1}$ and $\tau_j = t_j - t_{j-1}$ along the directions Ox and Ot, respectively in rectangle $\bar{Q} = \{0 \le x \le d, 0 \le t \le T\}$. The grid steps depend on the subscripts i and j. The steps should satisfy the normalization conditions $\sum_{i=1}^{n} h_i = d$ and $\sum_{j=1}^{m} \tau_j = T$.

In a similar way the boundary-value problem for the diffusion equation can be replaced by the difference equation using the pattern consisting of six grid points

$$(x_{i-1}, t_j), (x_i, t_j), (x_{i+1}, t_j), (x_{i-1}, t_{j+1}), (x_i, t_{j+1}), (x_{i+1}, t_{j+1}).$$

The diffusion equation (50) is replaced by the difference equation

$$\frac{v_i^{j+1} - v_i^j}{\tau_{j+1}} = \frac{D}{2\hat{h}_i}\left(\frac{v_{i+1}^{j+1} - v_i^{j+1}}{h_i} - \frac{v_i^{j+1} - v_{i-1}^{j+1}}{h_{i-1}} + \frac{v_{i+1}^j - v_i^j}{h_i} - \frac{v_i^j - v_{i-1}^j}{h_{i-1}}\right), \quad (53)$$

$$i = 1, 2, \ldots, n-1, \; j = 0, 1, \ldots, m-1,$$

where $\hat{h}_i = (h_i + h_{i-1})/2$.

The initial condition (51) is changed by

$$v_i^0 = \varphi(x_i), \; i = 0, 1, \ldots, n \quad (54)$$

and the boundary conditions (52)

$$\frac{v_1^j - v_0^j}{h_1} = 0, \; v_n^j = \psi(t_j), \; j = 1, 2, \ldots, m. \quad (55)$$

The difference scheme (53)–(55) is implicit and two-layer.

It is easy to show that the difference equation (53) provides an approximation for the diffusion equation (50) of the order $O(\tau_{\max}^2 + h_{\max})$ if the partial derivatives of function $u(t, x)$ satisfy the inequalities

$$\max\left|\frac{\partial^3 u(x,t)}{\partial t^3}\right|, \; \max\left|\frac{\partial^3 u(x,t)}{\partial x^3}\right| = c < \infty$$

in the region $\bar{Q}$, where

$$\tau_{\max} = \max\{\tau_j, \; j = 1, 2, \ldots, m\}, \; h_{\max} = \max\{h_i, \; i = 1, 2, \ldots, n\}.$$

The initial condition (51) is approximated by (54) exactly and the error of the approximation for the boundary condition (55) is $O(h_1)$.

8.1 The Calculation of a Solution

The procedure of the calculation is very similar to that of the Crank–Nicolson difference scheme (42)–(44). The calculation of the difference scheme (53)–(55) is performed in the series one grid layer after another one. At first the solution on the zero layer at $t = 0$ using the initial condition (54)

$$v_i^0 = \varphi(x_i),\ i = 0, 1, \dots, n$$

is calculated.

Now, suppose, the solution v_i^j is calculated at $j - th$ layer. Thus, the values v_i^j for $i = 0, 1, \dots, n$ are known. We proceed to solve the difference equation (53) subject to value v_i^{j+1} at $j + 1 - th$ layer. The coefficients of the difference equation (53) and the left-side boundary condition (55) tend to infinity if $h_i, \tau_j \to 0$. This can cause serious difficulties in the calculation. For this reason the difference equations are multiplied by the step of the grid τ_{j+1} and the boundary condition by the step of the grid h_1. This yields

$$v_i^{j+1} = v_i^j + \frac{\tau_{j+1} D}{2\hat{h}_i}\left(\frac{v_{i+1}^{j+1} - v_i^{j+1}}{h_i} - \frac{v_i^{j+1} - v_{i-1}^{j+1}}{h_{i-1}} + \frac{v_{i+1}^j - v_i^j}{h_i} - \frac{v_i^j - v_{i-1}^j}{h_{i-1}}\right), \tag{56}$$

$$i = 1, 2, \dots, n - 1$$

and the boundary condition for $j + 1 - th$ layer

$$v_0^{j+1} = v_1^{j+1},\ v_n^{j+1} = \psi(t_{j+1}),\ j = 1, 2, \dots, m - 1. \tag{57}$$

The difference equation together with the boundary condition establish the system of $n + 1$ linear equations with $n + 1$ unknown values

$$v_i^{j+1},\ i = 0, 1, \dots, n.$$

Let us rewrite this system in a form, suitable for the solution by the flow variant of the elimination method. For this purpose we introduce the notations

$$s_i = v_i^{j+1},\ i = 0, 1, \dots, n,$$

$$a_i = \frac{\tau_{j+1} D}{2\hat{h}_i h_{i-1}},\ b_i = \frac{\tau_{j+1} D}{2\hat{h}_i h_i},$$

$$c_i = 1 + \frac{\tau_{j+1} D}{h_i h_{i-1}},\ i = 1, 2, \dots, n - 1,$$

$$\beta_1 = 1,\ \beta_2 = 0,\ \mu_1 = 0,\ \mu_2 = \psi(t_j),$$

$$f_i = -v_i^j\left(1 - \frac{\tau_j + 1D}{h_i h_{i-1}}\right) - b_i v_{i+1}^j - c_i v_{i-1}^j,$$

$$i = 1, 2, \ldots, n-1.$$

With the aid of these notations the system of the linear equations (56), (46) takes the form

$$s_0 = \beta_1 s_1 + \mu_1, \tag{58}$$

$$a_i s_{i-1} - c_i s_i + b_i s_{i+1} = f_i,\ i = 1, 2, \ldots, n-1, \tag{59}$$

$$s_m = \beta_2 s_{m-1} + \mu_2. \tag{60}$$

The first and the last equations consist of two unknown values, and the remaining equations have three unknown values. As a result, the system of linear equations is with the three-diagonal matrix

$$B = \begin{pmatrix} 1 & -1 & 0 & \ldots & 0 & 0 & 0 & \ldots & 0 & 0 & 0 \\ a_1 & -c_1 & b_1 & \ldots & 0 & 0 & 0 & \ldots & 0 & 0 & 0 \\ \ldots & \ldots & \ldots & \ldots & \ldots & \ldots & \ldots & \ldots & \ldots & \ldots & \ldots \\ 0 & 0 & 0 & \ldots & a_i & -c_i & b_i & \ldots & 0 & 0 & 0 \\ \ldots & \ldots & \ldots & \ldots & \ldots & \ldots & \ldots & \ldots & \ldots & \ldots & \ldots \\ 0 & 0 & 0 & \ldots & 0 & 0 & 0 & \ldots & a_{n-1} & -c_{n-1} & b_{n-1} \\ 0 & 0 & 0 & \ldots & 0 & 0 & 0 & \ldots & 0 & 0 & 1 \end{pmatrix}.$$

Now the flow variant method can be applied to the solution of the system (53)–(60) as well as for the calculation of values $s_i = v_i^{j+1}$ for $i = 0, 1, \ldots, n$. The count of the difference solution on $j-th$ layer is finished. We will continue in a similar way up to last layer $n-th$. The method of the solution of the Crank–Nicolson difference scheme is finished.

8.2 *The Convergence and the Stability*

The Crank–Nicolson difference scheme (53)–(55) is unconditionally stable. It turns out that there is no restriction on the ratio between the steps of the grid τ_j and h_i. The steps can be taken without any restraints. It is a very important and useful feature of the Crank–Nicolson difference scheme.

The difference scheme (53)–(55) converges and the error of the difference solution is of order $O(\tau^2 + h)$.

9 The Explicit Difference Scheme in the Cylindrical Coordinates

Now we consider the diffusion equation in the cylindrical coordinates. The cylindrical coordinates are applied when the region of investigation has the cylindrical form or the region of investigation is approximated by a set of the cylinders.

The diffusion equation

$$\frac{\partial u}{\partial t} = D\left(\frac{\partial^2 u}{\partial x^2} + \frac{\partial^2 u}{\partial y^2} + \frac{\partial^2 u}{\partial z^2}\right) \tag{61}$$

takes the form in the cylindrical coordinate system (r, ϕ, z)

$$\frac{\partial u}{\partial t} = D\frac{1}{r}\frac{\partial}{\partial r}(r\frac{\partial u}{\partial r})$$

in the case when the solution $u = u(r,t)$ depends neither on z nor on ϕ, that is, in the case of the axial symmetry. Thus, we will investigate the diffusion process in the region $Q = \{0 < r < R,\ 0 < t \le T\}$.

The boundary-value problem for the diffusion equation takes the form

$$\frac{\partial u}{\partial t} = D\frac{1}{r}\frac{\partial}{\partial r}(r\frac{\partial u}{\partial r}),\ (r,t) \in Q, \tag{62}$$

$$u(r,0) = \varphi(r), 0 < r \le R, \tag{63}$$

$$\frac{\partial u(0,t)}{\partial r} = 0,\ u(R,t) = \psi(t),\ t, \in [0,T]. \tag{64}$$

Here $\varphi(r)$, $\psi(t)$ are the given functions.

We proceed as usual and introduce on the segment $0 \le r \le R$ the equidistant grid

$$\bar{Q}_{h,\tau} = \{(r_i, t_j), r_i = ih,\ t_j = j\tau,\ i = 0,1,\ldots,n,\ j = 0,1,\ldots,m,$$

$$hn = R,\ m\tau = T\}.$$

Still using the framework of this section, a difference scheme of the boundary-value problem for the diffusion equation can be obtained.

The diffusion equation is approximated by the difference equation

$$\frac{v_i^{j+1} - v_i^j}{\tau} = \frac{D}{hr_i}\left(r_{i+1/2}\frac{v_{i+1}^j - v_i^j}{h} - r_{i-1/2}\frac{v_i^j - v_{i-1}^j}{h}\right), \tag{65}$$

$$i = 1,2,\ldots,n-1,\ j = 0,1,\ldots,m-1$$

with the initial

$$v(r_i, 0) = \varphi(r_i),\ i = 0, 1, \ldots, n \tag{66}$$

and boundary conditions

$$\frac{v_1^j - v_0^j}{h} = 0,\ v_n^j = \psi(t_j),\ j = 1, 2, \ldots, m, \tag{67}$$

where $r_{i\pm 1/2} = (i \pm 1/2)h$. This scheme is explicit, two layer.

In a similar way one can show that the error of the approximation for the diffusion equation (62) by the difference equation (65) is $O(\tau + h^2)$ if the partial derivatives of function $u(t, x)$ satisfy the inequalities

$$\max |\frac{\partial^2 u(r,t)}{\partial t^2}|,\ \max |\frac{\partial^4 u(r,t)}{\partial r^4}| = c < \infty$$

in the region of the investigation. The error of the approximation of the left-side boundary condition (64) by (67) is $O(h)$. The error of the approximation of the right-side boundary and initial (63) conditions equal zero.

9.1 The Calculation of a Solution

The explicit difference scheme (65)–(67) is simple and has convenient algorithm of the calculation and the programming. The calculation of the difference scheme (65)–(67) is performed in the series one grid layer after another. At first the solution on the zero layer at $t = 0$ using initial condition (66)

$$v_i^0 = \varphi(r_i),\ i = 0, 1, \ldots, n$$

is calculated.

Now, suppose, solution v_i^j is calculated at $j - th$ layer. Thus, the values v_i^j for $i = 0, 1, \ldots, n$ are known. We proceed to solve the difference equation (65) subject to value v_i^{j+1} at $j + 1 - th$ layer. We have

$$v_i^{j+1} = v_i^j + \frac{\tau D}{h r_i}\left(r_{i+1/2}\frac{v_{i+1}^j - v_i^j}{h} - r_{i-1/2}\frac{v_i^j - v_{i-1}^j}{h} \right),$$

$$i = 1, 2, \ldots, n-1.$$

All values of function v_i^j on the right side of the equality are known. So one can calculate the values v_i^{j+1}, $i = 1, 2, \ldots, n-1$. The values v_0^{j+1} and v_n^{j+1} are calculated using the boundary conditions (67)

$$v_0^{j+1} = v_1^{j+1},\ v_n^{j+1} = \psi(t_j).$$

Further analogously we will find the solution v_i^j on the next layer and will continue up to the last layer $t = t_m$. The calculation of a solution of the explicit difference scheme is finished.

9.2 *The Convergence and Stability*

The difference scheme (65)–(67) has the following condition of the stability:

$$D\frac{\tau}{h^2} \leq 1/6. \tag{68}$$

It turns out that there is a restriction on the ratio between the steps of the grid τ and h.

The difference scheme (65)–(67) converges and the error of the difference solution is $O(\tau + h)$.

10 The Crank–Nicolson Difference Scheme in the Cylindrical Coordinates

The Crank–Nicolson difference scheme is among the most usable difference methods for the solution of the boundary-value problem for the diffusion equation in the cylindrical coordinates. The main reasons are:

no restraints on the ratio between the steps of the grid τ and h. The steps can be taken without any restraints. It is a very important and useful feature of the Crank–Nicolson difference scheme.

The scheme is economical and can be applied to the flow variant of the elimination method.

Now we will continue with the approximation of the boundary-value problem (62)–(64). In a similar way the diffusion equation (62) can be substituted by the following difference equation

$$\begin{aligned}\frac{v_i^{j+1} - v_i^j}{\tau} = \frac{D}{2r_i}\Bigg(& r_{i+1/2}\frac{v_{i+1}^{j+1} - v_i^{j+1}}{h} - r_{i-1/2}\frac{v_i^{j+1} - v_{i-1}^{j+1}}{h} \\ &+ r_{i+1/2}\frac{v_{i+1}^j - v_i^j}{h} - r_{i-1/2}\frac{v_i^j - v_{i-1}^j}{h}\Bigg)/h, \end{aligned} \tag{69}$$

$$i = 1, 2, \ldots, n-1,\ j = 0, 1, \ldots, m-1.$$

The initial (63) and boundary (64) conditions are approximated by (66) and (67). This difference scheme is implicit, two-layer.

In a similar manner one can show that the error of the approximation of the diffusion equation (62) by the difference equation is $O(\tau^2 + h^2)$ if the partial derivatives of function $u(t, x)$ satisfy the inequalities

$$\max |\frac{\partial^3 u(r,t)}{\partial t^3}|, \max |\frac{\partial^4 u(r,t)}{\partial r^4}| = c < \infty$$

in the region of the investigation. It was shown that the error of the approximation of the left-side boundary condition (67) is $O(h)$. The error of the approximation of the right-side boundary and initial (67) conditions equal zero.

10.1 The Calculation of a Solution

The procedure of the calculation is very similar to that of the implicit difference scheme (19)–(21) and the Crank–Nicolson scheme for the diffusion equation in one-dimensional case (42)–(44). The calculation of the Crank–Nicolson difference scheme (69), (66), (67) is performed in the series one grid layer after another. At first the solution on the zero layer at $t = 0$ using initial condition (66)

$$v_i^0 = \varphi(r_i),\ i = 0, 1, \ldots, n$$

is calculated.

Now let the solution v_i^j be calculated at $j - th$ layer. Thus, the values v_i^j for $i = 0, 1, \ldots, n$ are known. We proceed to solve the difference equation (69) the subject to value v_i^{j+1} at $j + 1 - th$ layer. The coefficients of the difference equation and the left-side boundary condition (71) tend to infinity if $h, \tau \to 0$. This can cause serious difficulties in the calculation. For this reason the difference equations are multiplied by the step of the grid τ and the boundary condition by the step of the grid h. This yields

$$v_i^{j+1} = v_i^j + \frac{\tau D}{2r_i}\Bigg(r_{i+1/2}\frac{v_{i+1}^{j+1} - v_i^{j+1}}{h} - r_{i-1/2}\frac{v_i^{j+1} - v_{i-1}^{j+1}}{h}$$
$$+ r_{i+1/2}\frac{v_{i+1}^{j} - v_i^{j}}{h} - r_{i-1/2}\frac{v_i^{j} - v_{i-1}^{j}}{h}\Bigg)/h, \tag{70}$$

$$i = 1, 2, \ldots, n-1$$

with the boundary condition

$$v_0^{j+1} = v_1^{j+1},\ v_n^{j+1} = \psi(t_{j+1}). \tag{71}$$

The difference equation together with boundary condition establish the system of n linear equations with n unknown values

$$v_i^{j+1},\ i = 0, 1, \ldots, n-1.$$

The difference equation (70) consists of three unknown values v_{i+1}^{j+1}, v_i^{j+1}, v_{i-1}^{j+1} and the left-side boundary condition equation (71) consists of two values v_1^{j+1}, v_0^{j+1}. The right-side boundary condition permits to calculate a value v_n^{j+1} immediately. In this case the linear system of the equations is three-diagonal.

The next step is a count of the values v_i^{j+1}, $i = 0, 1, \ldots, n-1$. For this objective we rewrite the system of linear equations in a new form. Let us introduce the notations

$$s_i = v_i^{j+1},\ i = 0, 1, \ldots, n-1,$$

$$a_i = \frac{\tau D r_{i-1/2}}{2h^2 r_i},\ b_i = \frac{\tau D r_{i+1/2}}{2h^2 r_i},\ c_i = 1 + \frac{\tau D}{h^2},\ i = 1, 2, \ldots, n-2,$$

$$f_i = -\frac{\tau D}{2r_i}\left(r_{i+1/2}\frac{v_{i+1}^j - v_i^j}{h} - r_{i-1/2}\frac{v_i^j - v_{i-1}^j}{h}\right)/h,\ i = 1, 2, \ldots, n-2,$$

$$\beta_1 = 1,\ \mu_1 = 0,$$

$$\beta_2 = \left(\frac{\tau D r_{n-3/2}}{2r_{n-1}h^2}\right) / \left(1 + \frac{\tau D}{h^2}\right),$$

$$\mu_2 = f_{n-1} / \left(1 + \frac{\tau D}{h^2}\right),$$

$$f_{n-1} = -\frac{\tau D}{2r_{n-1}}\left(r_{n-1/2}\frac{v_n^j - v_{n-1}^j}{h} - r_{n-3/2}\frac{v_{n-1}^j - v_{n-2}^j}{h} + r_{n-1/2}\frac{\psi(t_{j+1})}{h}\right)/h.$$

In the sequel the system (70), (71) derives the form

$$s_1 = \beta_1 s_0 + \mu_1,$$

$$a_i s_{i-1} - c_i s_i + b_i s_{i+1} = f_i,\ i = 1, 2, \ldots, n-2,$$

$$s_{n-1} = \beta_2 s_{n-2} + \mu_2.$$

Now the flow variant method can be applied to the solution of this system of the linear equations. So the values

$$v_i^{j+1} = s_i,\ i = 0, 1, \ldots, n-1,$$

will be calculated. In similar way we proceed to solve the solution for next layer $j+2$ and will continue up last layer $j = m$. The calculation of the solution of the Crank–Nicolson difference scheme is finished.

10.2 *The Convergence and Stability*

The Crank–Nicolson difference scheme (69), (66), (67) is unconditionally stable. It turns out that there is no restriction on the ratio between the steps of the grid τ and h. The steps can be taken without any restraints. It is a very important and useful feature of the Crank–Nicolson difference scheme.

The difference scheme converges and the error of the difference solution is of order $O(\tau^2 + h)$.

11 The Discontinuous Diffusion Coefficient

In practice the region of the investigation is non-homogeneous. This is due to the discontinuity of the media. Usually the different media possess different coefficients of the diffusion. Consequently it is important to investigate various types of the problem.

Now we will investigate the boundary-value problem for the diffusion equation with discontinuous diffusion coefficient (a point of the discontinuity equals $x = d_1$)

$$\frac{\partial u}{\partial t} = D\frac{\partial^2 u}{\partial x^2} \tag{72}$$

in a rectangle $Q = \{0 < x < d,\, 0 < t \le T\}$
with the initial condition for $t = 0$

$$u(x, 0) = \varphi(x),\ x \in [0, d] \tag{73}$$

and the boundary condition

$$u(0, t) = \psi_1(t),\ u(d, t) = \psi_2(t),\ t \in [0, T]. \tag{74}$$

The diffusion coefficient is equal to

$$D = \begin{cases} D_1, \text{ if } 0 \le x < d_1 \\ D_2, \text{ if } d_1 < x \le d. \end{cases}$$

The point d_1 is a point of the discontinuity for the diffusion coefficient D. The solution of the problem (72)–(74) satisfies the matching conditions

$$u(d_1 - 0, t) = u(d_1 + 0, t),\ t \in [0, T], \tag{75}$$

$$D_1\frac{\partial u(d_1 - 0, t)}{\partial x} = D_2\frac{\partial u(d_1 + 0, t)}{\partial x},\ t \in [0, T]. \tag{76}$$

The first matching condition follows from the continuity of a concentration $u(x, t)$ and the second one follows from the continuity of a flow $D\partial u(x, t)/\partial x$.

The condition (76) can be obtained using the balance equation. For this purpose we integrate Eq. 72 in the segment $[d_1 - \epsilon, d_1 + \epsilon]$, where $\epsilon > 0$ is a small number. It follows that

$$\int_{d_1-\epsilon}^{d_1+\epsilon} \frac{\partial u(x,t)}{\partial t} dx = \int_{d_1-\epsilon}^{d_1+\epsilon} D\frac{\partial^2 u(x,t)}{\partial x^2} dx.$$

The simplest integration gives

$$\int_{d_1-\epsilon}^{d_1+\epsilon} \frac{\partial u(x,t)}{\partial t} dx = D_2\frac{\partial u(d_1+\epsilon,t)}{\partial x} - D_1\frac{\partial u(d_1-\epsilon,t)}{\partial x}.$$

From it when $\epsilon \to 0$ we obtain the matching condition

$$D_1\frac{\partial u(d_1-0,t)}{\partial x} = D_2\frac{\partial u(d_1+0,t)}{\partial x}, \quad t \in [0,T].$$

The boundary-value problem for the diffusion equation with a discontinuous diffusion coefficient (72)–(74) can be written in a more simple form. This form is more convenient for the numerical solution because the segment $[0,d]$ is divided into two segments $[0,d_1]$ and $[d_1,d]$ with continuous diffusion coefficient. Instead of one function $u(x,t)$ we introduce two functions $u_1(x,t)$ and $u_2(x,t)$. The function $u_1(x,t)$ satisfies:

$$\frac{\partial u_1}{\partial t} = D_1\frac{\partial^2 u_1}{\partial x^2}, \tag{77}$$

in a rectangle $Q_1 = \{0 < x < d_1, 0 < t \leq T\}$
with the initial condition for $t = 0$

$$u_1(x,0) = \varphi(x),\ 0 \leq x \leq d_1 \tag{78}$$

and the boundary condition

$$u_1(0,t) = \psi_1(t),\ \ 0 < t \leq T. \tag{79}$$

The function $u_2(x,t)$ satisfies:

$$\frac{\partial u_2}{\partial t} = D_2\frac{\partial^2 u_2}{\partial x^2}, \tag{80}$$

in a rectangle $Q_2 = \{d_1 < x < d,\ 0 < t \leq T\}$
with the initial condition for $t = 0$

$$u_2(x,0) = \varphi(x),\ d_1 \leq x \leq d \tag{81}$$

and the boundary condition

$$u_2(d,t) = \psi_2(x),\ 0 < t \le T. \tag{82}$$

The lacking boundary conditions are substituted by the matching conditions at the point d_1 of the discontinuity for the diffusion coefficient

$$u_1(d_1,t) = u_2(d_1,t),\ 0 < t \le T, \tag{83}$$

$$D_1 \frac{\partial u_1(d_1,t)}{\partial x} = D_2 \frac{\partial u_2(d_1,t)}{\partial x},\ 0 < t \le T. \tag{84}$$

12 The Explicit Difference Scheme

Now we will substitute the boundary-value problems for the diffusion equation with a continuous diffusion coefficient (77)–(84) by the explicit difference schemes (the same type as (10)–(12)). Let the grids

$$\bar{Q}_{1,h_1,\tau} = \{(x_i,t_j) \in \bar{Q}_1\},\ \bar{Q}_{2,h_2,\tau} = \{(x_i,t_j) \in \bar{Q}_2\}$$

with the steps h_1 ($h_1 n_1 = d_1$), h_2 ($h_2 n_3 = d - d_1$), $n_2 = n_1 + n_3$ and τ along the directions Ox and Ot, respectively in the rectangles $\bar{Q}_1$, $\bar{Q}_2$.

Let (x_i,t_j) be a fixed point of the grid inside the rectangles Q_1 or Q_2 and we replace the diffusion equations (77), (80) by the explicit difference equations

$$\frac{v_{1,i}^{j+1} - v_{1,i}^{j}}{\tau} = D_1 \frac{v_{1,i+1}^{j} - 2v_{1,i}^{j} + v_{1,i-1}^{j}}{h_1^2}, \tag{85}$$

$$i = 1,2,...,n_1 - 1,\ j = 1,2,...,m-1,$$

$$\frac{v_{2,i}^{j+1} - v_{2,i}^{j}}{\tau} = D_2 \frac{v_{2,i+1}^{j} - 2v_{2,i}^{j} + v_{2,i-1}^{j}}{h_2^2}, \tag{86}$$

$$i = n_1+1, n_1+2, ..., n_2 - 1,\ j = 1,2,...,m-1.$$

The initial conditions are

$$v_{1,i}^{0} = \varphi(x_i),\ i = 0,1,...,n_1,\ v_{2,i}^{0} = \varphi(x_i),\ i = n_1, n_1 - 2, ..., n_2 \tag{87}$$

and the boundary –

$$v_{1,0}^{j} = \psi_1(t_j),\ j = 1,2,...,m,\ v_{1,n_2}^{j} = \psi_2(t_j),\ j = 1,2,...,m. \tag{88}$$

The matching conditions are replaced

$$v_{1,n_1}^j = v_{2,n_1}^j, \; D_1 \frac{v_{1,n_1}^j - v_{1,n_1-1}^j}{h_1} = D_2 \frac{v_{2,n_1+1}^j - v_{2,n_1}^j}{h_2}, \; j = 1, 2, ..., m. \quad (89)$$

The difference equations (85), (86) together with the initial (87), the boundary (88) and the matching conditions (89) analogously form the difference scheme (85)–(89).

The error of an approximation of the differential equations (77), (80) by difference equations (85)–(86) is of order $O(\tau + h_1^2)$ and $O(\tau + h_2^2)$, the initial and boundary conditions are replaced exactly. The second matching condition has the error of the approximation $O(h_1 + h_2)$, the first one is replaced exactly.

12.1 *The Calculation of a Solution*

The calculation of the solution of the difference scheme (85)–(89) is performed in the series one grid layer after another (similar to the case of the explicit difference scheme). The distinction lies in the calculation at the point of the discontinuity of the diffusion coefficient D ($x = d_1$ or $x = x_{n_1}$). Let the values $v_{1,i}^j$, $i = 0, 1, \ldots, n_1 - 1$, and $v_{2,i}^j$, $i = n_1 - 1, n_1 - 2, \ldots, n_2$ be calculated. Then, in order to calculate the values v_{1,n_1}^j, v_{2,n_1}^j we apply the matching conditions (89). This yields

$$v_{1,n_1}^j = v_{2,n_1}^j = \frac{D_1 v_{1,n_1-1}^j h_2 + D_2 v_{2,n_1+1}^j h_1}{D_1 h_2 + D_2 h_1}.$$

Thus, the solution of the difference scheme has been calculated on layer $t = t_j + 1$.

Further, analogously we will find the solution v_i^j on the next layer and will continue up to the last layer $t = t_m$. The calculation of the solution of the difference scheme is fished.

13 The Crank–Nicolson Difference Scheme

We now will continue with an approximation of the boundary-value problem for the diffusion equation with discontinuous diffusion coefficient. The Crank–Nicolson difference scheme is applied to the solution of (77)–(84). Therefore, it follows that

$$\frac{v_{1,i}^{j+1} - v_{1,i}^j}{\tau} = \frac{D_1}{2}\left(\frac{v_{1,i+1}^j - 2v_{1,i}^j + v_{1,i-1}^j}{h_1^2} + \frac{v_{1,i+1}^{j+1} - 2v_{1,i}^{j+1} + v_{1,i-1}^{j+1}}{h_1^2}\right), \quad (90)$$

$$i = 1, 2, ..., n_1 - 1, \; j = 1, 2, \ldots, m - 1,$$

$$\frac{v_{2,i}^{j+1} - v_{2,i}^{j}}{\tau} = \frac{D_2}{2} \left(\frac{v_{2,i+1}^{j} - 2v_{2,i}^{j} + v_{2,i-1}^{j}}{h_2^2} + \frac{v_{2,i+1}^{j+1} - 2v_{2,i}^{j+1} + v_{2,i-1}^{j+1}}{h_2^2} \right), \tag{91}$$

$$i = n_1 + 1, n_1 + 2, ..., n_2 - 1, \; j = 1, 2, \ldots, m - 1.$$

The initial conditions are

$$v_{1,i}^0 = \varphi(x_i), \; i = 0, 1, \ldots, n_1, \; v_{2,i}^0 = \varphi(x_i), \; i = n_1 + 1, n_1 + 2, \ldots, n_2 \tag{92}$$

and the boundary –

$$v_{1,0}^j = \psi_1(t_j), \; j = 1, 2, \ldots, m, \; v_{2,n_2}^j = \psi_2(t_j), \; j = 1, 2, \ldots, m. \tag{93}$$

The matching conditions are replaced

$$v_{1,n_1}^j = v_{2,n_1}^j, \tag{94}$$

$$D_1\left(\frac{v_{1,n_1}^j - v_{1,n_1-1}^j}{h_1} + \frac{v_{1,n_1}^{j+1} - v_{1,n_1-1}^{j+1}}{h_1}\right) = D_2\left(\frac{v_{2,n_1+1}^j - v_{2,n_1}^j}{h_2} + \frac{v_{2,n_1+1}^{j+1} - v_{2,n_1}^{j+1}}{h_2}\right), \tag{95}$$

$$j = 1, 2, \ldots, m.$$

The Crank–Nicolson difference scheme (90)–(95) is aggregated with the difference equations (90), (91) together with the initial (92), the boundary (93) and the matching conditions (94), (95).

The error of an approximation of the differential equations (77), (80) by the difference equations (90), (91) is of order $O(\tau^2 + h_1^2)$ and $O(\tau^2 + h_2^2)$ respectively, the initial and the boundary conditions are replaced exactly. The second matching condition has an error of the approximation $O(h_1 + h_2)$, the first one is replaced exactly.

13.1 The Calculation of a Solution

The procedure of the calculation is very similar to that of the Crank–Nicolson difference scheme (42)–(44). The main distinction is the point of discontinuity. The calculation of the Crank–Nicolson difference scheme (90)–(95) is done in the series one grid layer after another. At first the solution on the zero layer at $t = 0$ using the initial condition (92)

$$v_{1,i}^0 = \varphi(x_i), \; i = 0, 1, \ldots, n_1, \; v_{2,i}^0 = \varphi(x_i), \; i = n_1 + 1, n_1 + 2, \ldots, n_2$$

is calculated.

Now let the solution $v_{1,i}^j$, $v_{2,i}^j$ be calculated at $j-th$ layer. Thus, the values $v_{1,i}^j$, $i=0,1,\ldots,n_1$ and $v_{2,i}^j$ for $i=n_1,n_1-1,\ldots,n_2$ are known. We proceed to solve the difference equations (90), (91) subject to value $v_{1,i}^{j+1}$, $v_{2,i}^{j+1}$ at $j+1-th$ layer. Multiplying by τ Eqs. (90), (91) we have

$$v_{1,i}^{j+1}=v_{1,i}^j+\tau\frac{D_1}{2}\left(\frac{v_{1,i+1}^j-2v_{1,i}^j+v_{1,i-1}^j}{h_1^2}+\frac{v_{1,i+1}^{j+1}-2v_{1,i}^{j+1}+v_{1,i-1}^{j+1}}{h_1^2}\right),\quad (96)$$

$$i=1,2,\ldots,n_1-1,$$

$$v_{2,i}^{j+1}=v_{2,i}^j+\tau\frac{D_2}{2}\left(\frac{v_{2,i+1}^j-2v_{2,i}^j+v_{2,i-1}^j}{h_2^2}+\frac{v_{2,i+1}^{j+1}-2v_{2,i}^{j+1}+v_{2,i-1}^{j+1}}{h_2^2}\right),\quad (97)$$

$$i=n_1+1,n_1+2,\ldots,n_2-1.$$

The boundary conditions (93) are for $j+1-th$ layer

$$v_{1,0}^{j+1}=\psi_1(t_{j+1}),\ v_{2,n_2}^{j+1}=\psi_2(t_{j+1}).\quad (98)$$

The matching conditions (94), (95) are written in the form

$$v_{1,n_1}^{j+1}=v_{2,n_1}^{j+1},\quad (99)$$

$$D_1(v_{1,n_1}^j-v_{1,n_1-1}^j+v_{1,n_1}^{j+1}-v_{1,n_1-1}^{j+1})$$
$$=\frac{D_2h_1}{h_2}(v_{2,n_1+1}^j-v_{2,n_1}^j+v_{2,n_1+1}^{j+1}-v_{2,n_1}^{j+1}).\quad (100)$$

The difference equations (96), (97) together with the boundary conditions (98) and the matching conditions (99), (100) comprise the system of n_2 linear equations with n_2 unknown values

$$v_{1,i}^j,\ i=1,2,\ldots,n_1,$$
$$v_{2,i}^j,\ i=n_1,n_1+1,\ldots,n_2-1.$$

The difference equations (96), (97) and the matching conditions (99), (100) consist of no more than three unknown values.

In this case the linear system of the equations is three-diagonal.

The next step is a count of the values $v_{1,i}^{j+1}$, $v_{2,i}^{j+1}$. For this purpose we rewrite the system of linear equations (96)–(100) in a new form and introduce the notations

$$a_i=b_i=\frac{\tau D_1}{h_1^2},\ c_i=1+\frac{\tau D_1}{h_1^2},\ i=2,3,\ldots,n_1-1,$$

$$a_i = b_i = \frac{\tau D_2}{h_2^2},\; c_i = 1 + \frac{\tau D_2}{h_2^2},\; i = n_1+1, n_1+2, \ldots, n_2-2,$$

$$a_{n_1} = D_1,\; c_{n_1} = D_1 + \frac{D_2 h_1}{h_2},\; b_{n_1} = \frac{D_2 h_1}{h_2},$$

$$\beta_1 = \frac{\tau D_1}{2(h_1^2 + \tau D_1)},\; \mu_1 = \frac{v_{1,1}^j (h_1^2 + \tau D_1) + (v_{1,2}^j + v_{1,0}^j + \psi_1(t_j))/2}{h_1^2 + \tau D_1}$$

$$\beta_2 = \frac{\tau D_2}{2(h_2^2 + \tau D_2)},$$

$$\mu_2 = \left(v_{2,n_2-1}^j + \frac{\tau D_2}{2h_2^2}\left(v_{2,n_2}^j - 2v_{2,n_2-1}^j + v_{2,n_2-2}^j + \psi_2(t_j)\right)\right)\frac{h_2^2 + \tau D_2}{h_2^2},$$

$$f_i = v_{1,i}^j + \tau\frac{D_1}{2}\left(\frac{v_{1,i+1}^j - 2v_{1,i}^j + v_{1,i-1}^j}{h_1^2}\right),$$

$$i = 2, 3, \ldots, n_1 - 1,$$

$$f_i = v_{2,i}^j + \tau\frac{D_2}{2}\left(\frac{v_{2,i+1}^j - 2v_{2,i}^j + v_{2,i-1}^j}{h_2^2}\right),$$

$$i = n_1+1, n_1+2, \ldots, n_2-2,$$

$$f_{n_1} = \frac{D_2 h_1}{h_2}(v_{2,n_1+1}^j - v_{2,n_1}^j) - D_1(v_{1,n_1}^j - v_{1,n_1-1}^j),$$

$$s_i = v_{1,i}^j,\; i = 1, 2, \ldots, n_1 - 1,\; s_i = v_{2,i}^j,\; i = n_1, n_1+1, \ldots, n_2-1.$$

Finally, the system takes the following form

$$s_1 = \beta_1 s_2 + \mu_1, \tag{101}$$

$$a_i s_{i-1} - c_i s_i + b_i s_{i+1} = f_i,\; i = 1, 2, \ldots, n-1, \tag{102}$$

$$s_{n_2-1} = \beta_2 s_{n_2-2} + \mu_2. \tag{103}$$

Now the flow variant method can be applied to the solution of the system (101)–(103). In this case we will find the solution of the Crank–Nicolson difference scheme (90)–(95) on the layer $t = t_{j+1}$. Further, analogously we will find the solution on next layer and will continue up to the last layer $t = t_m$.

So the solution of the scheme was converted to the system of linear equations on every layer with the three-diagonal matrixes. The obtained system of the linear equations were solved by the flow variant method. The distinction lies in the calculation at the point of the discontinuity of the diffusion coefficient D ($x = d_1$, $x = x_{n_1}$). The method of the solution of the Crank–Nicolson difference scheme is finished.

The Difference Schemes for the Reaction–Diffusion Equations

Contents

1 The Boundary-Value Problem for the System of Reaction–Diffusion Equations 293
2 The Explicit Difference Scheme 295
2.1 The Calculation of a Solution 296
2.2 The Convergence and the Stability 297
3 The Non-linear Crank–Nicolson Type Difference Scheme 298
3.1 Calculation of a Solution 299
3.2 The Convergence and the Stability 303
4 The Linear Crank–Nicolson Type Difference Scheme 303
4.1 The Calculation of a Solution 304
4.2 The Convergence and the Stability 307
5 Law of Conservation of Mass 307
6 The Alternating Directions Method 309
6.1 Calculation of a Solution 311
6.2 The Convergence and the Stability 312
7 The Explicit Method for the Multidimensional Problems 312
7.1 The Calculation of a Solution 314
7.2 The Convergence and the Stability 315

This chapter is devoted to various difference approximations of the reaction–diffusion equations. The difference technique, developed in a previous chapter, is employed for the construction of the difference scheme for the system of the reaction–diffusion equations. The main subject of investigation is the system of two nonlinear reaction–diffusion equations in one and two dimensional in space cases.

1 The Boundary-Value Problem for the System of Reaction–Diffusion Equations

The main subject of investigation is the nonlinear system of two partial differential equations

$$\frac{\partial S}{\partial t} = D_S \Delta S - \frac{V_{max} S}{K_M + S}, \tag{1}$$

R. Baronas et al., *Mathematical Modeling of Biosensors*, Springer Series on Chemical Sensors and Biosensors 9, DOI 10.1007/978-90-481-3243-0_11,

$$\frac{\partial P}{\partial t} = D_P \Delta P + \frac{V_{max} S}{K_M + S}, \tag{2}$$

where Δ is the Laplace operator

$$\Delta S = \frac{\partial^2 S}{\partial x^2} + \frac{\partial^2 S}{\partial y^2} + \frac{\partial^2 S}{\partial z^2},$$

$$\Delta P = \frac{\partial^2 P}{\partial x^2} + \frac{\partial^2 P}{\partial y^2} + \frac{\partial^2 P}{\partial z^2},$$

where D_S, D_P, are diffusion coefficients, V_{max} is the maximal enzymatic rate and K_M is the Michaelis constant. The functions $S = S(x, y, z, t)$, $P = P(x, y, z, t)$ are usually concentrations of the substrate and the product respectively, x, y, z are spatial coordinates and t is the time. The system of Eqs. 1, 2 is being investigated in some region $Q = \Omega \times [0, T]$. The three-dimensional open region Ω is usually an area of the biosensor reaction and diffusion. The value T is the duration of a biosensor action. The system of the Eqs. 1, 2 possesses an infinite number of solutions. In order to separate a unique solution it is necessary to add the initial and boundary conditions. The initial conditions describe the start state (at time $t = 0$) of a biosensor and are formulated in the region Ω. The boundary condition describes the state of the solution on the region Ω border. At first we will investigate a simple example of the region (parallelepiped) $\Omega = \{(x, y, z) : 0 < x < d_x, 0 < y < d_y, 0 < z < d_z\}$, where d_x, d_y, d_z are the size of the action area.

The initial conditions for $t = 0$ are

$$S(x, y, z, 0) = \varphi_S(x, y, z), \;\; P(x, y, z, 0) = \varphi_P(x, y, z), \;\; (x, y, z) \in \bar{\Omega} \tag{3}$$

and the boundary conditions are

$$S(x, y, z, t) = \psi_S(x, y, z), \; P(x, y, z, t) = \psi_P(x, y, z), \; (x, y, z) \in \partial\Omega, t \in [0, T], \tag{4}$$

where $\varphi_S(x, y, z)$, $\varphi_P(x, y, z)$, $\psi_S(x, y, z)$, $\psi_P(x, y, z)$ are given functions,

$$\bar{\Omega} = \{(x, y, z) : 0 \le x \le d_x, 0 \le y \le d_y, 0 \le z \le d_z\}$$

a closed region and $\partial\Omega$ is a boundary of the region Ω (a surface of the parallelepiped).

The system of the differential equations (1), (2) together with the initial (3) and boundary (4) conditions is named the boundary-value problem. It is worth noticing, that the boundary condition (4) belongs to a particular example and in next paragraphes we will investigate the other types. Further we will assume, that the solution of the boundary-value problem (1)–(4) exists, is unique and sufficiently smooth (the functions S, P have the smooth partial derivatives with respect to x, y, z, t of the necessary order).

2 The Explicit Difference Scheme

We start with the construction of a difference scheme for the reaction–diffusion processes in a biosensor in one-dimensional case (the time t and one space dimension x). The explicit difference scheme is popular because of the programming simplicity. We will consider the boundary-value problem for the system of two non-linear reaction–diffusion equations

$$\frac{\partial S}{\partial t} = D_P \frac{\partial^2 S}{\partial x^2} - \frac{V_{max} S}{K_M + S}, \tag{5}$$

$$\frac{\partial P}{\partial t} = D_P \frac{\partial^2 P}{\partial x^2} + \frac{V_{max} S}{K_M + S} \tag{6}$$

in a rectangle $Q = \{0 < x < d, 0 < t \leq T\}$ with the initial conditions for $t = 0$

$$S(x,0) = \varphi_S(x),\ P(x,0) = \varphi_P(x),\ x \in [0,d] \tag{7}$$

and the boundary conditions

$$\frac{\partial S(0,t)}{\partial x} = 0,\ S(d,t) = \psi_S(t),\ P(0,t) = 0,\ \frac{\partial P(d,t)}{\partial x} = 0,\ t \in [0,T], \tag{8}$$

where $\varphi_S(x)$, $\varphi_P(x)$ and $\psi_S(t)$ are the given functions and the diffusion coefficients $D_S > 0$, $D_P > 0$.

These boundary conditions are being applied when there is the flux of the substrate S on the right-side of the boundary condition (point $x = d$) and the product P on the left-side of the boundary condition (point $x = 0$). Jointly there is no flux of the substrate on the left-side of the boundary condition and the product on the right-side of the boundary condition.

Now the difference technique, developed in a previous chapter, is employed for the construction of the difference schemes for the system of the reaction–diffusion equations. So we have

$$\frac{S_i^{j+1} - S_i^j}{\tau} = D_S \frac{S_{i+1}^j - 2S_i^j + S_{i-1}^j}{h^2} - \frac{V_{max} S_i^j}{K_M + S_i^j}, \tag{9}$$

$$\frac{P_i^{j+1} - P_i^j}{\tau} = D_P \frac{P_{i+1}^j - 2P_i^j + P_{i-1}^j}{h^2} + \frac{V_{max} S_i^j}{K_M + S_i^j}, \tag{10}$$

$$i = 1,2,\ldots,n-1,\ j = 0,1,\ldots,m-1$$

with the initial conditions

$$S_i^0 = \varphi_S(x_i),\ P_i^0 = \varphi_P(x_i),\ i = 0,1,\ldots,n \tag{11}$$

and the boundary conditions

$$\frac{S_1^j - S_0^j}{h} = 0,\ S_n^j = \psi_S(t_j),\ P_0^j = 0,\ \frac{P_n^j - P_{n-1}^j}{h} = 0,\ j = 1, 2, \ldots, m. \quad (12)$$

The error of the approximation can be calculated in a similar way, i.e. as in the case of the boundary-value problem for the diffusion equation. If the partial derivatives of the functions $S(x,t)$ and $P(x,t)$ satisfy the estimations

$$\max|\frac{\partial^2 S(x,t)}{\partial t^2}|,\ \max|\frac{\partial^2 P(x,t)}{\partial t^2}| = c < \infty$$

and

$$\max|\frac{\partial^4 S(x,t)}{\partial x^4}|,\ \max|\frac{\partial^4 P(x,t)}{\partial x^4}| = c < \infty$$

in the region of the investigation $\bar{Q}$ then the order of the approximation for the system of the differential equations (5), (6) by the system of the difference equations (9), (10) equals $O(\tau + h^2)$ as $\tau, h \to 0$. The order of the approximation of the boundary conditions (8) by the difference boundary conditions (12) is of $O(h)$ (where there is no flux). The remainder boundary conditions and the initial conditions (7) are replaced exactly.

The difference scheme (9)–(12) is explicit and two-layer.

2.1 *The Calculation of a Solution*

The calculation of the solution of the difference scheme (9)–(12) is similar to the calculation of the solution of the explicit scheme for boundary-value problem (10)–(12) in a previous chapter. It is worth emphasizing that the difference equation (9) is independent of the difference equation (10) and may be calculated separately. It seems clear that the values S_i^j can be calculated first and all values P_i^j,

$$i = 0, 1, \ldots, n,\ j = 0, 1, \ldots, m$$

later. But this way is not optimal from the point of view of the computer memory. This case demands a computer memory for the reserving at least $(n+1)(m+1)$ values S_i^j. If we calculate the values S_i^j and P_i^j simultaneously, then it is enough to reserve

$$2(n+1)$$

values S_i^j and P_i^j, on two layers $t = t_j, t = t_{j+1}$. It turns out that for the calculation of the final values S_i^m and P_i^m it is more convenient to solve Eqs. 9, 10 simultaneously.

The calculation is performed in series one grid layer after another. At first a solution on the zero layer at $t = 0$ using the initial conditions (11)

$$S_i^0 = \varphi_S(x_i),\ P_i^0 = \varphi_P(x_i),\ i = 0, 1, \ldots, n$$

is calculated

Now let the solutions S_i^j and P_i^j be calculated at $j-th$ layer. Thus, the values S_i^j and P_i^j for $i = 0, 1, \ldots, n$ are known. We proceed to solve the difference equations (9), (10) subject to values S_i^{j+1} and P_i^{j+1} at $j+1-th$ layer. We have

$$S_i^{j+1} = S_i^j + \tau\left(D_S\frac{S_{i+1}^j - 2S_i^j + S_{i-1}^j}{h^2} - \frac{V_{max}S_i^j}{K_M + S_i^j}\right),$$

$$P_i^{j+1} = P_i^j + \tau\left(D_P\frac{P_{i+1}^j - 2P_i^j + P_{i-1}^j}{h^2} + \frac{V_{max}S_i^j}{K_M + S_i^j}\right),$$

$$i = 1, 2, \ldots, n-1,\ j = 0, 1, \ldots, m-1.$$

All values of the functions S_i^j and P_i^j are known on the right side of the equality. So we can calculate the values S_i^{j+1} from the first equation for $i = 1, 2, \ldots, n-1$ and then, consequently, P_i^{j+1} from the second equation for $i = 1, 2, \ldots, n-1$. The values S_0^{j+1}, S_n^{j+1} and P_0^{j+1}, P_n^{j+1} are calculated using boundary conditions (12) and

$$S_0^{j+1} = S_1^{j+1},\ S_n^{j+1} = \psi_S(t_{j+1}),\ P_0^{j+1} = 0,\ P_n^{j+1} = P_{n-1}^{j+1}.$$

Further, analogously, we will find the solutions S_i^{j+2} and P_i^{j+2} on the next layer and will continue up the last layer $t = t_m$.

2.2 *The Convergence and the Stability*

The explicit method is conditionally stable and converges with the rate $O(\tau + h^2)$. The differential equations are nonlinear and it is possible to formulate the sufficient stability conditions

$$\frac{\tau \max\{D_S, D_P\}}{h^2} \le 1/4,\ \frac{\tau V_{\max}}{K_M} \le 1/2.$$

3 The Non-linear Crank–Nicolson Type Difference Scheme

Now we will present the second difference scheme (a non-linear Crank–Nicolson type) for the solution of the reaction–diffusion processes in a biosensor in one-dimensional case. We will consider the boundary-value problem for the system of the same reaction–diffusion equations

$$\frac{\partial S}{\partial t} = D_S \frac{\partial^2 S}{\partial x^2} - \frac{V_{max} S}{K_M + S}, \tag{13}$$

$$\frac{\partial P}{\partial t} = D_P \frac{\partial^2 P}{\partial x^2} + \frac{V_{max} S}{K_M + S} \tag{14}$$

in a rectangle $Q = \{0 < x < d,\ 0 < t \leq T\}$ with the initial conditions for $t = 0$

$$S(x, 0) = \varphi_S(x),\ P(x, 0) = \varphi_P(x),\ x \in [0, d] \tag{15}$$

and the boundary conditions (differ from (8))

$$S(0, t) = 0,\ S(d, t) = \psi_S(t),\ P(0, t) = 0,\ P(d, t) = 0, \in [0, T], \tag{16}$$

where $\varphi_S(x)$, $\varphi_P(x)$ and $\psi_S(t)$, are the given functions and the diffusion coefficients $D_S > 0$, $D_P > 0$.

The boundary conditions (16) are being applied when there is the flux of the substrate and the product on both sides of the boundary (points $x = 0, x = d$.)

Now the difference technique, developed in a previous chapter for the Crank–Nicolson difference scheme (42)–(44), is employed for the construction of the difference scheme for the boundary-value problem for the system of the reaction–diffusion equations (13)–(16). We have

$$\begin{aligned} &\frac{S_i^{j+1} - S_i^j}{\tau} = \\ &\frac{D_S}{2}\left\{\frac{S_{i+1}^{j+1} - 2S_i^{j+1} + S_{i-1}^{j+1}}{h^2} + \frac{S_{i+1}^j - 2S_i^j + S_{i-1}^j}{h^2}\right\} \\ &- \frac{V_{max}(S_i^j + S_i^{j+1})/2}{K_M + (S_i^j + S_i^{j+1})/2}, \end{aligned} \tag{17}$$

$$\begin{aligned} &\frac{P_i^{j+1} - P_i^j}{\tau} = \\ &\frac{D_P}{2}\left\{\frac{P_{i+1}^{j+1} - 2P_i^{j+1} + P_{i-1}^{j+1}}{h^2} + \frac{P_{i+1}^j - 2P_i^j + P_{i-1}^j}{h^2}\right\} \\ &+ \frac{V_{max}(S_i^j + S_i^{j+1})/2}{K_M + (S_i^j + S_i^{j+1})/2}, \end{aligned} \tag{18}$$

$$i = 1, 2, \dots, n-1,\ j = 0, 1, \dots, m-1.$$

The initial and boundary conditions are replaced as follow with

$$S_i^0 = \varphi_S(x_i),\ P_i^0 = \varphi_P(x_i),\ i = 0, 1, \dots, n, \tag{19}$$

$$S_0^j = 0,\ S_n^j = \psi_S(t_j),\ P_0^j = 0,\ P_n^j = 0,\ j = 1, 2, \dots, m. \tag{20}$$

The error of an approximation can be calculated in a similar way, i.e. as in the case of the boundary-value problem for the diffusion equation. If the partial derivatives of the functions $S(x,t)$ and $P(x,t)$ satisfy the estimations

$$\max\left|\frac{\partial^3 S(x,t)}{\partial t^3}\right|,\ \max\left|\frac{\partial^3 P(x,t)}{\partial t^3}\right| = c < \infty$$

and

$$\max\left|\frac{\partial^4 S(x,t)}{\partial x^4}\right|,\ \max\left|\frac{\partial^4 P(x,t)}{\partial x^4}\right| = c < \infty$$

in the region of the investigation $\bar{Q}$ then the order of the approximation for the system of the differential equations (13), (14) by the system of the difference equations (17), (18) equals $O(\tau^2 + h^2)$ as $\tau,\ h \to 0$. The initial (19) and the boundary (20) conditions are replaced exactly.

The difference scheme is implicit and two-layer.

3.1 Calculation of a Solution

The calculation of the solution of the difference scheme (17)–(20) is similar to the calculation of the solution of the implicit scheme for boundary-value problem (42)–(44) in a previous chapter. But there is an essential difference – a non-linear term. The nonlinearity appears in the fraction

$$\frac{V_{max}(S_i^j + S_i^{j+1})/2}{K_M + (S_i^j + S_i^{j+1})/2}$$

by containing the unknown value on the upper layer S_i^{j+1} in the denominator. For the numerical realization of this equation it is essential to employ an iterative method of the solution.

The calculation is performed in the series one grid layer after another. At first the solution on the zero layer at $t = 0$ using initial conditions (19)

$$S_i^0 = \varphi_S(x_i),\ P_i^0 = \varphi_P(x_i),\ i = 0, 1, \dots, n$$

is calculated.

Suppose the solutions S_i^j and P_i^j are calculated at j-th layer. Thus, the values S_i^j and P_i^j for $i = 0, 1, \dots, n$ are known. We proceed to solve the difference equations

(17), (18) subject to values S_i^{j+1} and P_i^{j+1} at $j+1$-th layer. We multiply these equations by τ and have

$$S_i^{j+1} = S_i^j + \tau\left(\frac{D_S}{2}\left\{\frac{S_{i+1}^{j+1} - 2S_i^{j+1} + S_{i-1}^{j+1}}{h^2} + \frac{S_{i+1}^j - 2S_i^j + S_{i-1}^j}{h^2}\right\} - \frac{V_{max}(S_i^j + S_i^{j+1})/2}{K_M + (S_i^j + S_i^{j+1})/2}\right),$$

$$P_i^{j+1} = P_i^j + \tau\left(\frac{D_P}{2}\left\{\frac{P_{i+1}^{j+1} - 2P_i^{j+1} + P_{i-1}^{j+1}}{h^2} + \frac{P_{i+1}^j - 2P_i^j + P_{i-1}^j}{h^2}\right\} + \frac{V_{max}(S_i^j + S_i^{j+1})/2}{K_M + (S_i^j + S_i^{j+1})/2}\right),$$

$$i = 1, 2, \ldots, n-1.$$

The boundary conditions at $j+1$-th layer are

$$S_0^{j+1} = 0,\ S_n^{j+1} = \psi_S(t_{j+1}),\ P_0^{j+1} = 0,\ P_n^{j+1} = 0.$$

Consequently, the boundary values S_0^{j+1}, S_n^{j+1}, P_0^{j+1}, P_n^{j+1} are known. This yields a non-linear system of the order $2n-2$ with respect to unknown values S_i^{j+1} and P_i^{j+1} for $i = 1, 2, \ldots, n-1$. Due to, for the solution of these non-linear equations we will introduce an iterative method

$$S_i^{(l+1)} = S_i^j + \tau\left(\frac{D_S}{2}\left(\frac{S_{i+1}^{(l+1)} - 2S_i^{(l+1)} + S_{i-1}^{(l+1)}}{h^2} + \frac{S_{i+1}^j - 2S_i^j + S_{i-1}^j}{h^2}\right) - \frac{V_{max}(S_i^j + S_i^{(l+1)})/2}{K_M + (S_i^j + S_i^{(l)})/2}\right),$$

$$P_i^{(l+1)} = P_i^j + \tau\left(\frac{D_P}{2}\left\{\frac{P_{i+1}^{(l+1)} - 2P_i^{(l+1)} + P_{i-1}^{(l+1)}}{h^2} + \frac{P_{i+1}^j - 2P_i^j + P_{i-1}^j}{h^2}\right\} + \frac{V_{max}(S_i^j + S_i^{(l+1)})/2}{K_M + (S_i^j + S_i^{(l)})/2}\right),$$

$$i = 1, 2, \ldots, n-1$$

where $S_i^{(l)}$, $P_i^{(l)}$, and $S_i^{(l+1)}$, $P_i^{(l+1)}$ are the iteration values for S_i^{j+1}, P_i^{j+1}. Here l-th and $l+1$-th are the numbers of the iterations. For $l = 0$ we take the initial values for the iterations from j-th layer

$$S_i^{(0)} = S_i^j,\ P_i^{(0)} = P_i^j,\ i = 1, 2, \ldots, n-1.$$

Now we will continue with the calculation of the iteration values $S_i^{(l+1)}$, $P_i^{(l+1)}$. It goes without saying that suitably chosen symbolism can simplify the development. Let us introduce $s_i = S_i^{l+1}$, $p_i = P_i^{l+1}$, then the iteration method can be written

$$a_i s_{i-1} - c_i s_i + b_i s_{i+1} = f_i, \tag{21}$$

$$\bar{a}_i p_{i-1} - \bar{c}_i p_i + \bar{b}_i p_{i+1} = \bar{f}_i, \tag{22}$$

$$i = 1, 2, \ldots, n-1,$$

where

$$a_i = b_i = \frac{\tau D_S}{2h^2},\ c_i = 1 + \frac{\tau D_S}{h^2},$$

$$\bar{a}_i = \bar{b}_i = \frac{\tau D_P}{2h^2},\ \bar{c}_i = 1 + \frac{\tau D_P}{h^2},$$

$$f_i = -S_i^j - \tau\left(D_S \frac{S_{i+1}^j - 2S_i^j + S_{i-1}^j}{2h^2} - \frac{V_{max} S_i^j}{K_M + S_i^j},\right),$$

$$\bar{f}_i = -P_i^j - \tau\left(D_P \frac{P_{i+1}^j - 2P_i^j + P_{i-1}^j}{2h^2} + \frac{V_{max} S_i^j}{K_M + S_i^j},\right)$$

for $i = 2, 3, \ldots, n-2$.

The values

$$a_1 = 0,\ b_1 = \frac{\tau D_S}{2h^2},\ c_1 = 1 + \tau\left(\frac{D_S}{h^2} + \frac{V_{max}}{2K_M + S_1^j + S_1^{(l)}}\right),$$

$$a_{n-1} = \frac{\tau D_S}{2h^2},\ b_{n-1} = 0,\ c_{n-1} = 1 + \tau\left(\frac{D_S}{h^2} + \frac{V_{max}}{2K_M + S_{n-1}^j + S_{n-1}^{(l)}}\right),$$

$$\bar{a}_1 = 0,\ \bar{b}_1 = \frac{\tau D_P}{2h^2},\ \bar{c}_1 = 1 + \tau\left(\frac{D_P}{h^2} - \frac{V_{max}}{2K_M + S_1^j + S_1^{(l)}}\right),$$

$$\bar{a}_{n-1} = \frac{\tau D_P}{2h^2},\ \bar{b}_{n-1} = 0,\ \bar{c}_{n-1} = 1 + \tau\left(\frac{D_P}{h^2} - \frac{V_{max}}{2K_M + S_{n-1}^j + S_{n-1}^{(l)}}\right),$$

$$f_1 = -S_1^j - \tau\left(D_S \frac{S_2^j - 2S_1^j + S_0^j}{2h^2} - \frac{V_{max} S_1^j}{2K_M + S_1^j + S_1^{(l)}}\right),$$

$$f_{n-1} = -S_{n-1}^j - \tau\Bigg(D_S \frac{S_n^j - 2S_{n-1}^j + S_{n-2}^j}{2h^2}$$
$$- \frac{V_{max} S_{n-1}^j}{2K_M + S_{n-1}^j + S_{n-1}^{(l)}} + \frac{D_S \psi_S(t_{j+1})}{2h^2}\Bigg),$$

$$\bar{f}_1 = -P_1^j - \tau\left(D_P \frac{P_2^j - 2P_1^j + P_0^j}{2h^2} - \frac{V_{max} S_1^j}{2K_M + S_1^j + S_1^{(l)}}\right),$$

$$\bar{f}_{n-1} = -P_{n-1}^j - \tau D_P \frac{P_n^j - 2P_{n-1}^j + P_{n-2}^j}{2h^2}$$
$$+ \frac{\tau V_{max} S_{n-1}^j}{2K_M + S_{n-1}^j + S_{n-1}^{(l)}}.$$

All values S_i^j, P_i^j and $S_i^{(l)}$, $P_i^{(l)}$ are known. Therefore, the coefficients of the systems (21), (22)

$$a_i, b_i, c_i, f_i, \bar{a}_i, \bar{b}_i, \bar{c}_i, \bar{f}_i$$

can be calculated.

The next step is the solution of this system (the calculation of the values s_i and p_i $i = 1, 2, \ldots, n-1$). As in the previous cases (the Crank–Nicolson difference schemes) we will apply the elimination method for the solution of the system of linear equations with three-diagonal matrixes (the flow variant of the elimination method). In preparation for this we can set up the system (21), (22) by

$$s_1 = \beta_1 s_2 + \mu_1,$$

$$a_i s_{i-1} - c_i s_i + b_i s_{i+1} = f_i,\ i = 2, 3, ..., n-2,$$
$$s_{m-1} = \beta_2 s_{m-2} + \mu_2,$$
$$p_1 = \bar{\beta}_1 p_2 + \bar{\mu}_1,$$
$$\bar{a}_i p_{i-1} - \bar{c}_i p_i + \bar{b}_i p_{i+1} = \bar{f}_i,\ i = 2, 3, ..., n-2,$$
$$p_{m-1} = \bar{\beta}_2 p_{m-2} + \bar{\mu}_2,$$

where

$$\beta_1 = b_1/c_1,\ \mu_1 = -f_1/c_1,$$
$$\beta_2 = a_{n-1}/c_{n-1},\ \mu_2 = (f_{n-1} - b_{n-1})\psi(t_{j+1})/c_{n-1},$$
$$\bar{\beta}_1 = \bar{b}_1/\bar{c}_1,\ \bar{\mu}_1 = -\bar{f}_1/\bar{c}_1,$$
$$\bar{\beta}_2 = \bar{a}_{n-1}/\bar{c}_{n-1},\ \bar{\mu}_2 = f_{n-1}/c_{n-1}.$$

For the solution of these systems we can apply the elimination method. Firstly, the values s_i are computed, and secondly, the values p_i are computed. Therefore, the next iteration of $S_i^{(l+1)}$ and $P_i^{(l+1)}$ is known. Usually, the iteration process is stopped, when the adjacent iteration values are suitably close or more exact

$$|S_i^{(l+1)} - S_i^{(l)}| \leq \varepsilon,$$

$$|P_i^{(l+1)} - P_i^{(l)}| \leq \varepsilon, i = 1, 2, \ldots, n-1,$$

where $\varepsilon > 0$ is a small number. When the last conditions are satisfied the iteration proses is stopped and we take

$$P_i^{j+1} = P_i^{(l+1)}, \; S_i^{j+1} = S_i^{(l+1)}, i = 1, 2, \ldots, n-1.$$

The values $S_0^{(l+1)}$, $S_n^{(l+1)}$ and $P_0^{(l+1)}$, $P_n^{(l+1)}$ are calculated using the boundary conditions and are equal to

$$S_0^{(l+1)} = 0, \; S_n^{(l+1)} = \psi_S(t_{j+1}), \; P_0^{(l+1)} = 0, \; P_n^{(l+1)} = 0.$$

Further, analogously, we will find the solutions S_i^j and P_i^j on the next layer and will continue up to the last layer $t = t_m$.

3.2 *The Convergence and the Stability*

The iterative method converges if the sufficient condition is satisfied

$$\frac{\tau V_{\max}}{K_M} \leq 1/2.$$

The non-linear Crank–Nicolson method is unconditionally stable and converges with the rate $O(\tau^2 + h^2)$. It is worth noticing that the stability of the difference scheme and the convergence of iterative method for the solution of the non-linear system are independent questions.

4 The Linear Crank–Nicolson Type Difference Scheme

In this section we will consider the linear Crank–Nicolson type difference scheme. The linearity of a difference scheme is achieved by a simplification of the approximation of nonlinear term $V_{max}S/(K_M + S)$. This scheme does not require an iterative processes for the calculations. For the purposes of the calculation it is sometimes convenient to apply the linear scheme.

We will again consider the boundary-value problem (5)–(8). Similar to previous section one can obtain a difference scheme

$$\frac{S_i^{j+1} - S_i^j}{\tau} = \frac{D_S}{2}\left\{\frac{S_{i+1}^{j+1} - 2S_i^{j+1} + S_{i-1}^{j+1}}{h^2} + \frac{S_{i+1}^j - 2S_i^j + S_{i-1}^j}{h^2}\right\} - \frac{V_{max} S_i^{j+1}}{K_M + S_i^j}, \tag{23}$$

$$\frac{P_i^{j+1} - P_i^j}{\tau} = \frac{D_P}{2}\left\{\frac{P_{i+1}^{j+1} - 2P_i^{j+1} + P_{i-1}^{j+1}}{h^2} + \frac{P_{i+1}^j - 2P_i^j + P_{i-1}^j}{h^2}\right\} + \frac{V_{max} S_i^{j+1}}{K_M + S_i^j}, \tag{24}$$

$$i = 1, 2, \ldots, n-1, \; j = 0, 1, \ldots, m-1.$$

The initial and the boundary conditions are replaced with

$$S_i^0 = \varphi_S(x_i), \; P_i^0 = \varphi_P(x_i), \; i = 0, 1, \ldots, n, \tag{25}$$

$$S_0^j = 0, \; S_n^j = \psi_S(t_j), \; P_0^j = 0, \; P_n^j = 0, \; j = 1, 2, \ldots, m. \tag{26}$$

The error of an approximation can be calculated in a similar way, i.e. as in the case of the boundary-value problem for the diffusion equation. The order of the approximation for the system of the differential equations (13), (14) by the system of the difference equations equals $O(\tau + h^2)$. The initial and the boundary conditions are replaced exactly.

The difference scheme is implicit and two-layer.

4.1 The Calculation of a Solution

The calculation of the solution of the difference scheme (23–(26) is similar to the calculation of the solution of the Crank–Nicolson scheme for the boundary-value problem (45)–(49) in a previous chapter. The calculation is performed in the series one grid layer after another one. At first the solution on the zero layer at $t = 0$ using initial conditions (25)

$$S_i^0 = \varphi_S(x_i), \; P_i^0 = \varphi_P(x_i), \; i = 0, 1, \ldots, n$$

is calculated.

Suppose, the solutions S_i^j and P_i^j are calculated at $j-th$ layer. Thus, the values S_i^j and P_i^j for $i = 0, 1, \ldots, n$ are known. We proceed to solve the difference equations (23), (24) subject to values S_i^{j+1} and P_i^{j+1} at $j+1$-th layer. We multiply these equations by τ and have

$$S_i^{j+1} = S_i^j + \tau\left(\frac{D_S}{2}\left(\frac{S_{i+1}^{j+1} - 2S_i^{j+1} + S_{i-1}^{j+1}}{h^2} + \frac{S_{i+1}^j - 2S_i^j + S_{i-1}^j}{h^2}\right) - \frac{V_{max} S_i^{j+1}}{K_M + S_i^j}\right), \tag{27}$$

$$P_i^{j+1} = P_i^j + \tau\left(\frac{D_P}{2}\left(\frac{P_{i+1}^{j+1} - 2P_i^{j+1} + P_{i-1}^{j+1}}{h^2} + \frac{P_{i+1}^j - 2P_i^j + P_{i-1}^j}{h^2}\right) + \frac{V_{max} S_i^{j+1}}{K_M + S_i^j}\right), \tag{28}$$

$$i = 1, 2, \ldots, n-1.$$

The boundary conditions at $j+1$-th layer are

$$S_0^{j+1} = 0,\ S_n^{j+1} = \psi(t_{j+1}),\ P_0^{j+1} = 0,\ P_n^{j+1} = 0.$$

This yields a linear system of the order $2n-2$ with respect to unknown values S_i^{j+1} and P_i^{j+1} for $i = 1, 2, \ldots, n-1$. Now we will continue with the calculation of the iteration values $S_i^{(j+1)}$, $P_i^{(j+1)}$. Suitably chosen symbolism may simplify the development. Let us introduce $s_i = S_i^{j+1}$, $p_i = P_i^{j+1}$, then the system of the Eqs. 28, 27 can be written

$$a_i s_{i-1} - c_i s_i + b_i s_{i+1} = f_i, \tag{29}$$

$$\bar{a}_i p_{i-1} - \bar{c}_i p_i + \bar{b}_i p_{i+1} = \bar{f}_i, \tag{30}$$

$$i = 1, 2, \ldots, n-1,$$

where

$$a_i = b_i = \frac{\tau D_S}{2h^2},\ c_i = 1 + \tau\left(\frac{D_S}{h^2} + \frac{V_{max}}{K_M + S_i^j}\right),$$

$$\bar{a}_i = \bar{b}_i = \frac{\tau D_P}{2h^2},\ \bar{c}_i = 1 + \tau\left(\frac{D_P}{h^2} - \frac{V_{max}}{K_M + S_i^j}\right),$$

$$f_i = -S_i^j - \tau D_S \frac{S_{i+1}^j - 2S_i^j + S_{i-1}^j}{2h^2},$$

$$\bar{f}_i = -P_i^j - \tau D_P \frac{P_{i+1}^j - 2P_i^j + P_{i-1}^j}{2h^2},$$

for $i = 2, 3, ..., n-2$.

The values

$$a_1 = 0,\ b_1 = \frac{\tau D_S}{2h^2},\ c_1 = 1 + \tau\left(\frac{D_S}{h^2} + \frac{V_{max}}{K_M + S_i^j}\right),$$

$$a_{n-1} = \frac{\tau D_S}{2h^2},\ b_{n-1} = 0,\ c_{n-1} = 1 + \tau\left(\frac{D_S}{h^2} + \frac{V_{max}}{K_M + S_i^j}\right),$$

$$\bar{a}_1 = 0,\ \bar{b}_1 = \frac{\tau D_P}{2h^2},\ \bar{c}_1 = 1 + \tau\left(\frac{D_P}{h^2} - \frac{V_{max}}{K_M + S_i^j}\right),$$

$$\bar{a}_{n-1} = \frac{\tau D_P}{2h^2},\ \bar{b}_{n-1} = 0,\ \bar{c}_{n-1} = 1 + \tau\left(\frac{D_P}{h^2} - \frac{V_{max}}{K_M + S_i^j}\right),$$

$$f_1 = -S_1^j - \tau D_S \frac{S_2^j - 2S_1^j + S_0^j}{2h^2},$$

$$f_{n-1} = -S_{n-1}^j - \tau D_S \frac{S_n^j - 2S_{n-1}^j + S_{n-2}^j + \psi_S(t_{j+1})}{2h^2},$$

$$\bar{f}_1 = -P_1^j - \tau D_P \frac{P_2^j - 2P_1^j + P_0^j}{2h^2},$$

$$\bar{f}_{n-1} = -P_{n-1}^j - \tau D_P \frac{P_n^j - 2P_{n-1}^j + P_{n-2}^j}{2h^2}.$$

All values S_i^j and P_i^j are known. Therefore, the coefficients of the systems (29), (30)

$$a_i,\ b_i,\ c_i,\ f_i,\ \bar{a}_i,\ \bar{b}_i,\ \bar{c}_i,\ \bar{f}_i$$

can be calculated.

The next step is the solution of this system (the calculation of the values s_i, $\bar{s}_i$ and p_i, $\bar{P}_i$ $i = 1, 2, \ldots, n-1$). As in the previous cases (the Crank–Nicolson difference schemes) we will apply the elimination method for the solution of the system of linear equations with three-diagonal matrixes (the flow variant of the elimination method). In preparation for this we may set up the system (29), (30) by

$$s_1 = \beta_1 s_2 + \mu_1,$$

$$a_i s_{i-1} - c_i s_i + b_i s_{i+1} = f_i,\ i = 2, 3, ..., n-2,$$

$$s_{m-1} = \beta_2 s_{m-2} + \mu_2,$$

$$p_1 = \bar{\beta}_1 p_2 + \bar{\mu}_1,$$
$$\bar{a}_i p_{i-1} - \bar{c}_i p_i + \bar{b}_i p_{i+1} = \bar{f}_i, \; i = 2, 3, ..., n-2,$$
$$p_{m-1} = \bar{\beta}_2 p_{m-2} + \bar{\mu}_2,$$

where

$$\beta_1 = b_1/c_1, \; \mu_1 = -f_1/c_1,$$
$$\beta_2 = a_{n-1}/c_{n-1}, \; \mu_2 = (f_{n-1} - b_{n-1})\psi(t_{j+1})/c_{n-1},$$
$$\bar{\beta}_1 = \bar{b}_1/\bar{c}_1, \; \bar{\mu}_1 = -\bar{f}_1/\bar{c}_1,$$
$$\bar{\beta}_2 = \bar{a}_{n-1}/\bar{c}_{n-1}, \; \bar{\mu}_2 = f_{n-1}/c_{n-1}.$$

For the solution of this system we can apply the elimination method (28)–(38). Further, analogously, we will find the solutions S_i^j and P_i^j on the next layer and will continue up to the last layer $t = t_m$.

4.2 The Convergence and the Stability

The linear Crank–Nicolson method is unconditionally stable and converges with the rate $O(\tau + h^2)$.

5 Law of Conservation of Mass

Now we will discuss conservation laws. They play an important role in many reaction–diffusion processes. Chemical processes in a biosensor obey some laws of conservation mass. Now we will consider an example of a simple conservation law for nonlinear reaction–diffusion problem

$$\frac{\partial S}{\partial t} = D_S \frac{\partial^2 S}{\partial x^2} - \frac{V_{max} S}{K_M + S}, \tag{31}$$

$$\frac{\partial P}{\partial t} = D_P \frac{\partial^2 P}{\partial x^2} + \frac{V_{max} S}{K_M + S} \tag{32}$$

in the region $Q = \{0<x<d, 0<t \le T\}$ with the initial

$$S(0, x) = 0, \; x \in [0, d), \; S(0, d) = S_0, \; P(0, x) = 0, \; x \in [0, d] \tag{33}$$

and the boundary conditions

$$S(0,t) = 0,\ S(d,t) = 0,\ P(0,t) = 0,\ \frac{\partial P(d,t)}{\partial x} = 0,\ t \in [0,T], \tag{34}$$

Now we will sum up this system of equations and after that we will integrate it in the region $Q_{\bar{t}} = \{0 \le x \le d, 0 \le t \le \bar{t} \le T\}$. Using the initial and boundary conditions, we obtain

$$\int_0^d (S(x,\bar{t}) + P(x,\bar{t}))dx =$$
$$\int_0^{\bar{t}} D_S \frac{\partial S(d,t)}{\partial x} dt - \int_0^{\bar{t}} \left(D_S \frac{\partial S(0,t)}{\partial x} + D_P \frac{\partial P(0,t)}{\partial x} \right) dt. \tag{35}$$

The term

$$\int_0^d (S(x,\bar{t}) + P(x,\bar{t}))dx$$

describes the quantity of the mass substrate $S(x,t)$ and the product $P(x,t)$ at the moment $t = \bar{t}$, the term

$$\int_0^{\bar{t}} D_S \frac{\partial S(d,t)}{\partial x} dt$$

describes the amount of the inflowing substrate mass, the term

$$\int_0^{\bar{t}} \left(D_S \frac{\partial S(0,t)}{\partial x} + D_P \frac{\partial P(0,t)}{\partial x} \right) dt$$

– the amount of the effluent substrate and the product mass.

The difference scheme, which expresses the conservation law on the grid, is said to be conservative or divergent. For the conservative scheme the relevant law of conservation in the entire grid domain $Q_{t_1,\tau h}$ follows as an algebraic corollary difference equation. The Crank–Nicolson non-linear and linear (17)–(20), (23)–(26) schemes are the examples of the conservative schemes. Here we will investigate only the non-linear Crank–Nicolson scheme (17)–(20)

$$\frac{S_i^{j+1} - S_i^j}{\tau} =$$
$$\frac{D_S}{2} \left\{ \frac{S_{i+1}^{j+1} - 2S_i^{j+1} + S_{i-1}^{j+1}}{h^2} + \frac{S_{i+1}^j - 2S_i^j + S_{i-1}^j}{h^2} \right\}$$

$$-\frac{V_{max}(S_i^j + S_i^{j+1})/2}{K_M + (S_i^j + S_i^{j+1})/2},$$

$$\frac{P_i^{j+1} - P_i^j}{\tau} =$$

$$\frac{D_P}{2}\left\{\frac{P_{i+1}^{j+1} - 2P_i^{j+1} + P_{i-1}^{j+1}}{h^2} + \frac{P_{i+1}^j - 2P_i^j + P_{i-1}^j}{h^2}\right\}$$

$$+\frac{V_{max}(S_i^j + S_i^{j+1})/2}{K_M + (S_i^j + S_i^{j+1})/2},$$

$$i = 1, 2, \dots, n-1,\ j = 0, 1, \dots, m-1.$$

with the initial and boundary conditions

$$S_i^0 = \varphi_S(x_i),\ P_i^0 = \varphi_P(x_i),\ i = 0, 1, \dots, n,$$

$$S_0^j = 0,\ S_n^j = S_0,\ P_0^j = 0,\ P_n^j = P_{n-1}^j,\ j = 1, 2, \dots, m.$$

Now we will sum up this system of equations and after that we will sum it with respect to $i,\ j$ in region $Q_{\bar{t},\tau h}$. After simple reconstruction one can obtain an equality

$$\sum_{i=1}^{n-1} h\left(S_i^{\bar{j}} + P_i^{\bar{j}}\right) =$$
$$\frac{1}{2}\left(\sum_{j=1}^{\bar{j}} hD_S\frac{S_n^j - S_{n-1}^j}{h} - \sum_{j=1}^{\bar{j}} h\left(D_S\frac{S_1^j - S_0^j}{h} + D_P\frac{P_1^j - P_0^j}{h}\right)\right), \tag{36}$$

where $\bar{t} = \bar{j}\tau$. This equality is a grid analogue of the conservation law (35). Consequently, the non-linear Crank–Nocolson difference scheme is conservative in accordance with the last grid analogue of the conservation law. The equality (36) guarantees that the sum of the substrate and the product will obey at every moment of the time $\bar{t}$ the law of conservation of mass.

6 The Alternating Directions Method

Now we start the investigation of the reaction–diffusion problem on the plane. The biosensor action is described by the boundary-value problem

$$\frac{\partial S}{\partial t} = D_S\left(\frac{\partial^2 S}{\partial x^2} + \frac{\partial^2 S}{\partial y^2}\right) - \frac{V_{max}S}{K_M + S}, \tag{37}$$

$$\frac{\partial P}{\partial t} = D_P\left(\frac{\partial^2 P}{\partial x^2} + \frac{\partial^2 P}{\partial y^2}\right) + \frac{V_{max}S}{K_M + S} \tag{38}$$

in the region $Q = \{(x, y) \in \Omega,\ 0 < t \le T)$, where $\Omega = \{0 < x < d_x, 0 < y < d_y\}$ with the initial conditions for $t = 0$

$$S(x, y, 0) = \varphi_S(x, y),\ P(x, y, 0) = \varphi_P(x, y),\ (x, y) \in \bar{\Omega} \tag{39}$$

and the boundary conditions

$$S(x, y, t) = \psi_S(x, y, t),\ P(x, y, t) = \psi_P(x, y, t),\ (x, y) \in \partial\Omega,\ t \in [0, T]. \tag{40}$$

Here $\varphi_S(x, y)$, $\varphi_P(x, y)$, $\psi_S(x, y, t)$, $\psi_P(x, y, t)$ are the given functions. The boundary $\partial\Omega$ consists of the boundary of the rectangular Ω. For the sake of simplicity we have taken the initial and boundary conditions of the general form.

The alternating directions method (the scheme ascribed to Peaceman and Rachford) belongs to the popular methods for the solution of two-dimensional in the space problems for the diffusion equations. One of the most important issues in the numerical methods is the well-founded choice of the computational algorithms, the realization of which requires a minimal execution time in giving an approximate solution with prescribed accuracy. The total number of the necessary arithmetic operations for doing so is the main characteristic of the algorithms in question, since, the other characteristics, such as the quality of the related software and the availability of an advanced-architecture computers are beyond our control. In view of this, the economy requirements become rather urgent and extremely important in the numerical solutions of the multidimensional problems arising time and again, particularly in domains with complicated geometry. The alternating direction is an economical scheme. The total definition of an economical scheme means that the amount of arithmetical operations, which is necessary for the calculation of the solution values is proportional to the number of the grid points. The main idea of the alternating direction method is the conversion of the two-dimensional in space problem into two one-dimensional problems. The present values S_{ik}^j, P_{ik}^j and S_{ik}^{j+1}, P_{ik}^{j+1} of this difference scheme are put together with intermediate value $S_{ik}^{j+1/2}$, $P_{ik}^{j+1/2}$ a formal treatment of which is the value of S, P at the moment $t = t_{j+1/2} = t_j + \tau/2$. The passage from the j-th layer to the $j + 1$-th the layer can be done in two steps with the appropriate spacing 0.5τ:

$$\frac{S_{ik}^{j+1/2} - S_{ik}^j}{0.5\tau} =$$

$$D_S\left(\frac{S_{i+1,k}^{j+1/2} - 2S_{ik}^{j+1/2} + S_{i-1,k}^{j+1/2}}{h_x^2} + \frac{S_{i,k+1}^j - 2S_{ik}^j + S_{i,k-1}^j}{h_y^2}\right) - \frac{V_{max}S_{ik}^{j+1/2}}{K_M + S_{ik}^j} \tag{41}$$

$$\frac{P_{ik}^{j+1/2} - P_{ik}^{j}}{0.5\tau} = D_P\left(\frac{P_{i+1,k}^{j+1/2} - 2P_{ik}^{j+1/2} + P_{i-1,k}^{j+1/2}}{h_x^2} + \frac{P_{i,k+1}^{j} - 2P_{ik}^{j} + P_{i,k-1}^{j}}{h_y^2}\right) + \frac{V_{max}S_{ik}^{j+1/2}}{K_M + S_{ik}^{j}} \tag{42}$$

$$\frac{S_{ik}^{j+1} - S_{ik}^{j+1/2}}{0.5\tau} = D_S\left(\frac{S_{i+1,k}^{j+1/2} - 2S_{ik}^{j+1/2} + S_{i-1,k}^{j+1/2}}{h_x^2} + \frac{S_{i,k+1}^{j+1} - 2S_{ik}^{j+1} + S_{i,k-1}^{j+1}}{h_y^2}\right) - \frac{V_{max}S_{ik}^{j+1}}{K_M + S_{ik}^{j}} \tag{43}$$

$$\frac{P_{ik}^{j+1/2} - P_{ik}^{j+1}}{0.5\tau} = D_P\left(\frac{P_{i+1,k}^{j+1/2} - 2P_{ik}^{j+1/2} + P_{i-1,k}^{j+1/2}}{h_x^2} + \frac{P_{i,k+1}^{j+1} - 2P_{ik}^{j+1} + P_{i,k-1}^{j+1}}{h_y^2}\right) + \frac{V_{max}S_{ik}^{j+1}}{K_M + S_{ik}^{j}}, \tag{44}$$

$i = 1, 2, \ldots, n_x - 1,\ k = 1, 2, \ldots, n_y - 1,\ j = 0, 1, \ldots, m - 1.$

Let us stress here that the first scheme is implicit along the direction x and it is explicit along the direction y, while the second one is implicit along the direction y and it is explicit in the direction x. Equations (41)–(44) are supplemented with the initial conditions

$$S_{ik}^{0} = S_0(x_i, y_k),\ P_{ik}^{0} = P_0(x_i, y_k),\ i = 0, 1, \ldots, n_x,\ k = 0, 1, \ldots, n_y \tag{45}$$

and the difference boundary conditions of special type

$$S_{ik}^{j+1} = S_0(x_i, y_k),\ P_{ik}^{j+1} = P_0(x_i, y_k) \tag{46}$$

for $k = 0$ and $k = n_y$, $i = 0, 1, \ldots, n_x$

$$S_{ik}^{j+1/2} = S_0(x_i, y_k),\ P_{ik}^{j+1/2} = P_0(x_i, y_k) \tag{47}$$

for $i = 0$ and $i = n_x$, $k = 0, 1, \ldots, n_y$.

6.1 Calculation of a Solution

The calculation is done in series one grid layer after another. Now let the solutions S_{ik}^{j} and P_{ik}^{j} be calculated at $j - th$ layer. We move along the rows $k = 1, 2, \ldots, n_y - 1$ to solve (41)–(42) by the elimination method, which permits us to determine the values $S_{ik}^{j+1/2}$ and $P_{ik}^{j+1/2}$ at all the nodes of the grid. After that we move along the columns $i = 1, 2, \ldots, n_x - 1$ in an attempt to solve (43), (44) and find the values S_{ik}^{j+1} and P_{ik}^{j+1}. In passing from the $(n+1)$th layer to the $(n+2)$th layer the same procedure is adopted, thus causing the alternating directions methods.

The algorithm just established is economical, because the number of arithmetical operations is proportional to the number of the nodes of the grid.

6.2 The Convergence and the Stability

The alternating directions method is unconditionally stable and converges with the rate $O(\tau + h^2)$.

7 The Explicit Method for the Multidimensional Problems

Now we start the investigation of the reaction–diffusion problem on the parallelepiped. The biosensor action is described by the boundary-value problem

$$\frac{\partial S}{\partial t} = D_S\left(\frac{\partial^2 S}{\partial x^2} + \frac{\partial^2 S}{\partial y^2} + \frac{\partial^2 S}{\partial z^2}\right) - \frac{V_{max} S}{K + S}, \tag{48}$$

$$\frac{\partial P}{\partial t} = D_P\left(\frac{\partial^2 P}{\partial x^2} + \frac{\partial^2 P}{\partial y^2} + \frac{\partial^2 P}{\partial z^2}\right) + \frac{V_{max} S}{K + S} \tag{49}$$

in the region $Q = \{(x, y, z) \in \Omega,\ 0 < t \le T)$, where $\Omega = \{0 < x < d_x, 0 < y < d_y, 0 < z < d_z\}$ with the initial conditions for $t = 0$

$$S(x, y, z, 0) = \varphi_S(x, y, z),\ P(x, y, z, 0) = \varphi_P(x, y, z),\ (x, y, z) \in \bar{\Omega} \tag{50}$$

and the boundary conditions

$$\begin{aligned} &S(x, y, z, t) = \psi_S(x, y, z, t),\ P(x, y, z, t) = \psi_P(x, y, z, t), \\ &(x, y, z,) \in \partial\Omega,\ t \in [0, T]. \end{aligned} \tag{51}$$

Here $\varphi_S(x, y, z,)$, $\varphi_P(x, y, z,)$, $\psi_S(x, y, z, t)$, $\psi_P(x, y, z, t)$ are the given functions. The boundary $\partial\Omega$ consists of the boundary of the parallelepiped Ω. For the sake of simplicity we have taken the initial and boundary conditions of the general form.

Now we will describe the explicit method for the solution of the system of the reaction–diffusion equations.

The explicit difference scheme for the solution of the system of the reaction–diffusion equations is popular among specialists. The main reason is the simplicity of the numerical realization. We introduce a grid with steps τ along the axis t and h_x, h_y, h_z along the axes x and y and z, respectively

$$\begin{aligned}
&\bar{Q}_{h_x,h_y,h_z,\tau} = \{(x_i,\ y_k,\ z_l,\ t_j) : x_i = ih_x,\ i = 0, 1, ..., n_x,\ h_x n_x = d_x;\\
&y_k = kh_y,\ k = 0, 1, ..., n_y,\ h_y n_y = d_y;\\
&z_l = lh_y,\ l = 0, 1, ..., n_z,\ h_z n_z = d_z;\ t_j = j\tau,\ j = 0, 1, ...m,\ \tau m = T\}.
\end{aligned}$$

Now the system of the reaction and the diffusion equations is approximated by the system of the explicit difference equations

$$\begin{aligned}
\frac{S^{j+1}_{i,k,l} - S^{j}_{i,k,l}}{\tau} = D_S\Bigg(&\frac{S^{j}_{i+1,k,l} - 2S^{j}_{i,k,l} + S^{j}_{i-1,k,l}}{h_x^2} + \frac{S^{j}_{i,k+1,l} - 2S^{j}_{i,k,l} + S^{j}_{i,k-1,l}}{h_y^2}\\
&+ \frac{S^{j}_{i,k,l+1} - 2S^{j}_{i,k,l} + S^{j}_{i,k,l-1}}{h_z^2}\Bigg) - \frac{V_{max} S^{j}_{i,k,l}}{K + S^{j}_{i,k,l}},
\end{aligned} \tag{52}$$

$$\begin{aligned}
\frac{P^{j+1}_{i,k,l} - P^{j}_{i,k,l}}{\tau} = D_P\Bigg(&\frac{P^{j}_{i+1,k,l} - 2P^{j}_{i,k,l} + P^{j}_{i-1,k,l}}{h_x^2} + \frac{P^{j}_{i,k+1,l} - 2P^{j}_{i,k,l} + P^{j}_{i,k-1,l}}{h_y^2}\\
&+ \frac{P^{j}_{i,k,l+1} - 2P^{j}_{i,k,l} + P^{j}_{i,k,l-1}}{h_z^2}\Bigg) + \frac{V_{max} S^{j}_{i,k,l}}{K + S^{j}_{i,k,l}},
\end{aligned} \tag{53}$$

$i = 1, 2, \ldots, n_x - 1$, $k = 1, 2, \ldots, n_y - 1$, $l = 1, 2, \ldots, n_z - 1$, $j = 0, 1, \ldots, m - 1$. The initial conditions are explicitly specified

$$\begin{aligned}
&S^{0}_{i,k,l} = \varphi_S(x_i, y_k, z_l),\ P^{0}_{i,k,l} = \varphi_P(x_i, y_k, z_l),\\
&i = 0, 1, \ldots, n_x,\ k = 0, 1, \ldots, n_y,\ l = 0, 1, \ldots, n_z
\end{aligned} \tag{54}$$

and the boundary conditions

$$S^{j}_{i,k,l} = \psi_S(x_i, y_k, z_l, t_j),\ P^{j}_{i,k,l} = \psi_P(x_i, y_k, z_l, t_j) \tag{55}$$

where points (x_i, y_k, z_l) belong to the boundary $\partial\Omega$ and

$$i = 0, n_x,\ k = 0, 1, \ldots, n_y,\ l = 0, 1, \ldots, n_z,$$

$$k = 0, n_y,\ i = 0, 1, \ldots, n_x,\ l = 0, 1, \ldots, n_z,$$

$$l = 0, n_z,\ i = 0, 1, \ldots, n_x,\ k = 0, 1, \ldots, n_y.$$

The system of the difference equations together with the initial and boundary conditions framed the explicit difference scheme (52)–(55). This scheme is two-layer one. The pattern contains seven points (x_{i+1}, y_k, z_l, t_j), (x_i, y_k, z_l, t_j), (x_{i-1}, y_k, z_l, t_j), (x_i, y_{k+1}, z_l, t_j), (x_i, y_{k-1}, z_l, t_j), (x_i, y_k, z_{l+1}, t_j), (x_i, y_k, z_{l-1}, t_j), (x_i, y_k, z_l, t_{j+1}). As in the previous one-dimensional space case we can prove that the error of the approximation is of order $O(\tau + h_x^2 + h_y^2 + h_z^2)$. Now we will continue with the calculation of the solution of the difference scheme (52)–(55).

7.1 The Calculation of a Solution

The calculation of the solution of the difference scheme (52)–(55) is similar to the calculation of the solution of the previous explicit schemes. The calculation is done in series one grid layer after another one. At first a solution on the zero layer at $t = 0$ using the initial conditions (54) is calculated.

Now let the solutions $S_{i,k,l}^j$ and $P_{i,k,l}^j$ be calculated at j-th layer. We proceed to solve the difference equations (52), (53) subject to values $S_{i,k,l}^{j+1}$ and $P_{i,k,l}^{j+1}$ at $j + 1$-th layer. We have

$$S_{i,k,l}^{j+1} = S_{i,k,l}^j + \tau D_S \left(\frac{S_{i+1,k,l}^j - 2S_{i,k,l}^j + S_{i-1,k,l}^j}{h_x^2} + \frac{S_{i,k+1,l}^j - 2S_{i,k,l}^j + S_{i,k-1,l}^j}{h_y^2} + \frac{S_{i,k,l+1}^j - 2S_{i,k,l}^j + S_{i,k,l-1}^j}{h_z^2} \right) - \tau \frac{V_{max} S_{i,k,l}^j}{K + S_{i,k,l}^j},$$

$$P_{i,k,l}^{j+1} = P_{i,k,l}^j + \tau D_P \left(\frac{P_{i+1,k,l}^j - 2P_{i,k,l}^j + P_{i-1,k,l}^j}{h_x^2} + \frac{P_{i,k+1,l}^j - 2P_{i,k,l}^j + P_{i,k-1,l}^j}{h_y^2} + \frac{P_{i,k,l+1}^j - 2P_{i,k,l}^j + P_{i,k,l-1}^j}{h_z^2} \right) + \tau \frac{V_{max} S_{i,k,l}^j}{K + S_{i,k,l}^j},$$

$i = 1, 2, \ldots, n_x - 1$, $k = 1, 2, \ldots, n_y - 1, l = 1, 2, \ldots, n_z - 1$.
All values of the functions $S_{i,k,l}^j$ and $P_{i,k,l}^j$ are known on the right side of the equality. So we can calculate the values $S_{i,k,l}^{j+1}$ from the first equation for

$$i = 1, 2, \ldots, n_x - 1,\ k = 1, 2, \ldots, n_y - 1,\ l = 1, 2, \ldots, n_z - 1$$

and then, consequently, $P_{i,k,l}^{j+1}$ from the second equation for

$$i = 1, 2, \ldots, n_x - 1,\ k = 1, 2, \ldots, n_y - 1,\ l = 1, 2, \ldots, n_z - 1.$$

The remaining values S_{ik}^{j+1} and P_{ik}^{j+1} on the boundary are calculated using the boundary conditions (55). Further, analogously, we will find the solutions $S_{i,k,l}^{j+2}$ and $P_{i,k,l}^{j+2}$ on the next layer and will continue up to the last layer $t = t_m$. The algorithm just established is economical, because the number of arithmetical operations is proportional to the number of the nodes of the grid.

7.2 *The Convergence and the Stability*

The explicit method is conditionally stable and converges with the rate $O(\tau + h_x^2 + h_y^2 + h_z^2)$. The differential equations are nonlinear and it is possible to formulate the sufficient stability conditions

$$\frac{\tau \max\{D_S, D_P\}}{\min\{h_x^2, h_y^2, h_z^2\}} \le 1/12, \quad \frac{\tau V_{\max}}{K_M} \le 1/2.$$

References

1. Adamatzky A, De Lacy Costello B, Asai T (2005) Reaction-diffusion computers, Elsevier, Amsterdam
2. Amatore C, Szunerits S, Thouin L, Warkocz J-S (2001) The real meaning of Nernst's steady diffusion layer concept under non-forced hydrodynamic conditions. A simple model based on Levich's seminal view of convection. J Electroanal Chem 500:62
3. Amatore C, Oleinick AI, Svir I (2006) Construction of optimal quasi-conformal mappings for the 2d-numerical simulation of diffusion at microelectrodes. Part 1: Principle of the method and its application to the inlaid disk microelectrode. J Electroanal Chem 597:69
4. Amatore C, Oleinick A, Klymenko OV et al (2008) Theory and simulation of diffusion-reaction into nano- and mesoporous structures. Experimental application to sequestration of mercury(II). Anal Chem 80:3229
5. Ames WF (1977) Numerical methods for partial differential equations, 2nd edn. Academic, New-York
6. Amine A, Kauffmann JM, Patriarche GJ (1991) Long-term operational stability of a mixed glucose oxidase-redox mediator-carbon paste electrode. Anal Lett 24:1293
7. Amine A, Kauffmann JM, Guilbault GG (1993) Characterization of mixed enzyme-mediator-carbon paste electrodes. Anal Lett 26:1281
8. Antiochia R, Lavagnini I, Magno F (2004) Amperometric mediated carbon nanotube paste biosensor for fructose determination. Anal Let 37:1657
9. Arshak K, Jafer E, McDonagh D (2007) Modeling and simulation of a wireless microsensor data acquisition system using PCM techniques. Simul Model Pract Th 15:764
10. Aris R (1975) The mathematical theory of diffusion and reaction in permeable catalysts. The theory of the steady state. Clarendon, Oxford
11. Artursson T, Eklöv T, Lundström I et al (2000) Drift correction for gas sensors using multivariate methods. J Chemometrics 14:711
12. Babushka I, Pager M, Vitasek E (1966) Numerical processes in differential equations. Wiley, Chichester
13. Bacha S, Bergel A, Comtat M (1995) Transient response of multilayer electroenzymic biosensors. Anal Chem 67:1669
14. Bacha S, Montagne M, Bergel A (1996) Modeling mass transfer with enzymatic reaction in electrochemical multilayer microreactors. AICHE J 42:2967
15. Bacon NC, Hall EAH (1999) A sandwich enzyme electrode giving electrochemical scavenging of interferents. Electroanal 11:749
16. Baeumner AJ, Jones C, Wong CY, Price A (2004) A generic sandwich-type biosensor with nanomolar detection limits. Anal Bioanal Chem 378:1587
17. Baldini F, Chester AN, Homola J, Martellucci S (2006) Optical chemical sensors. Springer, Amsterdam
18. Bakhvalov NS, Panasenko GP (1989) Homogenization: averaging processes in periodic media. Kluwer, Dordrecht
19. Bard AJ, Faulkner LR (2001) Electrochemical methods. Fundamentals and applications, 2nd edn. Wiley, New York

20. Barnaby W (1997) Biological weapons: an increasing threat. Med Confl Surviv 13:301
21. Baronas R, Ivanauskas F, Kulys J (1998) Modeling of a microreactor on heterogeneous surface and an influence of geometry to microreactor operation. Nonlinear Anal Model Contr 3:19
22. Baronas V, Ivanauskas F, Kulys J (1999) Modeling a biosensor based on the heterogeneous microreactor. J Math Chem 25:245
23. Baronas R, Ivanauskas F, Survila A (2000) Simulation of electrochemical behavior of partially blocked electrodes under linear potential sweep conditions. J Math Chem 27:267
24. Baronas R, Ivanauskas F, Kulys J (2002) Modelling dynamics of amperometric biosensors in batch and flow injection analysis. J Math Chem 32:225
25. Baronas R, Christensen J, Ivanauskas F, Kulys J (2002) Computer simulation of amperometric biosensor response to mixtures of compounds. Nonlinear Anal Model Contr 7:3
26. Baronas R, Ivanauskas F, Kulys J (2003) Computer simulation of the response of amperometric biosensors in stirred and non stirred solution. Nonlinear Anal Model Contr 8:3
27. Baronas R, Ivanauskas F, Kulys J (2003) The influence of the enzyme membrane thickness on the response of amperometric biosensors. Sensors 3:248
28. Baronas R, Ivanauskas F, Kulys J, Sapagovas M (2003) Modeling of amperometric biosensors with rough surface of the enzyme membrane. J Math Chem 34:227
29. Baronas R, Kulys J, Ivanauskas F (2004) Modeling amperometric enzyme electrode with substrate cyclic conversion. Biosens Bioelectron 19:915
30. Baronas R, Ivanauskas F, Kulys J (2004) The effect of diffusion limitations on the response of amperometric biosensors with substrate cyclic conversion. J Math Chem 35:199
31. Baronas R, Ivanauskas F, Maslovskis R, Vaitkus P (2004) An analysis of mixtures using amperometric biosensors and artificial neural networks. J Math Chem 36:281
32. Baronas R, Kulys J, Ivanauskas F (2004) Mathematical model of the biosensors acting in a trigger mode. Sensors 4:20
33. Baronas R, Ivanauskas F, Kulys J, Sapagovas M (2004) Computational modeling of a sensor based on an array of enzyme microreactors. Nonlinear Anal Model Contr 9:203
34. Baronas R, Ivanauskas F, Kulys J (2005) Modelling biosensors with perforated membrane. Lith Math J 45(spec issue):449
35. Baronas R, Kulys J, Ivanauskas F (2006) Computational modeling of biosensors with perforated and selective membranes. J Math Chem 39:345
36. Baronas R, Ivanauskas F, Kulys J (2006) Mathematical modeling of biosensors based on an array of enzyme microreactors. Sensors 6:453
37. Baronas R, Ivanauskas F, Kaunietis I, Laurinavicius V (2006) Mathematical modeling of plate-gap biosensors with an outer porous membrane. Sensors 6:727
38. Baronas R, Ivanauskas F, Maslovskis R et al (2007) Locally weighted neural networks for an analysis of the biosensor response. Kybernetika 43:21
39. Baronas R, Ivanauskas F, Kulys J (2007) Computational modeling of the behaviour of potentiometric membrane biosensors. J J Math Chem 42:321
40. Baronas R, Gaidamauskaite E, Kulys J (2007) Modeling a peroxidase-based optical biosensor. Sensors 7:2723
41. Baronas R, Ivanauskas F, Kulys J (2007) Computational modeling of amperometric enzyme electrodes with selective and perforated membranes. In: Simos TE, Maroulis G (ed) Computation in modern science and engineering: proceedings of the international conference on computational methods in science and engineering 2007 (ICCMSE 2007), vol 2, parts A and B. AIP Press, pp 457–460
42. Baronas R, Kulys J (2008) Modeling amperometric biosensors based on chemically modified electrodes. Sensors 8:4845
43. Bartlett PN, Whitaker RG (1987) Electrochemical immobilization of enzymes. Part 1. Theory. J Electroanal Chem 224:27
44. Bartlett PN, Pratt KFE (1993) Modelling of processes in enzyme electrodes. Biosens Bioelectron 8:451

45. Bartlett PN, Pratt KFE (1995) Theoretical treatment of diffusion and kinetics in amperometric immobilized enzyme electrodes Part I: Redox mediator entrapped within the film. J Electroanal Chem 397:61
46. Bartlett PN, Birkin PR, Wallace ENK (1997) Oxidation of β-nicotinamide adenine dinucleotide (NADH) at poly(aniline)-coated electrodes. J Chem Soc, Faraday Trans 93:1951
47. Bergel A, Comtat M (1984) Theoretical evaluation of transient responses of an amperometric enzyme electrode. Anal Chem 56:2904
48. Bertram R, Pernarowski M (1998) Glucose diffusion in pancreatic islets of Langerhans. Biophys J 74:1722
49. Bieniasz LK, Britz D (2004) Recent developments in digital simulation of electroanalytical experiments. Polish J Chem 78:1195
50. Bindra DS, Zhang Y, Wilson GS et al (1991) Design and in vitro studies of a needle-type glucose sensor for subcutaneous monitoring. Anal Chem 63:1692
51. Blaedel WJ, Kissel TR, Boguslaski RC (1972) Kinetic behavior of enzymes immobilized in artificial membranes. Anal Chem 44:2030
52. Blaedel WJ, Boguslaski RC (1978) Chemical amplification in analysis: a review. Anal Chem 50:1026
53. Bosch ME, Sánchez AJR, Rojas FS, Ojeda CB (2007) Recent development in optical fiber biosensors. Sensors 7:797
54. Boujtita E, el Murr N (2006) Biosensors for analysis of ethanol in food. J Food Sci 60:201
55. Briggs GE, Haldane JBS (1925) A note on the kinetics of enzyme action. Biochem J 19:338
56. Britz D (2005) Digital simulation in electrochemistry, 3rd edn. Springer, Berlin
57. Bro R (2003) Multivariate calibration: What is in chemometrics for the analytical chemist? Anal Chim Acta 500:185
58. Bruice TC (2006) Computational approaches: reaction trajectories, structures, and atomic motions. Enzyme reactions and proficiency. Chem Rev 106:3119
59. Buerk DG (1995) Biosensors: theory and applications. CRC Press, Lancaster
60. Cambiaso A, Delfino L, Grattarola M et al (1996) Modeling and simulation of a diffusion limited glucose biosensor. Sensor Actuat B-Chem 33:203
61. Carr PW (1977) Fourier analysis of the transient response of potentiometric enzyme electrodes. Anal Chem 49:799
62. Carr PW, Bowers LD (1980) Immobilized enzymes in analytical and clinical chemistry. Wiley, New York
63. Castillo J, Blöchl A, Dennison S et al (2005) Glutamate detection from nerve cells using a planar electrodes array integrated in a microtiter plate. Biosens Bioelectron 20:2116
64. Cenas NK, Kulys JJ (1981) Biocatalytic oxidation of glucose on the conductive charge transfer complexes. Bioelectrochem Bioenerg 8:103
65. Chaplin MF, Bucke C (1990) Enzyme technology. Cambridge University Press, Cambridge
66. Chaubey A, Malhotra BD (2002) Mediated biosensors. Biosens Bioelectron 17:441
67. Chen LC, Tseng KS, Ho KC (2006) General kinetic model for amperometric sensors based on Prussian blue mediator and its analogs: Application to cysteine detection. Electroanal 18:1313
68. Choi MMF (2004) Progress in enzyme-based biosensors using optical transducers. Microchimica Acta 148:107
69. Clarc LC, Lyons C (1962) Electrode system for continuous monitoring in cardiovascular surgery. Ann N Y Acad Sci 102:29
70. Coche-Guerente L, Labbe P, Mengeaud V (2001) Amplification of amperometric biosensor responses by electrochemical substrate recycling. 3. Theoretical and experimental study of the phenol-polyphenol oxidase system immobilized in Laponite hydrogels and layer-by-layer self-assembled structures. Anal Chem 73:3206
71. Corcuera JRD, Cavalieri R, Powers J, Tang J (2004) Amperometric enzyme biosensor optimization using mathematical modeling. In: Proceedings of the 2004 ASAE/Csae Annual International Meeting, Paper No. 047030 American Society of Agricultural Engineers, Ottawa
72. Cornish-Bowden A (2004) Fundamentals of enzyme kinetics, 3rd edn. Portland Press, London
73. Crank J (1975) The mathematics of diffusion, 2nd edn. Clarendon, Oxford

74. Della Ciana L, Bernacca G, Bordin F et al (1995) Highly sensitive amperometric measurement of alkaline phosphatase activity with glucose oxidase amplification. J Electronal Chem 382:129
75. Deslous C, Gabrielli C, Keddam M et al (1997) Impedance techniques at partially blocked electrodes by scale deposition. Electrochim Acta 42:1219
76. Devaux R, Bergel A, Comtat M (1995) Mass transfer with chemical reaction in thin-layer electrochemical reactors. AICHE J 41:1944
77. Devlin JP (ed) (1997) High throughput screening. Marcel Dekker, New York
78. Diamond D (ed) (1998) Principles of chemical and biological sensors. Chemical Analysis: A Series of Monographs on Analytical Chemistry and Its Applications. Wiley-Interscience, New York
79. Dirks JL (1996) Diagnostic blood analysis using point-of-care technology. AACN Clin Issues 7:249
80. Dixon M (1953) The determination of enzyme inhibitor constants. Biochem J 55:170
81. Dixon M, Webb EC, Thorne CJR, Tipton KF (1979) Enzymes, 3rd edn. Longman, London
82. Dohnal M (1992) Qualitative partial differential equations and their realistic applications. Comput Ind 20:209
83. Dormieux L, Lemarchand E (2001) Homogenization approach of advection and diffusion in cracked porous material. J Eng Mech ASCE 127:1267
84. Eggenstein C, Borchardt M, Diekmann C et al (1999) A disposable biosensor for urea determination in blood based on an ammonium-sensitive transducer. Biosens Bioelectron 14:33
85. Eggins BR (2002) Chemical sensors and biosensors. Analytical techniques in the sciences. Wiley, Chichester
86. Ehrfeld W, Hessel V, Lwe H (2000) Microreactors: new technology for modern chemistry. Wiley-VCH, New York
87. Ferreira LS, Souza MBD, Trierweiler JO et al (2003) Aspects concerning the use of biosensors for process control: experimental and simulation investigations.Comp Chem Eng 27:1165
88. Forrow NJ, Sanghera GS, Walters SJ (2002) The influence of structure in the reaction of electrochemically generated ferrocenium derivatives with reduced glucose oxidase. J Chem Soc, Dalton Trans, 3187
89. Forrow NJ, Bayliff SW (2005) A commercial whole blood glucose biosensor with a low sensitivity to hematocrit based on an impregnated porous carbon electrode. Biosens Bioelectron 21:3581
90. Fraser DM (ed) (1997) Biosensors in the body: continuous in vivo monitoring. Wiley, Chichester
91. Frew JE, Hill HO (1987) Electrochemical biosensors. Anal Chem 59:933A
92. Fuhrmann B, Spohn U (1998) An enzymatic amplification flow injection analysis (FIA) system for the sensitive determination of phenol. Biosens Bioelectron 13:895
93. Gaidamauskaite E, Baronas R Modeling a peroxide-based fluorescent biosensor. In: Louca LS, Chrysanthou Y, Oplatkova Z, Al-Begain K (eds) Proceedings, 22nd European Conference on modeling and Simulation ECMS 2008, 3–6 June 2008, Nicosia, Cyprus. EMCS 2008, Nicosia 2008
94. Gajovic N, Warsinke A, Huang T et al (1999) Characterization and mathematical modeling of a bienzyme electrode for L-malate with cofactor recycling. Anal Chem 71:4657
95. Garboczi EJ (1990) Permeability, diffusivity and microstructural parameters: a critical review. Cem Concr Res 20:591
96. Gorton L (1995) Carbon paste electrodes modified with enzymes, tissues, and cells. Electroanal 7:23
97. Gueshi T, Tokuda K, Matsuda H (1978) Voltammetry at partially covered electrodes. Part I. Chronopotentiometry and chronoamperometry at model electrodes. J Electroanal Chem 89:247
98. Gufreund H (1995) Kinetics for the life sciences. Cambridge University Press, Cambridge
99. Guilbault GG (1970) Enzymatic methods of analysis. Pergamon, Oxford
100. Guilbault GG (1980) Analytical uses of immobilized enzymes. Marcel Dekker, New York

101. Hall DL, McMullen SAH (2004) Mathematical techniques in multisensor data fusion. Artech House Information Warfare Library, 2nd edn. Artech House Inc, Norwood
102. Hameka HF, Rechnitz GA (1983) Theory of the biocatalytic membrane electrode. J Phys Chem 87:1235
103. Hameka HF, Rechnitz GA (1981) Steady-state theory of biocatalytic membrane electrodes. Anal. Chem. 53:1586
104. Ha J, Engler CR, Lee SJ (2008) Determination of diffusion coefficients and diffusion characteristics for chlorferon and diethylthiophosphate in Ca-Alginate gel beads. Biotechnol Bioeng 100:698
105. Hale PD, Lan HL, Boguslavsky LI et al (1991) Amperometric glucose sensors based on ferrocene-modified poly(ethylene oxide) and glucose oxidase. Anal Chim Acta 251:121
106. Harsanyi G (2000) Sensors in biomedical applications: fundamentals, technology and applications. CRC Press, New York
107. Hart JP, Crew A, Crouch E et al (2004) Some recent designs and developments of screen-printed carbon electrochemical sensors/biosensors for biomedical, environmental, and industrial analyses. Anal Lett 37:789
108. Hassan MM, Atiqullah M, Beg SA, Chowdhury MHM (1995) Analysis of non-isothermal tubular reactor packed with immobilized enzyme systems. Chem Eng J Biochem Eng J 58:275
109. Henri V (1902) Théorie générale de l'action de quelques diastases. Compt Rend Hebd Acad Sci Paris 135:916
110. Higham NJ (2002) Accuracy and stability of numerical algorithms, 2nd edn. SIAM, Philadelphia
111. Hobbs DW (1999) Aggregate influence on chloride ion diffusion into concrete. Cem Concr Res 29:1995
112. Ikeda T (1995) Enzyme-modified electrodes with bioelectrocatalytic function. Bunsuki Kagaku 44:333
113. Iliev I, Atanasov P, Gamburzev S et al (1992) Transient response of electrochemical biosensors with asymmetrical sandwich membranes. Sensor Actuat B-Chem 8:65
114. Ivanauskas F, Kaunietis I, Laurinavicius V et al (2005) Computer simulation of the steady state currents at enzyme doped carbon paste electrode. J Math Chem 38:355
115. Ivanauskas F, Baronas R, Kulys J (2005) Mathematical modeling of biosensors with perforated and selective membranes. Rakenteiden Mekaniikka - J Struct Mech 38:63
116. Ivanauskas F, Baronas R (2008) Modeling an amperometric biosensor acting in a flowing liquid. Int J Numer Meth Fluids 56:1313
117. Ivanauskas F, Baronas R (2008) Numerical simulation of a plate-gap biosensor with an outer porous membrane. Simul Model Pract Th 16:962
118. Ivanauskas F, Kaunietis I, Laurinavicius V et al (2008) Apparent Michaelis constant of the enzyme modified porous electrode. J Math Chem 43:1516
119. Jobst G, Moser I, Urban G (1996) Numerical simulation of multi-layered enzymatic sensors. Biosens Bioelectron 11:111
120. Jochum P, Kowalski BR (1982) A coupled two-compartment model for immobilized enzyme electrodes. Anal Chim Acta 144:25
121. Kalnin JR, Kotomin EA, Maier J (2002) Calculations of the effective diffusion coefficient for inhomogeneous media. J Phys Chem Solids 63:449
122. Kernevez JP (1980) Enzyme mathematics. Studies in mathematics and its applications. Elsevier, Amsterdam
123. Knopf GK, Bassi AS (2007) Smart biosensor technology. CRC Press, New York
124. Kohen A, Klinman JP (1999) Hydrogen tunneling in biology. Chem Biol 6(7):191
125. Kulys J, Kadziauskiene K (1978) Bioelectrocatalysis. Lactate-oxidizing electrode. Doklady Akademii Nauk SSSR 239:636
126. Kulys J, Samalius AS, Svirmickas G-JS (1980) Electron exchange between the enzyme active center and organic metal. FEBS Lett 114:7
127. Kulys J, Kadziauskiene K (1980) Yeast BOD sensor. Biotechnol Bioeng 22:221
128. Kulys JJ (1981) Analytical systems based on immobilized enzymes. Mokslas, Vilnius (in Russian)

129. Kulys J (1981) The development of new analytical systems based on biocatalysts. Anal Lett 14(B6):377
130. Kulys JJ (1981) Development of new analytical systems based on biocatalysers. Enzyme Microb Technol 3:344
131. Kulys J, Razumas V (1983) Biocatalysis in electrochemistry of organic compounds. Mokslas, Vilnius (in Russian)
132. Kulys JJ, Cenas NK (1983) Oxidation of glucose oxidase from penicillium vitale by one- and two-electron acceptors. Biochim Biophys Acta 744:57
133. Kulys JJ, Sorochinskii VV, Vidziunaite RA (1986) Transient response of bienzyme electrodes. Biosensors 2:135
134. Kulys J, Razumas V (1986) Bioamperometry. Mokslas, Vilnius (in Lithuanian)
135. Kulys J, Schmid RD (1990) A sensitive enzyme electrode for phenol monitoring. Anal Lett 23:589
136. Kulys JJ, Vidziunaite RA (1990) Amperometric enzyme electrodes with chemically amplified response. In: Wise DL (ed) Bioinstrumentation. Butterworths, Boston, pp 1263–1283
137. Kulys J, Schmid RD (1991) Bienzyme sensors based on chemically modified electrodes. Biosens Bioelectron 6:43
138. Kulys J (1991) Biosensors based on modified electrodes. In: Turner APF (ed) Advances in biosensors, vol 1. JAI Press, London Greenwich Connecticut, pp 107–124
139. Kulys J, Hansen HE, Buch-Rasmussen T et al (1994) Glucose biosensor based on the incorporation of Meldola blue and glucose oxidase within carbon paste. Anal Chim Acta 288:193
140. Kulys J, Hansen HE (1994) Carbon-paste biosensors array for long-term glucose measurement. Biosen Bioelectr 9:491
141. Kulys J, Hansen HE (1995) Long-term response of an integrated carbon paste based glucose biosensor. Anal Chim Acta 303:285
142. Kulys J (1999) The carbon paste electrode encrusted with a microreactor as glucose biosensor. Biosens Bioelectron 14:473
143. Kulys J, Krikstopaitis K, Ziemys A (2000) Kinetics and thermodynamics of peroxidase- and laccase-catalyzed oxidation of N-substituted phenothiazines and phenoxazines. J Biol Inorg Chem 5:333
144. Kulys J, Vidziunaite R (2003) Amperometric biosensors based on recombinant laccases for phenols determination. Biosens Bioelectron 18:319
145. Kulys J (2004) Modeling trienzyme biosensor at internal diffusion limitation. Anal Model Contr 9:139
146. Kulys J, Tetianec L (2005) Synergistic substrates determination with biosensors. Biosens Bioelectron 21:152
147. Kulys J (2005) Kinetics of biocatalytical synergistic reactions. Nonlinear Anal Model Contr 10:223
148. Kulys F, Baronas R (2006) Modeling of amperometric biosensors in the case of substrate inhibition. Sensors 6:1513
149. Kulys J, Tetianec L (2006) Modeling of amperometric biosensors in the case of substrate inhibition. Sensor Actuat B-Chem 113:755
150. Kwan RCH, Hon PYT, Mak WC et al (2006) Biosensor for rapid determination of 3-hydroxybutyrate using bienzyme system. Biosens Bioelectron 21:1101
151. Lammertyn J, Verboven P, Veraverbeke EA et al (2006) Analysis of fluid flow and reaction kinetics in a flow injection analysis biosensor. Sensor Actuat B-Chem 114:728
152. Laurinavicius VA, Kulys JJ, Gureviciene VV, Simonavicius KJ (1989) Flow through and cateter biosensors with an extended concentration range. Biomed Biochem Acta 48:905
153. Laurinavicius V, Razumiene J, Kurtinaitiene B et al (2002) Bioelectrochemical application of some PQQ-dependent enzymes. Bioelectrochem 55:29
154. Laurinavicius V, Razumiene J, Ramanavicius A, Ryabov AD (2004) Wiring of PQQ-dehydrogenases. Biosens Bioelectron 20:1217
155. Lawrence NS, Deo RP, Wang J (2004) Biocatalytic carbon paste sensors based on a mediator pasting liquid. Anal Chem 77:3735

156. Leatherbarrow RJ, Edwards PR (1999) Analysis of molecular recognition using optical biosensors. Curr Opin Chem Biol 3:544
157. Lemke K (1988) Mathematical simulation of an amperometric enzyme-substrate electrode with a pO2 basic sensor. Part 2. Mathematical simulation of the glucose oxidase glucose electrode. Med Biol Eng Comput 26:533
158. Levich VG (1962) Physicochemical hydrodynamics. Prentice-Hall, London
159. Liang JF, Li YT, Yang VC (2000) Biomedical application of immobilized enzymes. J Pharm Sci 89:979
160. Lide DR (ed) (2007-2008) Handbook of chemistry and physics, 88th edn. CRC Press, Boca Raton
161. Ligler FS, Taitt CR (2002) Optical biosensors: present and future Elsevier, Amsterdam
162. Lim T-C (2006) Application of binomial coefficients in representing central difference solution to a class of PDE arising in chemistry. J Math Chem 39:177
163. Llinas JR, Ruiz JM (1986) Multivariate analysis of chemical data sets with factor methods. In: Vemin G, Chanon M (eds) Computer aids to chemistry. Wiley, New York
164. Lyons MEG, Greer JC, Fitzgerald CA et al (1996) Reaction/diffusion with Michaelis-Menten kinetics in electroactive polymer films. Part 1. The steady-state amperometric response. Analyst 121:715
165. Lyons MEG, Bannon T, Hinds G, Rebouillat S (1998) Reaction/diffusion with Michaelis-Menten kinetics in electroactive polymer films. Part 2. The transient amperometric response. Analyst 123:1947
166. Lyons MEG, Murphy J, Rebouillat S (2000) Theoretical analysis of time dependent diffusion, reaction and electromigration in membranes. J Solid State Electrochem 4:458
167. Lyons MEG (2001) Mediated electron transfer at redox active monolayers. Sensors 1:215
168. Lyons ME (2006) Modeling the transport and kinetics of electroenzymes at the electrode/solution interface. Sensors 6:1765
169. Lobanov AV, Borisov IA, Gordon SH et al (2001) Analysis of ethanol-glucose mixtures by two microbial sensors: application of chemometrics and artificial neural networks for data processing. Biosens Bioelectron 16:1001
170. Luckarift HR (2008) Silica-immobilized enzyme reactors. J Liq Chromatogr R T 31:1568
171. Mackey D, Killard AJ, Ambrosi A, Smyth MR (2007) Optimizing the ratio of horseradish peroxidase and glucose oxidase on a bienzyme electrode: Comparison of a theoretical and experimental approach. Sensor Actuat B-Chem 122:395
172. Magner E (1998) Trends in electrochemical biosensors. Analyst 123:1967
173. Malinauskas A, Kulys J (1978) Alcohol, lactate and glutamate sensors based on oxidoreductases with regeneration of nicotinamide adenine dinucleotide. Anal Chim Acta 98:31
174. Malkavaara P, Aln R, Kolehmainen E (2000) Chemometrics: an important tool for the modern chemist, an example from wood-processing chemistry. J Chem Inf Comput Sci 40:438
175. Manz A, Graber N, Widmer HM (1990) Miniaturized total chemical analysis systems: a novel concept for chemical sensing. Sensor Actuat B-Chem 1:244
176. Mak KKW, Wollenberger U, Scheller F, Renneberg R (2003) An amperometric bi-enzyme sensor for determination of formate using cofactor regeneration. Biosens Bioelectron 18:1095
177. Martens H, Ns T (1989) Multivariate calibration. Wiley, Chichester
178. Marti S, Roca M, Andres J et al (2004) Theoretical insights in enzyme catalysis. Chem Soc Rev 33:98
179. Matuszewski W, Trojanowicz M (1988) Graphite paste-based enzymatic glucose electrode for flow-injection analysis. Analyst 113:735
180. Mell LD, Maloy T (1975) A model for the amperometric enzyme electrode obtained through digital simulation and applied to the immobilized glucose oxidase system. Anal Chem 47:299
181. Mell LD, Maloy T (1976) Amperometric response enhancement of the immobilized glucose oxidase enzyme electrode. Anal Chem 48:1597
182. Mello LD, Kubota LT (2002) Review of the use of biosensors as analytical tools in the food and drink industries. Food Chem 77:237
183. Merino S, Grinfeld M, McKee S (1998) A degenerate reaction diffusion system modeling an optical biosensor. Z Angew Math Phys 49:46

184. Meyerhoff ME, Duan CM, Meusel M (1995) Novel nonseparation sandwich-type electrochemical enzyme immunoassay system for detecting marker proteins in undiluted blood. Clin Chem 41:1378
185. Michaelis L, Menten ML (1913) Die Kinetik der Invertinwirkung. Biochem Z 49:333
186. Miscoria SA, Barrera GD, Rivas GA (2005) Enzymatic biosensor based on carbon paste electrodes modified with gold nanoparticles and polyphenol oxidase. Electroanal 17:1578
187. Mitchell A, Griffits D (1980) The finite difference methods in partial differential equations. Wiley, New York
188. Mizutani F, Yamanaka T, Tanabe Y, Tsuda K (1985) An enzyme electrode for L-lactate with a chemically amplified response. Anal Chim Acta 177:153
189. Mizutani F, Yabuki S, Okuda A, Katsura T (1991) Glucose-sensing electrode based on carbon paste containing ferrocene and polyethylene glycol-modified enzyme. Bull Chem Soc Jpn 64:2849
190. Mizutani F (1999) Application of enzyme-modified electrodes to biosensors. Bunseki Kagaku 48:809
191. Moreira JE, Midkiff SP, Gupta M et al (2000) Java programming for high performance numerical computing. IBM Systems J 39:21
192. Morton KW (1995) Numerical solution of convection-diffusion problems. Chapman/CRC, London
193. Mulchandani A, Rogers K (eds) (1998) Enzyme & microbial biosensors: techniques and protocols. Methods in biotechnology. Humana Press, Totowa, New Jersey
194. Mullen WH, Keedy FH, Churchouse SJ, Vadgama PM (1986) Glucose enzyme electrode with extended linearity: application to undiluted blood measurements. Anal Chim Acta 183:59
195. Murray RW (1980) Chemically modified electrodes. Accts Chem Res 13:135
196. Nakamura H, Karube I (2003) Current research activity in biosensors. Anal Bioanal Chem 377:446
197. Nakamoto T, Hiramatsu H (2002) Study of odor recorder for dynamical change of odor using QCM sensors and neural network. Sens Actuators B-Chem 85:98
198. Naujikas R, Malinauskas A, Ivanauskas F (2007) Modeling of electrocatalytic processes at conducting polymer modified electrodes. J Math Chem 42:1069
199. Nernst W (1904) Theorie der Reaktionsgeschwindigkeit in heterogenen Systemen. Z Phys Chem 47:52
200. Nistor C, Rose A, Wollenberger U et al (2002) A glucose dehydrogenase biosensor as an additional signal amplification step in an enzyme-flow immunoassay. Analyst 127:1076
201. Özisik MN (1980) Heat conduction. Wiley, New York
202. Ojeda CB, Rojas FS (2006) Recent development in optical chemical sensors coupling with flow injection analysis. Sensors 6:1245
203. Pandey PC, Kayastha AM, Pandey V (1992) Amperometric enzyme sensor for glucose based on graphite paste-modified electrodes. Appl Bioch Biotech 33:139
204. Passaro VMN, Dell'olio F, Casamassima B, Leonardis FD (2007) Guided-wave optical biosensors. Sensors 7:508
205. Pfeiffer D, Scheller FW, Setz K, Schubert F (1993) Amperometric enzyme electrodes for lactate and glucose determinations in highly diluted and undiluted media. Anal Chim Acta 281:489
206. Pickup JC, Thévenot DR (1993) European achievements in glucose sensor research. In: Turner APF (ed) Advances in Biosensors, Supplement 1. JAI Press, London, pp 201–225
207. Popovtzer R, Neufeld T, Ron EZ et al (2006) Electrochemical detection of biological reactions using a novel nano-bio-chip array. Sensor Actuat B-Chem 119:664
208. Popovtzer R, Natan A, Shacham-Diamand Y (2007) Mathematical model of whole cell based bio-chip: An electrochemical biosensor for water toxicity detection. J Electroanal Chem 602:17
209. Press WH, Flannery BP, Teukolsky SA, Vetterling WT (1993) *Numerical Recipes in C: The Art of Scientific Computing* Cambridge University Press, Cambridge

210. Rahamathunissa G, Rajendran L (2008) Application of He's variational iteration method in nonlinear boundary value problems in enzyme-substrate reaction diffusion processes: part 1. The steady-state amperometric response. J Math Chem 44:849
211. Razumas VJ, Kulys JJ, Malinauskas AA (1980) Kinetic amperometric determination of hydrolase activity. Anal Chim Acta 117:387
212. Reder S, Dieterle F, Jansen H et al (2003) Multi-analyte assay for triazines using cross-reactive antibodies and neural networks. Biosens Bioelectron 19:447
213. Renedo OD, Alonso-Lomillo MA, Martinez MJA (2007) Recent developments in the field of screen-printed electrodes and their related applications. Talanta 73:202
214. Reyes DR, Iossifidis D, Auroux PA, Manz A (2002) Micro total analysis systems. 1. Introduction, theory, and technology. Anal Chem 74:2623
215. Rickus JL (2005) Impact of coenzyme regeneration on the performance of an enzyme based optical biosensor: a computational study. Biosens Bioelectron 21:965
216. Richtmayer R, Morton K (1957) Difference methods for initial-value problems. An introduction. Interscience, New York
217. Rodriguez-Mozaz S, Marco MP, de Alda MJL, Barcelo D (2004) Biosensors for environmental applications: Future development trends. Pure Appl Chem 76:723
218. Rogers KR (1995) Biosensors for environmental applications. Biosens Bioelectron 10:533
219. Rong ZM, Cheema U, Vadgama P (2006) Needle enzyme electrode based glucose diffusive transport measurement in a collagen gel and validation of a simulation model. Analyst 131:816
220. Ruzicka J, Hansen EH (1988) Flow injection analysis. Wiley, New York
221. Sakura S, Buck RP (1992) Amperometric processes with glucose oxidase embedded in the electrode. Bioelect Bioenerg 28:387
222. Samarskii AA (2001) The theory of difference schemes. Marcel Dekker, New York-Basel
223. Sanz V, de Marcos S, Galbán J (2007) Direct glucose determination in blood using a reagentless optical biosensor. Biosens Bioelectron 22:2876
224. Sapelnikova S, Dock E, Solna R et al (2003) Screen-printed multienzyme arrays for use in amperometric batch and flow systems. Anal Bioanal Chem 376:1098
225. Schachl K, Turkušić E, Komersová A et al (2002) Amperometric determination of glucose with a carbon paste biosensor. Collect Czech Chem Commun 67:302
226. Scheller FW, Pfeiffer D (1978) Enzyme electrodes. Z Chem 18:50
227. Scheller F, Renneberg R, Schubert F (1988) Coupled enzyme reactions in enzyme electrodes using sequence, amplification, competition, and antiinterference principles. In: Mosbach K (ed) Methods in enzymology, vol 137 Academic, New-York pp 29–43
228. Scheller F, Schubert F, Pfeiffer D et al (1989) Research and development of biosensors. A review. Analyst 114:653
229. Scheller F, Schubert F (1992) Biosensors. Elsevier, Amsterdam
230. Schöning MJ (2005) "Playing around" with field-effect sensors on the basis of EIS structures, LAPS and ISFETs. Sensors 5:126
231. Schubert F, Kirstein D, Schröder KL, Scheller F (1985) Enzyme electrodes with substrate and co-enzyme amplification. Anal Chim Acta 169:391
232. Schulmeister T, Scheller F (1985) Mathematical treatment of concentration profiles and anodic current for amperometric enzyme electrodes. Anal Chim Acta 170:279
233. Schulmeister T, Scheller F (1985) Mathematical description of concentration profiles and anodic currents for amperometric two-enzyme electrodes. Anal Chim Acta 171:111
234. Schulmeister T (1987) Mathematical treatment of concentration profiles and anodic current of amperometric enzyme electrodes with chemically amplified response. Anal Chim Acta 201:305
235. Schulmeister T (1990) Mathematical modeling of the dynamic behavior of amperometric enzyme electrodes. Selective Electrode Rev 12:203
236. Schulmeister T, Pfeiffer D (1993) Mathematical modelling of amperometric enzyme electrodes with perforated membranes. Biosens Bioelectron 8:75
237. Schulmeister T, Rose J, Scheller F (1997) Mathematical modelling of exponential amplification in membrane-based enzyme sensors. Biosens Bioelectron 12:1021

238. Senda M, Ikeda T, Miki K, Hiasa H (1886) Amperometric biosensors based on a biocatalyst electrode with entrapped mediator. Anal Sci 2:501
239. Smith JM, Szathmary E (1996) On the likelihood of habitable worlds. Nature 384:107
240. Somasundrum M, Aoki K (2002) The steady-state current at microcylinder electrodes modified by enzymes immobilized in conducting or non-conducting material. J Electroanal Chem 530:40
241. Song M-J, Yun D-H, Jin J-H et al (2006) Steady state kinetics of cyclic conversions of substrate in amperometric bienzyme sensors. Jpn J Appl Phys 45:7197
242. Sorochinskii VV, Kurganov BI (1996) Amperometric biosensors with a laminated distribution of enzymes in their coating. Steady state kinetics. Biosens Bioelectron 11:45
243. Sorochinskii VV, Kurganov BI (1996) Steady state kinetics of cyclic conversions of substrate in amperometric bienzyme sensors. Biosens Bioelectron 11:225
244. Sorochinskii VV, Kurganov BI (1996) Diffusion-kinetic theory of stationary behaviour of amperometric bienzyme electrodes. Biosens Bioelectron 11:709
245. Sorochinskii VV, Kurganov BI (1997) Theoretical principles of the application of potentiometric enzyme electrodes. Appl Biochem Micro 33:116
246. Spichiger-Keller UE (1998) Chemical sensors and biosensors for medical and biological applications. Wiley-VCH, New York
247. Stamatin I, Berlic C, Vaseashta A (2006) On the computer-aided modeling of analyte-receptor interactions for an efficient sensor design. Thin Solid Films 495:312
248. Stefan R-I, Bokretsion RG, van Staden JF, Aboul-Enein HY (2003) Simultaneous determination of creatine and creatinine using amperometric biosensors. Talanta 60:1223
249. Stefano LD, Arcari P, Lamberti A et al (2007) DNA optical detection based on porous silicon technology: from biosensors to biochips. Sensors 7:214
250. Streffer K, Kaatz H, Bauer CG et al (1998) Application of a sensitive catechol detector for determination of tyrosinase inhibitors. Anal Chim Acta 362:81
251. Suzuki H (2000) Advances in the microfabrication of electrochemical sensors and systems. Electroanal 12:703
252. Suzuki H, Arakawa H, Karube I (2001) Fabrication of a sensing module using micromachined biosensors. Biosens Bioelectron 16:725
253. Svancara I, Vytras K, Barek J, Zima J (2001) Carbon paste electrodes in modern electroanalysis. Crit Rev Anal Chem 31:311
254. Takoh K, Ishibashi T, Matsue T, Nishizawa M (2005) Localized chemical stimulation of cellular micropatterns using a porous membrane-based culture substrate. Sensor Actuat B-Chem 108:683
255. Tang LX, Koochaki ZB, Vadgama PM (1990) Composite liquid membrane for enzyme electrode construction. Anal Chim Acta 232:357
256. Treloar PH, Christie IM, Vadgama PM (1995) Engineering the right membranes for electrodes at the biological interface; solvent cast and electropolymerised. Biosens Bioelectron 10:195
257. Tudorache M, Bala C (2007) Biosensors based on screen-printing technology, and their applications in environmental and food analysis. Anal Bioanal Chem 388:565
258. Turner APF, Karube I, Wilson GS (1987) Biosensors: fundamentals and applications. Oxford University Press, Oxford
259. Urban PL, Goodall DM, Bruce NC (2001) Biosensor microsystems. Sens Update 8:189
260. Urban PL, Goodall DM, Bruce NC (2006) Enzymatic microreactors in chemical analysis and kinetic studies. Biotechnol Adv 24:42
261. Updike SJ, Hicks GP (1967) The enzyme electrode. Nature 214:986
262. Vermeir S, Nicolai BM, Verboven P et al (2007) Microplate differential calorimetric biosensor for ascorbic acid analysis in food and pharmaceuticals. Anal Chem 79:6119
263. Vo-Dinh T (2003) Biomedical photonics handbook. CRC Press, New York
264. Vojinovic V, Esteves FMF, Cabral JMS, Fonseca LP (2006) Bienzymatic analytical microreactors for glucose, lactate, ethanol, galactose and l-amino acid monitoring in cell culture media. Anal Chim Acta 565:240
265. Wanekaya AK, Chen W, Mulchandani A (2008) Recent biosensing developments in environmental security. J Environ Monitor 10:703

266. Wang J, Liu J, Cepra G (1997) Thermal stabilization of enzymes immobilized within carbon paste electrodes. Anal Chem 69:3124
267. Wang J (2000) Analytical electrochemistry, 2nd edn. Wiley, New-York
268. Wang J (2001) Glucose biosensors: 40 years of advances and challenges. Electroanal 13:983
269. Wang YM, Pao CV (2006) Time-delayed finite difference reaction-diffusion systems with nonquasimonotone functions. Numerishe Mathematik 103:485
270. Weibel MK, Bright HJ (1971) The glucose oxidase mechanism. J Biol Chem 246:2734
271. Whitaker S (1999) The method of volume averaging. Kluwer, Boston
272. Wightman RM, Wipf DO (1989) Voltammetry at ultramicroelectrodes. In: Bard AJ (ed) Electroanalytical chemistry, vol 15. Marcel Dekker, New York, pp 267–353
273. Wilson R, Turner APF (1992) Glucose oxidase: an ideal enzyme. Biosens Bioelectron 7:165
274. Wollenberger U, Schubert F, Pfeiffer D, Scheller FW (1993) Enhancing biosensor performance using multienzyme systems. Trends Biotechnol 11:255
275. Wollenberger U, Lisdat F, Scheller FW (1997) Frontiers in biosensorics 2. Practical applications. Birkhauser Verlag, Basel
276. Wu X, Detzel CJ, Van Wie BJ, Haarsma SJ, Kidwel DA (2004) Model-based optimization of a conductive matrix enzyme electrode. Biotechnol Bioeng 88:135
277. Wu BL, Zhang GM, Zhang Y, Shuang SM, Choi MMF (2005) Measurement of glucose concentrations in human plasma using a glucose biosensor. Anal Biochem 340:181
278. Wu B, Wang Y, Li J, Song Z, Huang J et al (2006) An optical biosensor for kinetic analysis of soluble interleukin-1 receptor I binding to immobilized interleukin-1α.Talanta 70:485
279. Zhang Q, Xu JJ, Chen HY (2006) Glucose microfluidic biosensors based on immobilizing glucose oxidase in poly(dimethylsiloxane) electrophoretic microchips. J Chromatogr A 1135:122
280. Zhao W, Xu JJ, Chen HY (2006) Electrochemical biosensors based on layer-by-layer assemblies. Electroanal 18:1737
281. Zhu ZQ, Zhang J, Zhu JZ (2005) An overview of Si-based biosensors. Sensor Lett 3:71
282. Zupan J, Gasteiger J (1999) Neural networks in chemistry and drug design, 2nd edn. Wiley-VCH, Weinheim
283. Xi Y, Bazant ZP (1999) Modeling chloride penetration in saturated concrete. J Mater Civil Eng 11:58
284. Yang HM (2000) Mathematical model for liquid-liquid phase-transfer catalysis. Chem Eng Comm 179:117
285. Yang XS (2004) Pattern formation in enzyme inhibition and cooperativity with parallel cellular automata. Parallel Comput 30:741
286. Ylilammi M, Lehtinen L (1988) Numerical analysis of a theoretical one-dimensional amperometric enzyme sensor. Med Biol Eng Comput 26:81
287. Yokoyama K, Koide S, Kayanuma Y (2002) Cyclic voltammetric simulation of electrochemically mediated enzyme reaction and elucidation of biosensor behaviors. Anal Bioanal Chem 372:248

Index

A

absorbance, 197
adenosine triphosphate, 14
amperometric biosensor, xvii, 34, 47
amperometric electrode, 5
amperometric transducer, xviii, 29
amperometry, 47
amplification, 16
analytical solution, 56
anodic current, 47, 116, 242
apparent bimolecular constant, 179
array of enzyme microreactors, 217
array of microreactors, 217
artificial neural network, 119
ATP, 14
averaged concentration, 207
averaging diffusion coefficient, 164

B

batch analysis, 72
Beer–Lambert law, 196
bienzyme biosensor, 12, 14
bienzyme electrode, 11
biocatalyst, 3, 6
biocatalytical biosensor, xvii
biocatalytical membrane, xvii, 5, 6
biocatalytical process, 4
biocatalytical reaction, xvii
biocatalytical scheme, xvii
biochemical oxygen demand biosensor, 30
biochemical recognition, 5
biological catalyst, 3
biosensor, 3
biosensor based on microorganisms, 30
biosensor calibration, 117
biosensor current, 15, 47, 116, 166, 181, 211, 222
biosensor current density, 47
biosensor potential, 94
biosensor resistance, 70
biosensor response, xvii, 6, 11, 14, 16, 22, 34, 36, 47, 92, 176, 181, 196, 211, 221, 232, 242
biosensor selectivity, 5, 173
biosensor sensitivity, 15, 17, 48, 52, 196, 197
Biot number, 156, 158, 162
BOD biosensor, 30
boundary conditions, 6, 11, 15, 19, 33, 36, 45, 181, 185, 195, 210, 221, 231, 240, 241

C

carbon paste, 204
carbon paste electrode, xviii, 204
carbon paste porous electrode, 228
Cardano formula, 24
catalytic activity, 3
catalytic constant, 179
CCE, 121
CCE mode, 127
CCE scheme, 121, 123
CE scheme, 124
CEC, 121
CEC mode, 126
CEC scheme, 121
chemical amplification, 81, 120
chemical conversion, 16
chemically modified electrode, xviii, 18, 178, 179
CME, 18, 178, 179
co-immobilized enzymes, 113
compound, 5
compound I, 193, 194, 199
consecutive substrate conversion, 10, 35, 121
consecutive substrates conversion, 12
consumption–regeneration cycle, 120
Cottrell equation, 4
CPC, 204

CPC-matrix, 214
CPE, 204, 208
creatine, 12
creatininase, 12
creatinine, 12
current density, 11, 36, 232
cyclic mediator conversion, 4
cyclic product conversion, xviii
cyclic substrate conversion, 35, 120
cyclic substrates conversion, 15, 35

D

Damköhler number, 46, 183
dialysis membrane, 237
diffusion layer, 155, 164, 174, 230
diffusion module, 13, 15, 18, 34, 46, 102, 161, 162, 183
dimensionless concentration, 46, 160, 184
dimensionless coordinate, 46, 184
dimensionless current, 48, 161, 186
dimensionless distance, 184
dimensionless maximal gradient, 48
dimensionless model, 46, 160, 184, 186
dimensionless parameter, 24, 46, 160, 184
dimensionless sensitivity, 48, 196, 197
dimensionless thickness, 184
dimensionless time, 160, 184

E

effective diffusion coefficient, 164, 207, 208
effective diffusion layer, 4
effective thickness, 196, 197
electrochemical amplification, 125
electrochemical conversion, 29
electrochemical reaction, 5
electrode coverage, 225
electrode potential, 29
electrode stability, 12
electronic signal, 5
enzymatic amplification, 121, 125
enzymatic reaction, 6
enzymatic substrates conversion, 120
enzymatic trigger reaction, 121
enzyme activity, 3
enzyme inhibition, 103
enzyme layer, xviii
enzyme membrane, xviii, 6
enzyme microreactor, 218, 223
enzyme peroxidase, 199
enzyme-substrate complex, 4
excitation light, 196
explicit finite difference scheme, 50
external diffusion layer, 149, 155, 158, 160, 174, 211
external diffusion limitation, xvii, 7, 21
extinction coefficient, 196

F

faradaic current, 5
Faraday constant, 47
ferrocyanide, 11
FIA, 73
finite difference technique, xviii
first-order reaction, 7
first-order reaction constant, 7
first-order reaction rate, 53
flow injection analysis, 73
fluorescence, 196, 197
fluorescence biosensor, 197
fluorescent, 196
four-compartment model, 176, 177

G

glucose biosensor, 9
glucose electrode, 12
glucose oxidase, 9, 10, 12, 14, 17, 205
glucose oxidation, 14
GO, 9, 204
governing equation, 44, 127, 179, 184, 185, 194, 206, 207, 219, 230, 239
governing equations, 46, 160, 161, 164, 174

H

half-time of steady state, 49
Hansenula anomala, 27
hexacyanoferrate, 17
hexokinase, 14, 15
hexokinase reaction, 15
homogenization process, 164, 174
homogenized biosensor, 207
homogenized MR, 207
hydrogen peroxide, 9, 11, 12, 16, 193–195, 198, 199

I

immobilized enzyme, 3
implicit finite difference scheme, 51, 105, 109, 126, 150
inert porous membrane, 163
initial boundary value problem, 35, 104, 109, 211, 222, 233
initial concentration, 4

initial conditions, 6, 28, 33, 36, 45, 180, 185, 195, 209, 220, 231, 240
injection analysis, 73
internal diffusion limitation, xvii, 7, 10, 12
ion-selective electrode, 5, 6, 29

L
light absorbance, 196

M
matching conditions, 142, 175, 180, 185, 195, 221, 232, 241
mathematical model, 239
maximal enzymatic rate, 4, 45, 66, 83, 90, 92, 104, 115, 129, 141, 148, 164, 206, 220, 231, 240
maximal enzyme rate, 21
maximal gradient, 48, 52
mediator, 179, 204
mediator reoxidation, 205
membrane biosensor, 44
membrane perforation, 245
membrane thickness optimization, 63
metabolite biosensor, 27
Michaelis constant, 4, 7, 45, 53, 65, 83, 92, 104, 115, 141, 148, 164, 174, 206, 220, 231, 240
Michaelis–Menten function, 56
Michaelis–Menten kinetics, xvii, 24
Michaelis–Menten scheme, 28, 104
microbial biosensor, 27
microbial cell, xvii, 27, 30
microreactor, xvii, 204
mixture of compounds, 114
model validation, 59, 152
mono-layer biosensor, 147
mono-layer enzyme biosensor, 222
MR, 204, 208
multi-component analysis, 119
multi-enzyme biosensor, xviii, 114
multi-layer approach, 140
multi-layer biosensor, 147
multi-layer enzyme membrane, 172, 237
multi-layer model, 140, 141
multienzyme system, xviii, 140

N
Nernst approach, 157, 221, 230
Nernst diffusion layer, xviii, 157, 158, 222, 234
Nernst equation, 34
Nernst layer, 24
non Michaelis–Menten kinetics, 23
non-leakage boundary condition, 45, 92, 150, 175, 210, 211, 241
nontransparent plate, 24
normalized steady state current, 225
numerical approximation, 143
numerical error, 57

O
one-layer model, 147
optical biosensor, xviii, 193
optical fibber, 5
optical system, 5
optical transducer, 6
outer membrane, 140, 230, 237
overall biosensor response, 116
oxidized compound, 5
oxygen electrode, 30
oxygen electrode current, 31

P
parallel substrate conversion, 35
parallel substrates conversion, 14
partial stationary current, 49
perforated membrane, xviii, 173, 174, 177, 238
perforated membrane openness, 245, 246
periodic media, 164, 174, 208, 230
permeability, 27, 168
peroxidase, 10, 194
peroxidase activity, 11
peroxidase-based optical biosensor, 193
phenol sensitive biosensor, 82
plate-gap biosensor, 228, 229
plate-gap model, 228
polyenzyme system, 22
porosity, 168, 208, 209, 213, 214
porous electrode, 228
porous membrane, 163, 164, 168, 230
potentiometric biosensor, xvii, 91, 149
potentiometric biosensor response, 34
potentiometric transducer, xviii, 29
product fluorescence, 197
product inhibition, xviii, 108, 109
product inhibition constant, 109
product regeneration, xviii

Q
QSS, 4
quantum yield, 196
quasi steady state, 44

quasi steady state approximation, 140
quasi-steady state, 4
quasi-steady state approach, 4

R
reduced compound, 5
reduced diffusion module, 184, 189
redundant information, 119
relative fluorescence units, 197
response amplification, 120, 125, 131–133
response stability, 63, 100
response time, 49, 96, 135
response time prolongation, 135
resultant relative output signal, 49
RFU, 197

S
sandwich-like multi-layer biosensor, xviii, 140
sarcosine, 12
saturating substrate concentration, 31, 33
saturation constant, 30
selective layer, 140
selective membrane, xviii, 140, 172, 173, 176, 237–239, 241
semipermeable membrane, 30
sequential injection analysis, 80
signal gain, 90
silica carrier, 204
stationary concentration, 57
stationary condition, 7
stationary current, 57
steady state, 47
steady state biosensor current, 47, 222
steady state concentration, 57
steady state condition, 7
steady state conditions, 10, 12, 29, 96
steady state current, 57, 183, 232, 236, 242
steady state fluorescence, 197
steady state response, 34
stoichiometric coefficient, 30
substrate conversion, 10, 23, 30
substrate cyclic conversion, 82
substrate inhibition, xviii, 103, 105
substrate inhibition constant, 104
synergistic reaction, 4
synergistic scheme, 17
synergistic substrates conversion, 16

T
three-compartment model, 166
three-layer biosensor, 146
three-layer model, 145, 176
time of maximal current, 77
tortuosity, 209, 213, 214
transducer, 5, 6
transducer selectivity, 5
trienzyme biosensor, 12
trigger mode, 121
trigger reaction, 121
trigger scheme, 120
triggering, 120, 121
two-compartment mathematical model, xviii
two-compartment model, 147, 149, 153, 160, 162, 166, 178, 222
two-enzyme biosensor, xviii, 113
two-phase composite, 208

Z
zero flux boundary condition, 150
zero-order reaction rate, 34, 56

About Authors

Romas Baronas was born in 1959 in Kybartai, Lithuania. He enrolled at the Vilnius University, where he studied applied mathematics and received his Ph.D. degree. Now he is a professor and serves as the head of the department of Software Engineering at Vilnius University. He also joined the Software Engineering Department at the Institute of Mathematics and Informatics. His research interests are in database systems, software engineering, and computational modeling of nonlinear phenomena in life sciences. He is a member of the editorial board of journals "Nonlinear Analysis: Modelling and Control", "The Open Information Systems Journal", "The Open Structural Biology Journal" and "International Journal of Artificial Intelligence". His e-mail address is: romas.baronas@mif.vu.lt and his Web-page can be found at http://www.mif.vu.lt/ baronas.

Feliksas Ivanauskas was born in Vitebsk district, Belarus, on July 24, 1945. He received the Ph.D. degree in mathematics from Moscow state university and the second degree (habilitation doctor) in mathematics from Institute of Mathematical Modeling, Russia AS (Moscow), in 1974 and 1992, respectively. Since 1972 he has been with Vilnius University; at present he is Head and Professor at Department of Computer Science II. In 1982–1983 he was as a Visiting Scientist at Oxford University, Great Britain and in 1980 and at Moscow University, Russia. He has published over 250 technical papers and holds two patents on solid-state sources of light. Currently his research group works on numerical methods for solution of partial differential equations and on mathematical modelling of problem in physics (characterization of light-emitting diodes, nonlinear optics), biochemistry (biosensors), chemistry (diffusion and reaction processes). Prof. Ivanauskas received the Lithuanian National Prize in Science 1995. He is a Corresponding Member of the Lithuanian Academy of Sciences, a President of the Lithuanian Association of Nonlinear Analysts, Vice President of Association of Information Technology, Telecommunications and Office equipment, a member of the Lithuanian Mathematical Society. Prof. F. Ivanauskas is a member of editorial board of four scientific journals. He has contributed to Lithuanian national and international conferences being a Member of Organizing and Program Committee.

Juozas Kulys received Ph.D. at the M. V. Lomonosov Moscow State University in 1970, and Habil. Dr. at the N. N. Semenov Chemical Physics institute in 1982. As of 1984 he is a professor of physical chemistry. From 1974 he works as the head of department of Enzyme Chemistry in the Institute of Biochemistry, Vilnius. As of 2001 he also works as the head of the department of Chemistry and Bioengineering of Vilnius Gediminas Technical University. He was awarded by the State Science Prize (1987), the Baltic Assembly Premium of Science (1995), National Science Premium (2002), was elected as Academician of the Lithuania Academy of Sciences (1996). He is a member of the editorial board of journals "Biosensors & Bioelectronics", "Bioelectrochemistry", "The Open Nanoscience Journal", "The Open Enzyme Inhibition Journal", "The Open Biotechnology Journal", "Biologija", "Nonlinear Analysis: Modelling and Control". The main scientific interests cover chemical problems of biocatalysis, biosensors.

FSC
www.fsc.org
MIX
Papier | Fördert gute Waldnutzung
FSC® C083411

Kolibri 360 GmbH
Ferdinand-Jühlke-Straße 7
99095 Erfurt, Deutschland
produktsicherheit@kolibri360.de